FLORA
of Surrey

FLORA OF SURREY

J. E. LOUSLEY

Based on Records collected by
THE SURREY FLORA COMMITTEE
1957–1974

with a Chapter on Geology and Soils by
A. J. STEVENS

DAVID & CHARLES
NEWTON ABBOT : LONDON
NORTH POMFRET (VT) : VANCOUVER

ISBN 0 7153 7048 0
Library of Congress Catalog Card Number 75-2923

Set in 10 on 12pt Times
and printed in Great Britain by
Redwood Burn Limited, Trowbridge & Esher
for David & Charles (Publishers) Limited
Brunel House   Newton Abbot   Devon

Published in the United States of America
by David & Charles Inc
North Pomfret   Vermont 05053   USA

Published in Canada
by Douglas David & Charles Limited
1875 Welch Street North   Vancouver   BC

Contents

J. E. Lousley: An Appreciation

Job Edward Lousley was born on 18 September 1907 and lived all his life in Surrey where he died very suddenly on 6 January 1976. His family originally came from Berkshire where his great-great-grandfather, Job Lousley (1790–1855), was a farmer and naturalist. J. E. Lousley was a banker by profession, working in the Trustee Department of Barclay's Bank where he held high office and from which he retired in 1967.

From his earliest days he developed an interest in natural history and started his first notebook on the subject at the age of eight. He possessed great powers of concentration, an outstanding memory and a tremendous capacity for work. His earliest papers were published in the *Journal of Botany* and elsewhere before the age of twenty-one and to aid his researches he prepared a card index of botanical references, culled from the literature on British botany to be found in the libraries of the Natural History Museum, Kew, and the South London Botanical Institute. He also set to work to make an herbarium and, by the time of his death, had amassed a collection of 25,000 sheets, not only of British plants, but also including thousands of sheets of alien species—one of his special interests. He was also interested in the varieties and forms of various taxa found only by careful observation; he found, on a site 100yd from his home, *Senecio londinensis*, a hybrid between *S. squalidus* and *S. viscosus*, hitherto unknown to science. Another special interest was the *Polygonaceae* which he studied with great thoroughness, publishing his results between 1938 and 1944 in the reports of the Botanical Society. At the time of his death he was preparing a handbook on this difficult family for the Society.

His eye was ever active and his observations, while working in London during World War II, led to the publication of accounts of the flora of many bomb sites and finally, with R. S. R. Fitter, to *The Natural History of the City* in 1953. In his own immediate neighbourhood he noticed many isolated trees of *Quercus petraea* surviving in gardens and parks and this led to papers establishing the fact that the great North wood of South London was composed mainly of this tree and not, as might be expected, of *Quercus robur*.

7

On visits to the Isles of Scilly he studied the flora assiduously and his *Flora of the Isles of Scilly* was published in 1972.

J. E. Lousley was a gifted broadcaster, lecturer and writer: his 'The Wild Flowers of Chalk and Limestone', published in the *New Naturalist Series* in 1950, has been reprinted several times, and his classes at Morley College and the City Literary Institute were always oversubscribed. He also led botanical parties abroad and careful preparatory work, together with his skill and knowledge, ensured their success. He was married in 1940, and his wife, Dorothy, shared his interests and often accompanied him on such expeditions at home and abroad.

Above all, his interest was the county of Surrey. No one had a more detailed knowledge of this region or a better memory for the odd corners where scarce, wild plants were to be found. Before any of these excursions he always did his 'homework' and so knew what to look for and where to find it. Many old Surrey records were traced in this way, particularly during World War II when travel was limited and he could not go further afield.

He had long been dissatisfied with the account of Surrey plants in the latter part of Salmon's *Flora*, and a new Flora was therefore in his mind from early days and, almost unconsciously, he accumulated records for it. Lousley was convener of the meeting of Surrey botanists held in January 1957 at which the Surrey Flora Committee was formed. He was elected chairman, a post which he held until 1972 when he was appointed editor and recorder following the untimely death of Dr D. P. Young. The long illness of Dr Young had delayed the project, but Lousley set to work with characteristic energy, setting himself a timetable to complete the task. Many expeditions were made, photographs taken, records checked, often alone or in the company of the Surrey Flora Committee on one of its excursions. He had approved the typescript as prepared for press only a few weeks before his death.

It is indeed tragic that he did not live to see its publication.

J. E. Lousley found time to undertake many other tasks. He served the Botanical Society of the British Isles as treasurer, secretary and as president, 1961–5. He was president of the London Natural History Society, 1963–4, and a vice-president of the Surrey Naturalists' Trust from its foundation in 1960. He was the first secretary of the Council for Nature and sat on the England Committee of the old Nature Conservancy. He found time to take a very active interest in the South London Botanical Institute, acting as president at the time of his death. He received the Bloomer Award of the Linnean Society of London in 1963. Nor did he neglect his own local affairs, for he served on the boards of managers of two local schools.

A great many botanists all over the world will be grateful for the benefit of the expert knowledge which he shared so freely; others owe to him their first introduction to the delights and pleasure of field botany, while all who knew him will miss his companionship, his kindness and his levelheaded enthusiasm.

C. T. Prime and E. C. Wallace

Surrey Flora Committee

Publisher's Note

The members of the Surrey Flora Committee and Mrs Dorothy Lousley wish to express their thanks to Dr C. T. Prime and Dr C. P. Petch who assumed responsibility for the *Flora*, completed the proof reading and saw the book through to publication. The publishers also are grateful to Dr Prime and Dr Petch and in addition they wish to thank the Botanical Society of the British Isles who greatly assisted publication by guaranteeing to purchase seven hundred and fifty copies of the book.

Preface

Surrey is one of the loveliest counties in England, with large areas of wild and unspoilt country and a great number of attractive wildflowers and ferns, some of them rare. It is a county where botanists and residents share their interest in flowers with visiting motorists and walkers every weekend. A modern *Flora* is much needed to take the place of the last which appeared over forty years ago.

This need I have tried to fill. The *Flora* is based on the records collected by the Surrey Flora Committee set up in 1957 and these records are shown on the 504 distribution maps and in the text. Since the flora is constantly changing I have included every species I can trace as having been established in Surrey at any time, but full details are given only for those found since 1957. Even in this short time some of the plants have become scarcer or even extinct, so a great deal of attention has been devoted to aspects of conservation. It is hoped that this book will serve as a reference book to indicate the plants at risk and their requirements.

All my life I have lived in the area covered by this book, and much of my experience of field botany was gained in the county. I hope that it will enable others to share the enjoyment I have had from Surrey's wildflowers. In any case I seem to have been destined to write about them. When C. E. Salmon died in 1930 with his *Flora of Surrey* partly in the press and the rest still to be written, I was one of those invited to complete it, but business commitments prevented my acceptance. Dr Donald Peter Young was appointed to edit the present work in 1967, but sadly he died in 1972 after a long illness. As my *Flora of the Isles of Scilly* was on the point of publication, I was able to take it on. The colour plates are published in memory of Donald Young and his scientific work on orchids and *Oxalis*, and the cost has been met by the Botanical Society of the British Isles.

Part One

Chapter One

Basis of the Flora

Surrey is a superb county for flowers. A wide variety of soils and habitats, often in wild and beautiful country, support a fine flora. Proximity to London has made it accessible, from the earliest days of the study of botany, to people who lived in the metropolis and recently the 3 million residents in the 'geographical county' have included a great many interested in flowers.

Surrey is often regarded as a dormitory area for London, and indeed it is. The north is over-run with a dense network of railways and houses with small gardens; large areas farther out are taken up by the 'stockbroker belt', where the gardens are larger and, although sometimes left as woodland, the loss to the public is equally great. That is not the Surrey of the naturalist and gives a totally misleading impression. Of the county (excluding the London Boroughs), 26,871 acres are commonland—that is, 5.82 per cent of the area—a higher proportion than in any other lowland county of England except Hampshire, and Hampshire has the New Forest. The area to which there is public access is over 36,000 acres. The protection given by the Surrey Hills Area of Outstanding Natural Beauty (covering 160 square miles), the Green Belt legislation (which now covers nearly the whole of the remaining county), the National Trust and other bodies, is likely to preserve a great part of Surrey for the naturalist and the walker (see Chapter Five). It is a county of open spaces and lovely flowers.

Definition of the Area

There are many different 'counties' of Surrey—so many that it is difficult to find two standard reference books which agree on the area. The present administrative county covers about 415,879 acres ($=653$ sq miles$=1,692$km^2); but, to save altering their records every time there is a boundary change, botanists use the boundaries as they were in 1859. This *Flora* covers the Watsonian vice-county of Surrey as shown on all the maps in its pages. The area is about 485,324 acres ($=758$ sq miles$=1,964$km^2).

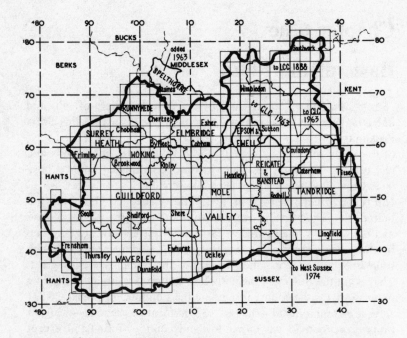

Fig 1 Administrative Areas of Surrey: The heavy outline encloses the
botanical county of Surrey (V-c 17) which is the subject of this
Flora. The present *administrative* council has lost the areas transferred to the LCC in 1888, to the GLC in 1963 and to West
Sussex in 1974. Spelthorne, formerly in Middlesex, was added to
the administrative county in 1963

The system of vice-counties was devised by H. C. Watson to provide
112 very roughly equal areas in England, Wales and Scotland to
record the geographical distribution of plants. Surrey was one of
these units—'V-c 17'—with boundaries as defined by him in 1859
(*Cybele Britannica* **4,** 139–42) and was essentially the old geographical
county. There were boundary changes under the Reform Acts of
1867 and 1885, and larger ones under the Local Government Act
of 1888, which set up the administrative county of London (the
LCC) and removed from Surrey administration the ancient parishes
of Wandsworth, Lambeth, Camberwell and Southwark and all parts
of the Hundred of Brixton. Surrey lost about 25,000 acres to London.
More recently the London Government Act, 1963, set up the Greater
London Council (the GLC) and made even greater changes. Surrey

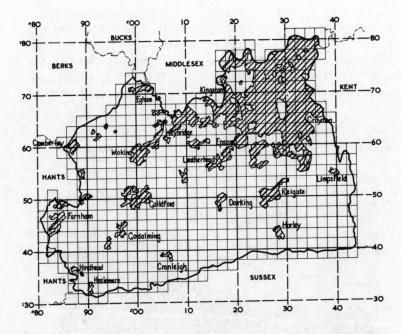

Fig 2 Built-up Areas: showing land most densely covered with houses and other buildings

lost to London the boroughs of Barnes, Beddington and Wallington, Kingston-upon-Thames, Maldon Combe, Mitcham, Richmond, Surbiton, Sutton and Cheam, Wallington, and the urban districts of Carshalton, Coulsdon and Purley, and Merton and Morden. Croydon, the only county borough, also became part of Greater London. On the other hand, the urban districts of Staines and Sunbury on Thames, on the other side of the Thames, were added to Surrey, so the net loss was only about 52 sq miles. Apart from these, there have been a few minor boundary changes, such as a slight variation in the boundary in the north which followed the Thames, and Dockenfield parish near Frensham, which has been disputed with Hampshire.

Topography

Surrey is a county of contrasts. There are two east-to-west ranges, rising to 868ft and 965ft respectively, with the Thames Valley to the north and the Weald Clay lowland to the south. The chalk escarpment enters the county from Kent, is at first wooded round Titsey and Botley Hills and soon reaches its highest point at 868ft above Oxted Chalkpit; past Godstone to Reigate Hill, 763ft, Colley Hill to Box Hill, 600ft. Then, the other side of the Mole Gap, past Ranmore Common to White Downs, 724ft, and Hackhurst Downs and Albury Downs, past Guildford to the Hog's Back, which attains its highest point, 504ft, at Monk's Hatch. The chalk provides

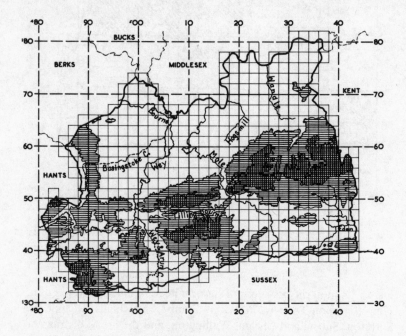

Fig 3 Physical Features: land over 250ft and 500ft, rivers and canals

wonderful scenery and flowers from one side of the county to the other.

The range to the south of the chalk, on Lower Greensand, is very different. It enters Surrey from Kent near Limpsfield Chart at 653ft, and attains 590ft at Tilburstow Hill south of Godstone; it then decreases in height to just beyond Reigate. Then there is a gap until Dorking where the range suddenly widens and soon attains 965ft at Leith Hill. This is the highest point in south-east England and from the top of the tower, at 1,000ft, there are fine views over the still wooded Weald to the south. The ridge extends through the Hurtwood, past Pitch Hill, 843ft, to the Wey gap, then Hascombe Hill, 645ft, and Hydon Ball, 586ft, to culminate in the Devil's Punch Bowl and Gibbet Hill, 895ft, and pass out of the county at Hindhead and Haslemere. The Lower Greensand is not rich in species, but provides a large area of grand country which is a valuable contrast to the soils over other formations.

To the south of these two ranges is the Weald Clay area, much of it under 200ft, and to the north the Thames Valley with its London Clay and Bagshot Beds. Even this area is seldom flat and has some interesting ridges—Chobham Ridges rise to over 400ft. In the north-east, London Clay forms hills round Norwood, while the Blackheath Pebble Beds of Addington Hills and Croham Hurst near Croydon have considerable botanical interest. The topography is further discussed in relation to the geology in Chapter Two.

Rivers and Canals

The rivers of Surrey provide an interesting series of habitats. The Thames forms the northern boundary for the whole of its length. It enters the county charged with calcium carbonate in solution and over the ages this has been deposited on the alluvial soils of the flood meadows on its banks. Runnymede, Chertsey and Ham, for example, all have calcicoles as a result. At its eastern end the Thames is tidal up as far as Teddington Lock. Unfortunately most of the maritime species which thrived on and by the tidal mud have now been lost owing to the embankments made necessary to prevent flooding. In the higher reaches the Thames had formerly a rich flora of fresh water aquatic species, but these are now rare or extinct, their destruction having been brought about mainly by fast-moving motor cruisers.

The next largest river is the Wey, which rises south of Haslemere just over the boundary in Sussex and, after a sweep through Hampshire, enters Surrey near Frensham. Its course from there until it enters the Thames near Weybridge, with its tributary the Tillingbourne, and the connected canals, provides one of our best habitats for aquatics. The National Trust owns the whole of the Wey Navigation, from Guildford to the Thames, and also the Godalming Navigation from Godalming to Guildford. The Wey is about 35 miles long and the Mole a little shorter. This also has its headwaters in Sussex, but gathers several branches near Horley to wander through the plain south of Reigate to Brockham and the Mole Gap, with its famous 'Swallow Holes' near Mickleham, to enter the

Thames near East Molesey. To the east of this there are several small streams draining into the Thames—the Hogsmill River from Ewell to Kingston, the Beverley Brook from Cheam to Barnes, and the River Wandle from Waddon to Wandsworth. For the most part their courses are cased in concrete, but there are still a few bits of marsh beside them. There is one more stream draining into the Thames —the Bourne—with two branches, one coming down from Chobham through Woodham, the other from Thorpe and Chertsey. The Bourne enters the Thames near Weybridge.

In the south-east the Eden Brook and River Eden drain the clays round Blindley Heath and Lingfield before crossing into Kent to flow into the Medway. Their wet pastures have an interesting flora, but have suffered severely from modern farming practices. In the south-west the River Arun rises near Haslemere, runs north round Chiddingfold and Dunsfold, and turns south to pass out of the county to drain through Sussex to the English Channel. The long-disused Wey and Arun Canal joined the Arun near the county boundary. This canal and the Basingstoke Canal and Wey Navigation are discussed later (p 44). They have made a great contribution to the aquatic flora of Surrey and provided routes by which the distribution of many aquatics has been extended.

Climate

Surrey, as might be expected from its position, is a little wetter, and warmer in winter, than the east of England, and a little drier, and colder in winter and hotter in summer than farther west. The climatic conditions are in no way extreme, though hard frosts are sufficiently frequent to prevent the persistence of tender species, except in specially sheltered spots, and the summers are seldom hot enough to encourage the germination and growth of plants requiring continental conditions. Wisley (05D5), from its geographical position and records as a horticultural centre, is probably the most suitable of the meteorological stations for which data is available, and this is set out in Table 1.

At the more frequently quoted Kew (17D3), the rainfall is less (23.95in), and it is a little warmer (eg only ninety ground frosts a year). Gatwick, farther south (24E1), is considerably wetter (30.29in), and sunnier (1,596 hours), but has more snow or sleet (twenty days). Rainfall in Surrey increases from east to south-west with elevation, so the wettest areas are likely to be on the high ground round Leith Hill and Hindhead. In spite of the importance of climate in con-

Table 1 Climate

(Station: Wisley, 150ft above mean sea level, with averages for periods stated.
Compiled from figures supplied by the Meteorological Office)

	Jan	Feb	Mar	Apr	May	Jun	Jul	Aug	Sep	Oct	Nov	Dec	Year
TEMPERATURE °C (1931–60)													
Daily Maximum	6.6	7.3	10.7	13.9	17.5	20.7	22.1	21.9	19.1	14.5	10.1	7.5	14.3
Daily Minimum	0.9	1.0	2.2	4.2	6.8	10.0	12.0	11.7	9.8	6.5	3.8	2.0	5.9
Mean	3.7	4.1	6.4	9.0	12.1	15.3	17.1	16.8	14.4	10.5	7.0	4.7	10.1
Extreme Max. }(1904–	15.0	17.8	22.8	27.8	31.7	33.3	33.3	35.6	33.3	28.3	20.0	15.0	35.6
Extreme Min. }1972)	−13.9	−15.0	−12.2	−4.4	−2.8	0.6	2.2	2.8	−1.7	−4.4	−8.3	−13.3	−15.0
SUNSHINE in hours (1931–60)													
Total	46	63	114	155	191	209	194	175	136	95	51	38	1,467
Daily Mean	1.49	2.22	3.69	5.16	6.18	6.96	6.25	5.66	4.52	3.07	1.17	1.24	4.02
RAINFALL in inches (1916–50)													
Total	2.43	1.80	1.63	2.02	1.86	1.67	2.63	2.17	1.94	2.53	2.82	2.53	26.03
RAIN DAYS (more than 0.01in)	17.5	13.6	13.1	14.5	12.6	11.1	13.3	13.1	13.0	15.3	16.3	17.6	170.9
AVERAGE NUMBER OF DAYS OF . . .													
Snow or Sleet	1.8	2.0	1.6	0.9	—	—	—	—	—	—	0.5	1.3	8.1
Snow lying (at 09.00 GMT)	1.2	1.3	0.6	0.3	—	—	—	—	—	—	—	1.5	4.9
Air Frost	14.2	11.9	9.3	3.3	0.7	—	—	—	0.1	0.8	6.2	10.4	56.9
Ground Frost	19.9	17.1	17.0	12.5	6.2	0.9	—	0.2	2.1	8.2	13.3	18.3	115.7
Fog at 09.00	2.3	2.3	0.7	0.7	—	—	—	0.5	2.1	3.2	3.2	3.8	18.8
EARTH TEMPERATURES °C (1931–60)													
1ft Depth	3.9	3.9	5.7	9.2	13.0	17.1	19.3	18.9	16.3	11.9	7.9	5.3	11.0
4ft Depth	6.5	5.7	6.2	8.3	11.0	14.3	16.8	17.8	16.9	14.2	10.8	8.2	11.4

trolling which species can grow in Surrey, the changes in general climate within the county are too small to produce obvious instances of their influence on the distribution of species. Differences in micro-climates are of far greater importance. There are marked differences between the floras of colder and wetter north-facing slopes and their sunnier south-facing counterparts. The same applies to walls; ferns are much more plentiful on those which face north and especially so on the north-facing brickwork of railways. Many species depend on micro-climates as, for example, *Dryopteris pseudomas* which favours moist and frost-free ghylls. But it is the nature of the soils rather than climate which controls the detailed distribution of species within the county.

Chapter Two

Geology and Soils by A. J. Stevens

The county of Surrey may be divided geologically, for the purposes of description, into three broad areas extending in a general sense east-west across the map. In the northern area (1) lie all the solid rocks which are younger than the Chalk. Area 2, a band of uneven width, but of a general triangular shape, is the outcrop of the Chalk while Area 3 is made up of the outcrops of the rocks older than the Chalk in that part of the Weald that is within the county limits. This broad division of the county into three is especially relevant botanically, because most of the more acid habitats occur in the northern and southern areas, while most of the calcareous ones are in the centre. There are, however, a great many exceptions to this general statement, as will appear later.

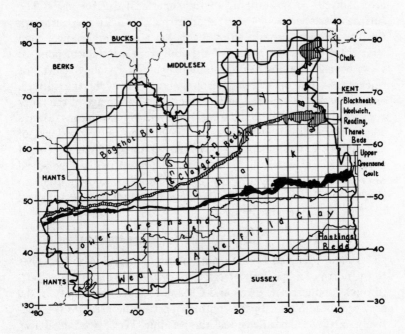

Fig 4 Solid Geology

Area 1, the area of those beds beneath which the Chalk dips to the north, contains the outcrops of the Reading, Woolwich and Thanet Beds, the London Clay, and the Bagshot, Bracklesham and Barton Beds, together with the complex patches of gravels and alluvium which in many places obscure them.

Area 2 is occupied by the outcrop of the Upper, Middle and Lower Chalks, but here too the surface is often made up of such superficial materials as Clay-with-Flints, flint-gravels and plateau gravels.

Area 3, where the present surface rocks have been exposed by the removal of the Chalk which formerly overlay them, is made up of a varied succession of clays, sandstones, sands and limestones. These are the Upper Greensand and Gault, Folkestone Beds, Sandgate and Bargate Beds, Hythe Beds, Atherfield and Weald Clays, and, in the extreme south-east, the Tunbridge Wells Sand, Wadhurst Clay and Ashdown Sand. In this area, the presence of alluvium and gravels is of relatively less importance, but even the Weald Clay is diversified by the irregular outcrop of sandstone and of the Paludina Limestone in some parts.

Thus in terms of solid geology some twenty different types of rocks are involved, to say nothing of the superficial drift deposits. The surface of the county is therefore an extremely varied one, although for the present purpose lithology is of more importance than stratigraphy; and the nature of the soil, from a botanical point of view, depends on whether the parent material is sandy, clayey or calcareous—and on the geomorphological history of the surface, and on the land use in the past—rather than on the age of the rocks concerned. The soil boundaries, however, by no means always coincide very closely with those of the geology since, on slopes of any steepness, soil creep will have moved some of the weathered parent material downhill. In addition, the maps of the Geological Survey generally indicate only those deposits more than 3ft in thickness, and a much smaller thickness than that may be sufficient to affect the soil character.

Area 1   North of the Chalk Outcrop

This area, occupied by the main outcrop of the Eocene beds, lies to the north of a line drawn from just north of Farnham in the west, through Leatherhead, Ewell and Cheam, to leave the county in the east near Addington. In the southern part of the area are the relatively narrow Thanet Beds, and the Reading, Woolwich and Blackheath Beds, succeeded to the north by the much greater width of the London Clay and Claygate Beds. The north-west of the county, adjoining the Berkshire and Hampshire borders, is covered by the irregular patches of the Bagshot, Bracklesham and Barton Beds, forming the heaths and commons of Bagshot and Chobham. The whole area is thus one of alternating clays, sands and pebble beds, from which calcareous materials are almost entirely absent. From a botanical point of view, the relatively simple 'solid' geological pattern is made more complicated by the frequent occurrence of areas of plateau gravels, alluvium and peat.

The oldest of the beds—the Thanet Beds—occur in general only as a narrow band bordering the Chalk as far west as a point between Horsley and East Clandon, although rather larger patches are to be found near Beddington and Waddon, and as outliers to the south on the back-slope of the Chalk downs, such as those at Walton and Headley Heaths. They consist of sands of a light colour, but may become clayey at the base, often with a layer of unworn flints derived from the Chalk below.

In the east of the county the Thanet Beds pass northwards under the estuarine loamy material of the Woolwich Beds but, somewhere in the neighbourhood of Sutton, the latter soon change laterally into the fluviatile clays of the Reading Beds. Thin layers of clay may, however, occur in the Woolwich Beds, while sandy material is not uncommon in the Reading Beds. To the east of Croydon the Woolwich Beds will often be found to be overlain by the pebble beds, laid down as great shingle banks, known as the Blackheath Beds. These form much of the escarpment of the Addington Hills and Croham Hurst.

A much greater area is occupied by the London Clay. This formation has a width of little more than a mile to the north of a line between Farnham and Guildford, but widens out between the Wey and the Mole, northwards to the neighbourhood of Ripley and Ockham, in the oakwood and commons belt and cultivated clay plain of north Surrey. Beyond the Mole it widens out even more under the southern suburbs of London. It is, however, often obscured not only by buildings but by stretches of river gravels, so that it is usually only on the lower slopes of gravel-capped hills that it appears at the surface. Made up of stiff, usually bluish-grey, clay which weathers to brown, with occasional concretions of clayey limestone, this formation is almost uniformly heavy except where sandy layers occur at the base.

At the top of the London Clay, and outcropping in the Claygate, Cobham, Walton and Oxshott districts, are the sandy transition beds to which the name of Claygate Beds has been given. They consist of well-defined alternating layers of sands and clays. Although the area they occupy is only small, they frequently make a marked contrast with the predominantly heavy clays of the London Clay to the south.

Most of the north-west of the county, as already mentioned, is covered by the Bagshot, Bracklesham and Barton Beds, in a series of flat-topped, gravel-capped hills. These beds are chiefly of fine-grained sands and pebble beds but, especially in the Bracklesham Beds, clay bands occur which give rise to local water-tables, often marked by small bogs on the hill-sides, with a considerable effect on the vegetation. Wide stretches of alluvium, laid down since Neolithic (sub-Boreal) times, also diversify the surface of this district, notably in the Blackwater valley, in the neighbourhood of Woking, Horsell and Chobham, and in the lower Mole valley.

Within this whole northern area, therefore, the soils display all types of texture from the very heavy to the very light, but all with a marked tendency towards acidity. The soils of both the London Clay and Reading Beds are heavy intractable clays which suffer water-logging in winter. These will frequently show the mottled rust colours, in the horizons affected by the fluctuations in the water-table, which are characteristic of gleyed soils. That is to say, while the soils are water-logged, iron compounds, toxic to some plants, are present in solution, but these are re-precipitated as rust-coloured mottlings when, in the summer, the water-table falls again. In the centre of the northern half of the county, the soils on the Claygate Beds, and those on the Woolwich Beds in the eastern half, containing as they do a much greater amount of sand, are less liable to water-logging.

In the north-west, the characters of the soils will depend on whether they are developed on the sandy Bagshot and Barton Beds or on the more loamy Bracklesham Beds, and also on the amount of gravel which is present above them. They will, however, all show to some extent a podzolic profile—that is to say, a thin peaty horizon will lie above a characteristic light-coloured layer of leached sands. In some places the re-deposition of humus and iron lower down may have produced a 'pan' which impedes the drainage, giving rise to patches of damp ground. It is on soils such as these that the dry and damp heaths of the Bagshot and Chobham district have developed.

Area 2   The Chalk Outcrop

Running the whole length of the county, from just west of Farnham to the Kent border between Addington and Tatsfield, the Chalk, with its bold south-facing scarp, forms one of the most conspicuous topographical features of Surrey. Although simple enough in solid stratigraphy, the surface is an extremely varied one. Its geomorphological development has been very complex and the stages of this development are marked by the occurrence of a very varied collection of superficial deposits, some of which are conspicuous and others less so.

Only examination of the ground can reveal this great variety, since the Geological Survey maps distinguish only between 'Netley Heath Deposits' and 'Deposits of Doubtful Age' on the one hand, and 'Clay-with-Flints' on the other. Even in the last-named, gravels and sands sometimes make up nearly all of the accumulations. For the present purpose, three different types of drift may be distinguished on the Chalk: Clay-with-Flints *sensu stricto* (above about 690ft); Plateau Drift, probably of Early Pleistocene age (at heights between about 600 and 690ft); and Sandy High-level Flint Gravels (at heights between about 490 and 600ft). When it is realised that any of these may have been transported, by hillwash or by periglacial sludging, to lower levels, there to be incorporated into the upper layers of the underlying beds in greater or lesser amounts, the complexity of the surface becomes apparent.

Nor is this all, since, on the northern slopes of the Downs, many patches of the Eocene beds have survived erosion and remain *in situ*, and even where this is not so, relics of these materials may have become incorporated in the soils overlying the Chalk. To this type of surface, representing the former landscape on which the Eocene beds were laid down, the name of 'Sub-Eocene Surface' has been given. Even the Chalk itself is variable, from the pure white limestone of the Upper Chalk, which occupies the greater part of the solid outcrop, to the marl beds of the Middle and Lower Chalks. All of this, together with the varied history of land use in the area, makes the Chalk Downs much more varied than might be supposed from a cursory examination of the geological map.

The soils developed on the Chalk uplands are consequently very diverse, and have as yet been only incompletely described. However, certain well-marked types may be distinguished, corresponding to the different parent materials. On the highest parts (above about 690ft) the soils are generally brown silty clay loams; a foot or two below these is the flinty clay, yellowish-red in colour, which may be regarded as the Clay-with-Flints *sensu stricto*. This occurs on relatively undisturbed surfaces where the Chalk has been weathered at depth. At rather lower elevations (down to about 600ft) the soils developed above the Plateau Drift contain more sandy material, together with some Eocene flint pebbles. Where the Early Pleistocene beach gravels and sands are present in quantity, the soils may show partial or complete podzolisation, and hence may be termed 'podzols' or 'podzolic brown earths.' The freer draining nature of these soils, under *Calluna* and *Pteridium*, form a strong contrast to the heavy soils of the Clay-with-Flints proper. Lower down on the spurs of the northern dry valleys, these soils may be more commonly developed on the Sandy High-level Flint Gravels, and may be termed 'sandy brown earths'. All three of these soil types are of an acid nature, with pH values as low as 4.5.

On the slopes and floors of the dry valleys and on the Sub-Eocene Surface occur the calcimorphic soils. In these, the mineral colour is grey or greyish-brown and the organic material colour is masked by the excess of chalk. They may be termed 'rendzinas' and their quality depends to a great extent on the angle of the slope and on the extent to which clayey hillwash has been incorporated into the soil profile. Only where little contamination by hillwash has taken place and where the soils have developed under woodland will the profile show the characters of a true dark-rendzina. These truly calcareous soils are of remarkably little extent on the chalk uplands.

Area 3   South of the Chalk Outcrop

To the south of a line running from near Dippenhall on the Hampshire border to near Tatsfield on the Kent boundary begins the varied assemblage of sands and clays of the Weald. The youngest of these beds, at the foot of the Chalk scarp, is the Albian of continental geologists, sometimes given the name of Selbornian in this country. This formation consists of the Gault Clay below, with the sands and hard sandstone of the Upper Greensand above. The first of these generally forms a conspicuous vale, sometimes called the Holmesdale, north of the Lower Greensand country between the Kent border and the Mole gap. Apart from an area near Westcott, this valley feature is less evident beyond Dorking, as the outcrop becomes narrower as far west as the western end of the Hog's Back.

Towards Farnham, the outcrop widens again, and south of this town, around Rowledge and Frensham, are some of the widest areas of this clay in the county. The Upper Greensand, wherever the hard 'malmstone' is present, forms a conspicuous bench feature at the foot of the Chalk scarp. This is narrow in the east of the county, but becomes wider between Merstham and Dorking. To the west of Gomshall, the outcrop is once more very narrow and is noticeable again only west of the fault near Whiteways End to the east of Farnham. Beyond Farnham, it forms the north side of the Wey valley and is increasing in width when the Hampshire border is reached. These two beds often form a sharp landscape contrast between the low-lying damp land of the Gault Clay and the more steeply sloping and somewhat drier land of the Upper Greensand.

The formation which emerges from beneath the foregoing is the Lower Greensand, and this forms a very wide area of country west of the Leith Hill salient, reaching a width of about 10 miles between the Wey valley and the Sussex boundary near Haslemere. East of Leith Hill, the outcrop is very much narrower and is only about 2 miles wide at the Kent border. Five different beds are involved in this formation, the lowest—the Atherfield Clay—being the most inconspicuous, and almost indistinguishable from the Weald Clay which it overlies. It forms the damp floor of the Peasemarsh valley to the east and west of Shalford.

Above this clay, the sandy Hythe Beds, with thick layers of hard siliceous chert, form the top part of the bold scarps of Leith Hill, Holmbury Hill, Hascombe and Gibbet Hill. To the north of this sandy country is the variable outcrop of the Bargate Beds, which occupies a large area only between Compton and Witley, to the west of Godalming, with an extension nearly a mile wide beyond Thursley to the Hampshire border near Churt. The shelly and calcareous sands and sandstones of this formation, wherever they occur in sufficient quantity, form country in sharp contrast to the mainly heath-covered landscape to the north and south of them. Of even more importance in the present connection is the fact that, by forming the slopes of the plateau-like country around Godalming, often with steep roadside banks, the Bargate Beds present calcareous habitats in an otherwise acid district. Where streams from these limestones run down on to the heaths, the modification of the soil-water produces conditions more suitable for plants of calcicole type.

The fourth of the Lower Greensand beds, in ascending order, are the clayey sands and loams of the Sandgate Beds. These occupy a considerable area of the commons between Witley and Puttenham, on Hydon Heath and on parts of Munstead Heath, and extend westward through Thursley to Churt. Further smaller patches occur in the east of the area, between Albury and Wotton, and between Reigate and Oxted. The Fuller's Earth of the Nutfield district may be regarded as being a clayey facies of this formation. The fifth and youngest part of the Lower Greensand is the Folkestone Beds, consisting of coarse, ferruginous quartz sand, with very little fine material. These beds, too, have their greatest surface extension in the west of the county, forming the commons of Frensham, Hankley and Crooksbury, and extending from there to the Kent border in a narrow band usually between half a mile to a mile in width. There are, however, some larger patches to the south, at Blackheath, and at Albury and Farley Heaths.

Except for a small area in the extreme south-east, the rest of Area 3 is occupied by the Weald Clay. The northern edge of the outcrop runs from the slopes of Gibbet Hill in the west along the south of the Hythe Beds scarp to below Leith Hill. Beyond, the outcrop widens to over 6 miles south of Dorking and is not much less at the Kent border. The Weald Clay is a non-calcareous silty clay, of a brown or blue colour weathering to yellow, and is often of a rather shaly nature with intercalations of sands and clay-ironstones. The landscape of low-swelling rises between the numerous streams is diversified by the occurrence in many places of outcrops of sandstones and the 'Paludina Limestone'. The sandstones form the high ground from between Newdigate and Charlwood along the east side of the valley of the Dean Oak Brook, and in the neighbourhood of Outwood, and here and there between Bletchingley and Crowhurst. They also reach the surface at numerous other places, especially around Alfold and Chiddingfold. The Paludina Limestone, too, appears as low ridges, notably near Charlwood, to the south of Earlswood, to the west and south of Outwood, and between Crowhurst and Lingfield. In the clay country south and east of Leith Hill, and in the Ockley district, there are further outcrops, and away to the west a long band of the limestone stretches towards Dunsfold village. The Weald Clay is also covered in places with wide patches of river gravels and 'head deposits', especially near Horley and in the Mole Valley, and farther to the west between Cranleigh and Dunsfold northwards along the Bramley Wey to Guildford.

In the extreme south-east of the county occur the sands, sandstones and variable bands of clay, known as the Hastings Beds. The upper layers of sand—the Tunbridge Wells Sand—are separated from the Ashdown Sand by the Wadhurst Clay and, although the whole area occupies no more than about 15 sq miles, the activity of streams and the occurrence of a number of faults have brought to the surface all of these beds. This higher country, culminating in the eminence of Dry Hill on the Kent boundary, forms a striking contrast with the damper landscape of the Weald Clay away to the north.

As a consequence of the great variety of rocks which appear at the surface in Area 3, the soils show correspondingly great differences, frequently over quite short distances. At the foot of the Chalk scarp, the more open-textured soils of the Upper Greensand contrast with the heavier, usually gleyed, soils of the Gault, besides the slope contrasts already mentioned. Both are somewhat calcareous, whether from the lime in the rocks themselves or from the down-wash from the Chalk above. Beyond the low-lying country of the Gault, the higher land of the Folkestone Beds forms one of the more striking transitions in the county. Besides the steeper slopes of the latter beds, emphasised where the presence of ferruginous 'carstone' has rendered them more resistant to erosion, the free-draining nature of these coarse sands has produced highly podzolised upper horizons supporting a notably acidiphilous flora. Both the Sandgate and Bargate Beds carry loamier and less strongly leached soils, the former ferruginous and the latter calcareous brown earths. They thus form a band of often enclosed and cultivated country, with the associated weed flora, between the heaths to the north and south of them.

On the high heathlands in the western half of the county the soils of the Hythe Beds are all podzolised to some extent. The variations between the acid iron-humus podzols at one extreme and brown earths at the other are considered to be the result of differences in past land use. On the more level stretches at lower elevations much of the land was formerly farmed, and here the soils are acid brown earths, partly reflecting perhaps the marling that was practised on these farmlands. On the higher unenclosed ground, much of which is under woodland, strongly developed podzols and podzolic brown earths are the characteristic soils.

Except for the upper 100ft or so, the scarps of the Hythe Beds salients are composed of the Atherfield and Weald Clays, but land-slipping has, together with hill-wash, carried much material from the sandstone capping down to lower levels. Podzolic soils are there-fore present some distance out on to the Weald Clay, merging into peaty gleys wherever depressions in the scarp have created conditions of impeded drainage. On the Weald Clay itself the soils are non-calcareous silty clays, often suffering winter waterlogging, except on the low divides between the streams, where brown earths occur which are gleyed only in the lower horizons.

Of more importance in the present connection are the great varia-tions in soil conditions within the Weald Clay area, produced by the materials other than clay already mentioned. No detailed studies have apparently been made of these variations, but they may be said to derive from two main causes. In the first place, the sandstones and limestones give rise to steeper slopes, and hence to better drainage;

and, secondly, the presence of calcareous beds within both the sandstones and limestones provide conditions for plant growth in strong contrast to those on the Weald Clay itself. Thus small areas of calcicole species may appear within the general background of more calcifuge types. Many of the lighter soils rest directly on a largely impermeable subsoil of clay, so that the pattern is very complex.

Contrasts of a somewhat similar nature, but on a much smaller scale, occur in the extreme south-east where the land is rising to the High Weald beyond the county limits. The reasons for these contrasts have already been mentioned. The soils of the Tunbridge Wells and Ashdown Sands are everywhere strongly leached, except where clay and silt intercalations weather into soils of a loamier character. Where the somewhat calcareous Wadhurst Clay comes to the surface in small areas there are heavier, poorly-drained soils, but since this formation also contains beds of sandstone and ironstone it is probable that here, within the small compass of this south-eastern district, there exist conditions for plant growth that are as variable as any within the county.

Further reading on Surrey's geology

Clayton, K. M. (ed.) (1964). *Guide to London Excursions*. 20th International Geographical Congress.

Dewey, H. and Bromehead, C. E. N. (1915). *The Country around Windsor and Chertsey*. Memoir of the Geological Survey, Sheet 269.

Dewey, H. and Bromehead, C. E. N. (1921). *The Country around South London*. Memoir of the Geological Survey, Sheet 270.

Dines, H. G. and Edmunds, F. H. (1929). *The Country around Aldershot and Guildford*. Memoir of the Geological Survey, Sheet 285.

Dines, H. G. and Edmunds, F. H. (1933). *The Country around Reigate and Dorking*. Memoir of the Geological Survey, Sheet 286.

Hall, A. D. and Russell, E. J. (1911). *A Report on the Agriculture and Soils of Kent, Surrey and Sussex*. Board of Agriculture, Misc. Publications No 11.

Thurrell, R. G., Worssam, B. C. and Edmunds, E. A. (1968). *The Country around Haslemere*. Memoir of the Geological Survey, Sheet 301.

Wooldridge, S. W. and Goldring, F. (1953). *The Weald*.

Wooldridge, S. W. and Hutchings, G. E. (1957). *London's Countryside*.

Chapter Three

The Study of the Flora

Surrey has been fortunate in the long series of able botanists who have recorded its flowers. The north of the county attracted attention at a very early date. This began in the fifteenth and sixteenth centuries when homes and gardens at Southwark and Lambeth, and estates at Mortlake, Kew and Richmond, were accessible by water while roads were impassable.

It started with **William Turner** (c1508–68), 'Father of English Botany' who, on return from exile for his religious views, became in 1547 physician to the Duke of Somerset at Syon House, and lived at Kew in a house belonging to his patron. Turner is credited with fifteen first records of Surrey plants in his *The Names of Herbs* (1548), including Yellow Loosestrife, *Lysimachia vulgaris*, from 'the Temes syde beside Shene', and Camomile, *Chamaemelum nobilis*, which 'groweth on Rychmund grene'. **John Gerarde** (1545–1612) gave 27 plants from Surrey in his famous *Herball* (1597), but some of these were errors. He seems to have botanised very little in the county, and the same is true of **Thomas Johnson** (1600–44) who added 9 species from Surrey in his greatly improved issues of the *Herball* (1633, 1636). **John Parkinson** (1567–1650) included 18 Surrey records in his *Theatrum Botanicum* (1640) and **William How** a few more in his *Phytologia Britannica* (1650), but the first really substantial list of Surrey plants was that of **Christopher Merrett** (1614–95) in his *Pinax Rerum Naturalium*. Here there are over a hundred printed records, and a further fifteen manuscript entries in the copy at the British Museum. Green Hound's-tongue, *Cynoglossum germanicum*, (plate 18), still grows in Norbury Park, near Leatherhead, and Burnet Rose, *Rosa pimpinellifolia*, on Barnes Common. The *Pinax*, of which nearly all the copies of the first print were destroyed in the Great Fire of London, was a compilation.

Merrett and other early writers employed as a paid collector **Thomas Willisel** (died 1675), an old Cromwellian soldier with a sharp eye for plants. His movements are not easy to trace, as his discoveries were often reported by his employers without acknowledgement, but he is known to have collected in Surrey and no doubt some of our first records were really Willisel's. **John Ray** (1627–1705), considered by some the greatest of English naturalists, does cite Willisel. Ray visited Charles Howard, who owned the manor at 'Darking' (now Dorking), and while he was there Willisel showed him Toothwort, *Lathraea squamaria*, in a shady lane, which may well be West Humble Lane where it still grows. Ray also saw Large Bitter Cress, *Cardamine amara*, 'in boggy and watery places near Dorking'. **John Evelyn** (1620–1706) was a contemporary famous for his diary and his work on trees and arboriculture. His family home, Wotton House, is now used by the Fire Service, but there are still some fine trees in the grounds, and his influence on their planting by other Surrey landowners was considerable. **John Aubrey** (1626–97), the eccentric and much-loved antiquary, included interesting and sometimes quaint plant records in his five-volumed *Natural History and Antiquities of Surrey* (1718–19).

Although not primarily botanists, John Evelyn and John Aubrey contributed specimens to the great herbarium amassed by **Charles Du Bois** (1656–1740), London merchant and treasurer of the East India Company, who lived at Mitcham. This included a good many specimens from Surrey, and especially from the Mitcham district (Druce, 1928). **John Martyn** (1699–1768) lived at Streatham from 1752 to 1761, while he was Professor of Botany at Cambridge, and made a number of interesting records. **William Curtis** (1746–99), an apothecary, closely associated with Chelsea Physic Garden, author of *Flora Londinensis* and founder of the *Botanical Magazine* (which is still being published), was the first to record and illustrate many plants from St George's Fields, Battersea Fields, and other places round London. A nurseryman, **James Dickson** (1738–1822), who died at Broad Green near Croydon, was the first to record *Cerastium pumilum* and *Minuartia hybrida*. He localised them both from 'near Croydon', but it is quite likely that he found them on Banstead Downs, where they still grow together.

From the early part of the nineteenth century the roads improved, the use of the Thames as a highway became relatively less important, and there was less emphasis on the extreme north of the county. The famous economist **John Stuart Mill** (1806–73) walked extensively in Surrey in the first half of his life and wrote his own 'Flora of Surrey' (*J. Bot.*, **42**, 297, 1904) and his records were freely used by Brewer. He was the first to report Orange Balsam, *Impatiens capensis*, which has since spread along rivers and canals to much of southern England, and his Medlar, *Mespilus germanica*, near Redhill has delighted several generations of botanists since. One of his walking companions was **William Pamplin** (1806–99), who became a bookseller and publisher and generously supported the study of field botany by continuing a useful periodical, the *Phytologist*, at a loss. His *Catalogue of the rarer plants of Battersea and Clapham* appeared when he was twenty-one. **Hewett Cottrell Watson** (1804–81) was the first botanist with a highly critical and challenging knowledge of British plants to reside in Surrey. He moved to Thames Ditton in 1835 and remained there for the rest of his life, finding time from his primary interest in phytogeography to collect and study, write on and distribute specimens of plants from many parts of Surrey. It is on Watson's division of Britain into 'vice-counties' for the recording of plant distribution that the boundaries of Surrey used in this *Flora* are based.

Daniel Cooper (1817–42) was a medical student when at the age of nineteen he published his *Flora Metropolitana* (1836) with forty-five pages listing in rather small type the plants of individual Surrey habitats. These extended as far out as Dorking, Guildford and Godalming and much of it was based on his own fieldwork—he claims to have visited Wimbledon Common thirty times in the previous three years. The book—a remarkable achievement for someone of his age—was reissued with a supplement in 1837. This dynamic young man took a leading part in the formation of the Botanical Society of London in 1836, and was the first Secretary, but in 1842 he joined the army as an assistant surgeon, and died the same year.

George Luxford (1807–54), a printer, produced his *Flora of Reigate* in 1838. This was very much a personal record of his own fieldwork, and the area covered was stretched to include records rather far from Reigate. Many of them, such as Marsh Helleborine, *Epipactis palustris*, from Box Hill where it still grows, are of considerable interest. **Alexander Irvine** (1793–1873), another walking companion of John Stuart Mill, lived for a time in Surrey when he was a schoolmaster at Albury, and later at Guildford. His records appear in many publications, but especially in his misleadingly entitled *London Flora* of 1838.

From this time onwards Surrey botanists became much more numerous, and only a few can be mentioned. **John Drew Salmon** (1802–59) of Thetford moved to Godalming in 1843 and, after producing a few papers on local plants, started to collect records for a *Flora* of the whole county. It was to be based on a geological division of the area (*Phyt.*, **4/1**, 558–66, 1851), but he died before his manuscript was completed. **James Alexander Brewer** (1818–86) was the first Secretary of the Holmesdale Club (later the Holmesdale Natural

History Club) founded in 1857 with its headquarters at Reigate. In the previous year he had produced the *New Flora of Reigate*, an advance on Luxford's book but a little lacking in personal observations. It fell on Brewer to complete Salmon's manuscript and edit the first *Flora of Surrey* for his Club. It appeared in 1863 and was one of the best county *Floras* published to date—clear with a very high standard of accuracy; many of the plants recorded can still be found in the localities he gives.

Soon after this **Arthur Bennett** (1843–1929) started to publish new records of Surrey plants in the *Proceedings of the Holmesdale Natural History Club*. Bennett lived all his life in Croydon where he had a business as builder and decorator started by his father, interrupted by several financial crises. Blessed with an incredibly good memory, he acquired a first-class knowledge of European botanical literature. He was accepted as a leading authority on the British flora and the authority on the Potamogetons of the world. Sadly, he spoilt his reputation by going on too long. I only knew him towards the end of his life when he had become confused and, although his memory of past events was wonderfully clear (and, I believe, accurate), his determinations of plants were by then unreliable. On one occasion when I called on him, he sent me a postcard immediately afterwards complaining that it was a long time since I last came to see him! (*J. Bot.*, **68,** 217–21, 1929). He handed over responsibility for a new *Flora* to W. H. Beeby in 1884.

In 1876 appeared the *New London Flora* by **Dr Eyre de Crespigny** (1821–95). This listed plants under localities in the same way as the earlier *Flora Metropolitana*, and both works are extremely valuable as records of the changes which have taken place in the habitats described. **William Whitwell** (1839–1920), on the staff of the Inland Revenue, was born in Surrey but spent much of his life elsewhere. His best find was Dwarf Milkwort, *Polygala amara*, which he discovered near Caterham in 1888. Houses have been built on the site, the only one in the county. At about this time **William Hadden Beeby** (1849–1910), who worked in a bank in London, was exceedingly active in field work, and intent on preparing a new *Flora*. He was an exceptionally accurate botanist, who made few mistakes, and it is unfortunate for us that by 1907 he was so absorbed in his work on the flora of Shetland that he handed his manuscript over to C. E. Salmon.

In 1885 **Edward Shearburn Marshall** (1858–1919) became curate at Witley, rising to vicar in the adjoining parish of Milford in 1890, where he remained until 1900. Marshall was another leading British field botanist with a good eye for critical plants, and while in Surrey he added substantially to scientific knowledge of the flora and especially of the genera *Epilobium, Hieracium* and *Rubus* (*J. Bot.*, **58**, 1–11, 1920). He helped **Stephen Troyte Dunn** (1869–1938) with his *Flora of South-West Surrey* (1893), a disappointing work overloaded with abbreviations. Dunn is better known as a botanist for his work on British alien plants, on India and the Far East.

During this period **Charles Edgar Salmon** (1872–1930), an architect of Reigate, was becoming known as a botanist. A meticulously careful worker with an eye to detail, and keenly interested in the smaller difficult groups, specialising in *Limonium*, he was a scholar through and through and his interests spilled out in the pages of his *Flora of Surrey*, published posthumously in 1931. As I remember Salmon, he was quiet, extremely shy and self-effacing, a typical Quaker whose main object was to get on with his botanical work without attracting attention and with great kindliness (*J. Bot.*, **68**, 50–3, 1930). **Charles Edward Britton** (1872–1944) had some of the same characteristics. Employed by the Post Office on work which involved night shifts, he lived at New Malden until he retired and in 1932 moved to Warlingham. I first met him in 1922 when he was spending most of his leisure in solitary field work, and making a deeper and deeper study of the small critical groups, especially *Centaurea* and then *Melampyrum*. His critical work suffered from lack of scientific training and of wider experience, and he found it difficult to accept the views of younger botanists, but he made an outstanding contribution to our records (*Rep. BEC*, **12**, 639–41, 1946). **Joanna Charlotte Davy**, later Lady Davy, (1865–1955) moved to Pyrford in 1909 and to West Byfleet in 1922; at both homes she entertained botanists freely, taking them to see Surrey plants and gaining their help in exchange. Her wonderfully keen eye for plants, and delight in training young botanists, had a part in many important finds— Miss R. M. Cardew's detection of *Holosteum umbellatum* in 1906; the rediscovery of *Carex filiformis* by Miss A. A. M. Tulk and Miss Gertrude Bacon (later Mrs Foggitt) in 1916, and the finding of *Luzula pallescens* on her own lawn in 1910. A remarkable woman, her influence was immense (*Proc. BSBI*, **2**, 190–2, 1956).

John Fraser (1854–1935) was a quiet, retiring, kindly Scot employed by Lord Avebury for his work on seedlings; he became editor of horticultural periodicals and books, and spent most of his Sundays in solitary botanising. He specialised on mints and willows and is the only man who has taken me botanising wearing a top hat, his normal headgear. He lived in one room at Kew, which served for the storage of his books and large herbarium, as well as his living requirements, and yet he found space for a spittoon beside the table where he worked. His botanical knowledge was very wide (*J. Bot.*, **73**, 81–3, 1935). **George Simonds Boulger** (1853–1922) had the rare ability to write for the masses without sacrifice of scientific accuracy and he is best known for his popular works. However, his main interest was research into the history of British botany and he wrote the excellent 'History of Botany in Surrey' for Salmon's *Flora*. This was set up in type before he died in 1922, and gives much fuller detail than is possible here about some of the early botanists. **William Harrison Pearsall** (1860–1936) was a schoolmaster in Dalton-in-Furness, specialising on aquatic plants, who retired to Kent in 1925. Unfortunately Pearsall was not familiar with the plants of southern England and many of his determinations are now regarded as errors. It was unfair to invite him to complete Salmon's *Flora* for the press, but he did his best in the difficult circumstances (*J. Bot.*, **74**, 352–3, 1936).

There are still a few recent workers to be mentioned, all very well known to me. **Randolph William Robbins** (1871–1941) of Limpsfield retired in 1935 as Renter Warden of Guy's Hospital. He was a founder-member of the London Natural History Society and took a very active part in their botanical work, contributing many records from south-east Surrey (*LN*, **21**, 2–11, 1942). **Edmund Browne Bishop** (1864–1947) was another of the keen LNHS botanists. A civil servant, he retired to Godalming about 1924 and recorded the plants of that district, specialising on roses. Some of the rare ones he found have not been noticed recently, but only a few of our workers would expect to recognise them in the field (*LN*, **27**, 112–13, 1949). **Arthur Langford Still** (1872–1944), an analytical chemist, was born at Addington and lived at Wallington. He followed Fraser as our authority on mints and was also keenly interested in sedges, and walked much of the county in search of them (*Proc. BSBI*, **12**, 651–3, 1946). On his death, **Rex Alan Henry Graham** (né Knowling) (1915–58) took over the study of mints and, although he did not live in the county until shortly before he died, he spent a lot of time botanising in Surrey after the war. (*Proc. BSBI*, **4**, 505–7, 1962).

Gerald Mortimer Ash (1900–59), known to his friends as 'Farmer Ash', farmed Lower Birtley Farm, Witley, and had an intimate knowledge of the local flora. He took up the study of willowherbs on Marshall's old stamping ground, and in 1931 discovered as new to Britain *Epilobium adenocaulon*, an alien from N. America which is now one of our most common willowherbs in Surrey and widespread in England (*Proc. BSBI*, **4,** 106–7, 1960). **Charles Baynard Tahourdin** (d. 1942) was another specialist—this time on orchids. He lived at Wallington and his kindness and encouragement, backed with sound logical advice arising from long experience as a solicitor, put several young botanists on the right road. By reporting individual plants or habitats year after year he made a useful contribution to our knowledge of the orchids of Surrey. **William Charles Richard Watson** (1885–1934) became for a time the accepted authority on the brambles of south-east England (*Proc. BSBI*, **1,** 556–61, 1955). He was helped by **Charles Avery** (1880–1960) who searched the Surrey Commons for *Rubi* (*Proc. BSBI*, **4,** 351, 1961) and by **John Ezra Woodhead** (1883–1967) who accompanied him when he revisited the bushes he regarded as of special interest (*Proc. BSBI*, **7,** 495–6, 1968).

   Herbert William Pugsley (1868–1947) lived in the same house at Wimbledon for over half a century and wrote several papers on the flora of Wimbledon Common and Wimbledon Park, but he is best known for his work on fumitories, eyebrights and hawkweeds. Of these difficult critical groups, in turn he wrote accounts which were in each case a great advance on anything already available, and in general they have stood the test of time. He also worked on orchids and Narcissi and many small critical groups. Meticulously careful and cautious, and generous with his knowledge, Pugsley was one of the greatest of Surrey botanists and we owe him a great deal (*Watsonia*, **1,** 124–30, 1949). **Iolo Aneurin Williams** (1890–1962), who lived at Kew, described *Bromus britannicus* as a grass new to science in 1929 (*J. Bot.*, **57,** 67–9, 1929) with seven Surrey localities. His name has had to be displaced by *B. lepidus*, Holmb., published only five years earlier. Williams made many Surrey records, but his greatest contribution to botany was the very high standard of reporting of botanical news in *The Times* during the quarter of a century he was on their staff (*Proc. BSBI*, **4,** 507–8, 1962).

   William Bartram Turrill (1890–1961) became Keeper of the Herbarium at the Royal Botanic Garden, Kew. His work on experimental taxonomy had a great influence on the study of British plants, and his experimental work on *Centaurea* had particular relevance to Surrey. As one of his students at Chelsea Polytechnic, where he lectured on ecology and genetics, I joined his parties to Richmond Park, where with some of his colleagues he was making detailed studies of the vegetation, and to Ham (*Proc. BSBI*, **5,** 194–6, 1963). **Noel Yvri Sandwith** (1901–65) was also on the staff of Kew Herbarium and lived at Kew. He did a lot of field work in the county, especially during the war years, but published very little. His knowledge of British plants was profound and placed freely at the disposal of his friends. Noel Sandwith was an outstanding example of a botanist who shunned publicity and yet made a far greater contribution than most of those better known. Much of his work on Surrey appears under the names of other botanists (*Proc. BSBI*, **6,** 418–22, 1967). **Edmund Frederick Warburg** (1908–66), a brilliant taxonomist, and one of the authors of the *Flora of the British Isles* (1952), the standard work used in the collection of records for this book, was the son of Sir Otto Warburg of Headley (*Proc. BSBI*, **6,** 207–8, 1966 and **7,** 67–9, 1967).

Finally, **Donald Peter Young** (1917–72), who was educated and lived in Croydon, became one of the most able amateur botanists of his day. He specialised on *Epipactis* and *Oxalis*, on which he became an acknowledged authority, and the accounts in this *Flora* are based on his determinations. He organised the recording in the Croydon area for the *Atlas of the British Flora* and, when he became Recorder and later Editor for the Surrey *Flora*, he carried out the work with the same care as he applied to everything he undertook. The editing of the data for the distribution maps was almost entirely Donald Young's work and, but for his sad illness followed by an early death, he would have written the text as well.

Transport

This review of the study of the flora of Surrey reflects not only the tremendous advance in scientific knowledge of British flowers, but also the extent to which it was dependent on transport. Whenever William Turner moved into Surrey away from the Thames, he would have been faced with discomforts and slow progress difficult for us to imagine. Donald Young, who favoured sports cars, had every corner of the county open to him with little loss of time. In between the first and last worker mentioned there is a steady sequence through the long period when horses and carriages were used on main roads, which gradually improved to turnpike standards, and horseback elsewhere, to the more numerous records from places accessible by railway. Brewer's *Flora* of 1863 is very largely based on places opened up by the railways shown on the map issued with it, and especially on the London and South-Western Railway and its branches. Salmon's *Flora* of 1931 had the advantage of an increased railway network with the gaps largely filled in with good roads and bus services. The workers for the present *Flora* had the use of good, though sometimes congested, roads to cover every part of Surrey.

This has made possible the much more even coverage of the county and the inclusion of realistic distribution maps.

Putting the Flora on Record

It is interesting to see how the work of the above-mentioned botanists, who were the most active and often famous contributors to our records, was collected for publication. This is the third book covering the flora of the whole of Surrey; each had more than one editor before publication, and all three differed in the way they divided up the county and in other respects.

The first was the *Flora of Surrey* published for the Holmesdale Natural History Club by J. A. Brewer in 1863. John Stuart Mill and H. C. Watson were generous contributors of records, and the *Flora* had a high standard of reliability well above the general level of *Floras* of the time. The records were arranged under nine divisions devised by J. D. Salmon in 1852. These were based primarily on soil; it was a good attempt to split up the county into natural divisions and revealed many soil preferences.

The second was C. E. Salmon's *Flora of Surrey* which appeared in 1931 after a long period of gestation. This commenced in 1879 when Arthur Bennett announced that he was preparing a new *Flora*. In March 1884 he handed the project over to W. H. Beeby, who described eight districts based on river basins; these were increased to ten by 1902 when he contributed the account of the botany of Surrey to the *Victoria County History*. This was a retrograde step. River basins were being used at the time as units for the study of large areas abroad, where mountain chains formed very real barriers to the movement of plants, but in an area the size of Surrey their boundaries are difficult to trace precisely in the field and they obscured the important influence of soils in explaining distribution.

On 28 November 1907, under pressure from E. S. Marshall and Arthur Bennett, Salmon wrote to Beeby agreeing to take over the preparation of the new *Flora*, and received the manuscript which was 'in an advanced state'. When Salmon died on New Year's Day, 1930, about two-thirds of the *Flora* was in type, with entries of extravagant length. The remainder was greatly shortened and heavily edited by W. H. Pearsall and published in 1931.

The first proofs were dated 1920, and at the time of Salmon's death the text to page 550 had been printed and the type broken up. A further forty-five pages were printed as prepared by Salmon; the remainder contains many misleading entries as Pearsall, though he did his best, was a Lakeland botanist with little experience of southern counties. I have made great efforts to trace Salmon's manuscript, but have failed to establish that it was returned to Mrs Salmon after publication. The book is now fifty years out of date and great changes have taken place in Surrey in that time.

The present work arose out of the Botanical Society of the British Isles' Map Scheme, of which I was Chairman. This set out to map the distribution of the flowering plants and ferns by 10km squares over the whole of the British Isles, and culminated in the publication of the *Atlas* in 1962 and the *Supplement* in 1968. The logical step after completing the general maps on the 10km² scale was to fill in the detail by mapping counties on smaller squares. On 26 January 1957 I convened, as BSBI Local Secretary and Recorder for Surrey, a meeting attended by thirty-eight members resident in the county. The purpose was to make arrangements for the collection of the records still required for the Maps Scheme, and to organise the collection of records and work on a new *Flora of Surrey*. A committee, to be called the Surrey Flora Committee, was appointed with myself as Chairman, Mrs B. Welch as Secretary, and O. V. Polunin, E. C. Wallace and D. P. Young as members. Mrs J. E. Smith, J. C. Gardiner and W. E. Warren were co-opted in 1962; and, at subsequent Annual General Meetings, R. Clarke was added in 1966, C. P. Petch in 1971 and C. T. Prime in 1972. With the exception of O. V. Polunin, who resigned in 1966, and D. P. Young who was appointed Recorder in 1963 and Editor in 1967, and died in 1972, there were no other changes in membership.

From the beginning the committee took the view that the *Flora* should be illustrated with as many maps as it was possible to publish on a sufficiently large scale for interpretation. A Field Card was printed listing 445 selected species and to this a further forty-five were added later. For the rarer species and those regarded as critical, members were asked to send in Individual Record Cards giving more information. Volunteers were recruited to take on individual tetrads (groups of four 1km squares) or groups of tetrads.

Mrs Welch acted as Secretary and Recorder until January 1963 when to our great regret she resigned on moving out of the county. It would be impossible to praise too highly the care, patience and accuracy she devoted to the plodding spadework of the early stages. Mrs J. E. Smith then became Secretary and Dr Young, Recorder. In 1967, as the work of collecting records dropped off, Dr Young was also appointed Editor, on the understanding that finishing off some of his work on *Oxalis* and *Epipactis* would at first have priority over the *Flora*. When he died five years later, after a long illness during which he put up a heroic struggle to work against failing strength, he had completed the transfer of the records on to the working copies of the maps. I was appointed Editor and Recorder in May 1972, and handed over the Chairmanship to Dr Prime at the AGM in April 1973.

The maps, which were prepared for the printer by Dr James Stevens are the work of a great many people, co-ordinated mainly by Mrs Welch and Dr Young. For the rest of the work I accept responsibility. My aim has been to produce an account of the present flora of the county based primarily on the records collected since 1950; but, since the flora is constantly changing, it is sometimes only possible to understand the present by reference to the past. In such cases, and for the sake of completeness in the case of the rare species, I have not hesitated to include some older records. It is hoped too that this will form a handbook for conservation, an aspect to which the Surrey Flora Committee, in association with the Surrey Naturalists' Trust, now devotes most of its effort.

Chapter Four

The Changing Flora

Surrey was a 'late developer' compared with other southern counties. By the time our first botanical records were printed by Turner and Gerard in the sixteenth century there were still large areas which showed little evidence of the hand of man.

The area known to the early botanists was the north of the county, but even here there were still large marshes and tracts of unculti- vated ground. The fine houses of London merchants, or the court officials concerned with the royal residences at Nonsuch, Richmond and across the Thames at Hampton Court, had started Surrey on the path of suburbanisation, and it was from these houses and palaces that much of the early botanical work was done. Farther out, the market towns of Croydon, Kingston-on-Thames and Darking (sic) were centres of agricultural and other activity. Towards the south-west the cloth industries of Guildford, Farnham and Godal- ming used the wool of large numbers of sheep which grazed on the chalk and other well-drained soils, and the woad grown on the hills, and the Fuller's Earth dug out at Nutfield. On the Greensand and Bagshot Beds there were great stretches of acid heathland and bogs; on the clay, and especially the great areas of Wealden Clay in the south, there was dense forest on deep sticky soils, with the few roads impassable in winter and often in summer also. The woods were being felled to make charcoal for gunpowder, and to provide fuel for the iron-smelting industry with its associated hammer ponds, and more locally for glassmaking. These activities were making some impact on the forest, but the roads remained bad for another 300 years—read what William Cobbett had to say about them in the 1820s in his *Rural Rides*. This was the background against which the evolution of our present flora has to be considered.

The main factor has been the development of transport. At first this was mainly by water, so that the main residential area was by the Thames, which provided a reliable highway until roads were improved. By 1653 the canalisation of the Wey from the Thames to Guildford had been completed, so that timber, grain, iron, gun- powder and market-garden produce were moved cheaply and the Guildford area developed. The middle decades of the eighteenth century saw the beginning of improved roads under Turnpike Trusts. Improved roads led to rapid changes in agriculture and in the areas round the market towns—changes which were accelerated with the construction of canals and railways. New plants moved along these routes and had a big influence on the present flora.

Tracks and Roads

Surrey's only ancient track of importance was the one along, or below, the chalk escarpment which extended from the Kent border to Farnham and became known as the Pilgrim's Way. The trampling and grazing on this ancient highway no doubt had its influence, but no evidence of this can be traced today. The other early roads were not of the type common in many flatter counties, where wheeled vehicles and riders trying to avoid the ruts made the road wider and wider, so that when hedges were planted to contain it, and it was later given a metalled surface, broad verges were left down each side. Most of the county's verges are narrow, and many were eliminated years ago when the asphalt was widened to cope with modern traffic.

Of greater botanical interest are some of the sunken lanes, formed when they sank deeper as rain washed the surface downhill. These have well-drained banks on which some of our rarest hawkweeds thrive on the Greensand. Some roads still have attractive shows of flowers or rare species like the orchid *Epipactis phyllanthes* on their verges, but the narrowness increases the dangers from road widening, machine cutting and toxic sprays. Fortunately the Surrey Naturalists' Trust keeps a record of botanically important stretches of roadsides, and the Surrey County Council cooperates in arranging for their roadmen to avoid them. In a few places, on chalk or clay-with-flints, roads are flanked with belts of beeches or other trees planted by landowners. These harbour orchids, rare grasses and other plants of interest.

While the wide hard-shoulder of the motorway often forms a continuous path for plants to spread across the county, this development is as yet too new to show any recorded results. Motorway construction has destroyed areas surprisingly large for its length, and further stretches are under way or planned.

Canals

The massive construction of canals in Britain towards the end of the eighteenth and beginning of the nineteenth centuries created a new habitat for water plants and one that was flanked by a towpath convenient for close observation. Very slow-moving water, constantly disturbed by barges, and with excessive growth controlled by management, provided conditions between those of ponds and rivers, but nearer the former. The alkalinity varied according to the source of the canal water, and in our Surrey canals was usually well on the basic side. These waterways provided, with the connecting rivers, a complicated network of continuous and exceptionally favourable aquatic habitats by which water plants could spread quickly from one place to another. Britain's canal system at its peak provided a continuous freshwater route from Lancashire and Yorkshire to the English Channel, and from Wales to the Wash and the Thames Estuary. Native plants spread to canals from ponds, lakes and rivers; alien plants spread along them from points of introduction.

In Surrey the construction of the Basingstoke Canal started in 1788 and was completed to the Hampshire border by 1796, giving Surrey the finest habitat for aquatics it had ever had. The canal entered the county near Aldershot, turned north through Ash Vale to Frimley Green, and then east through Brookwood, Woking and Byfleet to join the Wey Navigation to Weybridge and the Thames. From the first the waterway had a difficult commercial history, and periods of activity alternated with periods of neglect. Light use of a canal by horse-drawn barges is ideal for plants, and the period when I knew it well, from 1925 to 1940, was one of the best. Barges were passing up and down from Aldershot to the Thames; excessive weed was cut; the pounds were dredged in turn to remove accumulated silt or, at longer intervals, drained and dug out. Thus there were always sections in various stages of recovery offering a wide range of conditions, with just enough barge traffic to keep the channel clear.

Pondweeds thrived best in the clear water for about two years after dredging or cleaning—*Potamogeton natans, lucens, gramineus* (very abundant), *alpinus, perfoliatus, obtusifolius,* and *pectinatus* and *Groenlandia densa* were amongst those found regularly. *Eleocharis acicularis* and Waterwort, *Elatine hexandra,* were rather local on the bottom in shallow water, and flowered only when the water-level fell. On the banks Cut-grass, *Leersia oryzoides,* grew in several places and a rare Horsetail, *Equisetum x litorale,* was quite plentiful near Sheerwater and elsewhere. The length of the canal in Surrey was ample for two full days of botanising, with a wonderful show of flowers and never a dull moment.

But during this time it was gradually falling into neglect, the revenue from traffic being insufficient to meet the heavy cost of maintenance. When failure to carry out major repairs stopped the through flow of water, the pounds became stagnant; some dried up, others became choked with coarse plants (plates 6, 7). The vegetation in many pounds is now that of a marsh rather than a canal, and parts have attracted the dumping of refuse. Nevertheless, interesting species are still to be found and especially about Ash Vale and Mychett Lake. Restoration of the through flow of water would soon restore the full botanical interest.

The important influence of the Basingstoke Canal on the distribution of aquatics is clearly shown on the maps. Good examples include *Rorippa amphibia*, M60,* *Myosoton aquaticum*, M78, and *Hydrocharis morsus-ranae* M372. Fringed Waterlily, *Nymphoides peltata* M246, probably an introduction in the Thames, has spread along the canal, as has the Water Fern, *Azolla filiculoides*, M19, from the places where they were introduced. *Eleocharis acicularis*, M431, and *Alisma lanceolatum*, M369, are more plentiful along the canal than elsewhere. The towpath also provided a continuous track for the distribution of plants whose seeds pass through the horse or adhere to its hooves. Forty years ago I noticed *Juncus tenuis*, M387, which has glutinous seeds, spreading along the towpath near Byfleet— when barge traffic ceased, the plant no longer spread.

The Wey and Arun Junction Canal, opened in 1816, was intended for through barge traffic from London via the Thames, Wey and Arun to the English Channel. The canal left the canalised Wey at Shalford and can still be traced through Bramley, Run Common and Laker's Green to the Sussex boundary near Sidney Wood. Traffic ceased in 1871, but for some time after that it was a well-known station for rare aquatics, such as the two hybrids of *Potamogeton lucens*, found in 1884 and 1886, and Cut-grass, *Leersia oryzoides*. Much of the course is now dry, but some stretches contain water and the influence of the canal on the distribution of water plants is still important. For example, Great Yellowcress, *Rorippa amphibia*, M60, Water Chickweed, *Myosoton aquaticum*, M78, *Myriophyllum alterniflorum* M182, and *Glyceria maxima*, M467, all show a string of localities along the route of the old canal, suggesting that they came from the river Wey when the water was continuous.

The Grand Surrey Canal ran from Rotherhithe to Croydon. It was opened in 1809, but the part near Croydon had a very short life; it was taken over by the London & Croydon Railway in 1836 and closed. By then it had built up a good flora and Daniel Cooper

---

*'M' numbers refer to distribution maps, pages 405–467.

published a list of the species growing by the canal between the Dartmouth Arms and Norwood in 1836 (*Fl. Metrop.*, 18–19). This includes our only record for *Sonchus palustris*, which was probably brought from the North Kent marshes by barges. All the canal south of Walworth and Peckham has been drained, apart from a small bit in a recreation ground, but the northern part, the Surrey Canal near Bermondsey, produced several recent records for this *Flora*.

Railways

The construction of railways destroyed many old habitats and created some new ones. They altered drainage, provided a route for the movement of plants as well as humans, and created new well-drained slopes—some facing the sun, others in cuttings which were shaded, cool and damp. Interesting examples arose out of the first stage of the London and Southampton Railway, soon renamed the London & South-Western Railway, which was opened from Nine Elms (near Vauxhall) to Woking Common, a run of 23 miles, in 1838. The route ran by the remains of Battersea Fields, across Wandsworth Common, along the south side of Lord Spencer's Park at Wimbledon, across Ditton Marsh and ended on Woking Heath. These were all good places for plants. Ballast for the track was dug on Wandsworth Common, and a ballast pit near the bridge where the Tooting Road crossed the line (Nat. Grid. 267743 approx.) produced a most remarkable flora, which was described at length by K. M'Ennes (*Phytol.*, **4**, 697, 1852). This included plants associated with sphagnum bogs, such as Marsh Clubmoss, *Lycopodiella inundata*, Round-leaved Sundew, *Drosera rotundifolia*, Lesser Skullcap, *Scutellaria minor*, Marsh Pennywort, *Hydrocotyle vulgare*, and Marsh Willowherb, *Epilobium palustre*. In the open water at the bottom of the pits there was Starfruit, *Damasonium alisma*, Fringed Waterlily, *Nymphoides peltata*, and other aquatics. On the dry exposed gravel, Sandspurrey, *Spergularia rubra*, Fenugreek, *Trifolium ornithopodioides*, Sticky Groundsel, *Senecio viscosus*, and many other species not otherwise known in the district. M'Ennes suggested that long-buried seeds had been dug up and, as evidence for this, pointed out that many strange plants had appeared on the Richmond line where some of the same ballast had been used. In fact, some of the seeds and fruits had been brought by air currents, others by horses and carts working in the pits, and by birds. It was a good example of the way interesting plants will arrive if suitable conditions are provided and of how the railways produced new botanical interest.

On the other hand this first bit of line provided many examples of the losses which were to follow widespread railway construction. One was fairly fully reported. For their first field meeting, members of the recently formed Botanical Society of London were quick to take advantage of the new railway soon after it opened on 21 May 1838. They travelled to Woking Common, on the site of the present Woking Station, which was then in the middle of extensive heathland. The interesting plants found included two Clubmosses, *Huperzia selago*, no longer known in the county, and *Lycopodium clavatum*, as Daniel Cooper, the Secretary, reported at their meeting on 3 August 1838 (*Proc. Bot. Soc. Lond.*, **1**, 74–6, 1839). This heathland flora was soon destroyed by the buildings which sprang up round the railway station—an early step in the loss of so much of the countryside to the homes of London commuters.

The well-drained slopes of railway cuttings, and their embankments of varying exposures, provide habitats which have proved valuable as sanctuaries. The fine display of Primroses on the railway banks near Surbiton station is an example. *Geranium pyrenaicum*, M99, and Ox-eye Daisies, *Leucanthemum vulgare*, are especially common in

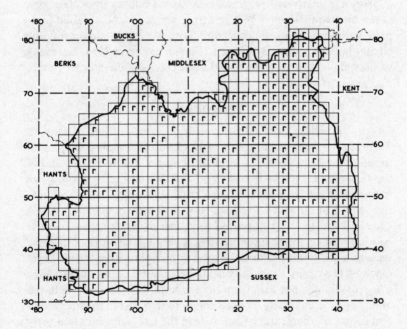

Fig 5 Railways: Tetrads with railways (1974)

such places, while Broad-leaved Everlasting Pea, *Lathyrus latifolius*, is one of the aliens very common on railway banks in Surrey. British Rail's Southern Region keeps the ballasted track on the electrified third-rail system almost free of weeds by using herbicides, but even so it has served as a continuous road for the spread of Oxford Ragwort, *Senecio squalidus*, M330, Sticky Groundsel, *S. viscosus*, M332, and a grass *Vulpia myuros*, M474. Small ferns find very suitable habitats on the north-facing brickwork of railway platforms, cuttings, viaducts and bridges. The commonest are Black Spleenwort, *Asplenium adiantum-nigrum*, Maidenhair Spleenwort, *A. trichomanes*, Wall-rue, *A. ruta-muraria*, Male Fern, *Dryopteris filix-mas*, and young plants of Broad Buckler-fern, *Dryopteris austriaca*. Less common are Rustyback, *Asplenium ceterach*, and Hart's-tongue, *A. scolopendrium*, while Maidenhair Fern, *Adiantum capillus-veneris*, *Cystopteris fragilis*, and *Gymnocarpium robertianum* occur rarely. The mortar between the bricks, permanently damp and shaded from the sun, provides ideal conditions for germination of the spores, and railway traffic probably plays a part in distributing the very light airborne spores.

Surrey has relatively few stretches of closed railway track. Our best is the one from Bramley to Cranleigh, which is now a public footpath with interesting flowers and ferns. Another, almost on the county boundary near Sydenham Hill and closed in 1954, has many woodland plants on the banks and a fine colony of Greater Horsetail, *Equisetum telmateia*.

Agriculture

Agriculture has never affected as high a proportion of land in Surrey as is usual in the south of England. The great commons of the Bagshot Beds, such as Chobham, Bagshot and Pirbright, and of the Greensand, such as Frensham, Thursley and the Hurtwood, are examples of the large areas of acid heathland which defied reclamation. Much of the south was covered with dense woodland on Weald Clay and had to await improved tools and methods before it could be reclaimed on anything more than a very limited scale. For these reasons a great part of Surrey was never brought within the medieval Open Field System. Eventually fields were carved out of the Wealden Clay woodlands, but seen from high ground, such as Leith Hill, the area still appears closely wooded owing to trees left in the hedgerows and copses. Parliamentary Enclosure in the late eighteenth and early nineteenth century only affected to any serious extent the north-eastern quarter of the county, where over 10 per cent of the land was enclosed, but much of this is now covered with houses. Hence the pattern of small fields with regular outlines of hedges, composed mainly of hawthorn, is less widespread in Surrey than elsewhere.

The big agricultural changes in the county have arisen from the agricultural revolution which started about 1949 and is still in course. At present less than two-fifths of the area of Surrey is used for agriculture. On the last available figures (1971) the area of crops and grass totalled 179,300 acres; the decrease from 283,934 acres in 1897 was mainly due to losses to housing and roads, especially in the London commuter area (Min. Agric. Fish & Food: *Agricultural Statistics*).

The great change in the last twenty-five years has been the replacement of the horse by the tractor. In 1897 well over 10,000 horses were used in agriculture in Surrey; in 1939–44 an average of just over 3,000; in 1947 2,112; by 1958 the number had fallen to 459 and the Ministry of Agriculture ceased to collect the figures thereafter. Each working horse required pasture and hay meadows for its support, so some of this land could be 'improved' and used for cattle to supply more milk. When it was drained, ploughed and resown, the associated wildflowers were destroyed. Since permanent pasture carried a characteristic flora of perennial wildflowers many of these, especially the Green-winged Orchid, *Orchis morio*, have become rare. Hay meadows dropped from 51,718 acres in 1936 to 21,199 in 1971 and much of the remainder had been resown. One of the worst sufferers has been *Oenanthe silaifolia*, a Water Dropwort, which has a life cycle attuned to that of the grass. It grows with the hay; its fruit ripens as the hay ripens, and it is cut with the hay. The rate of loss of this species in a few years has been staggering.

The use of the tractor for ploughing has had an even greater effect on arable weeds. When horses were used, their numbers were restricted by the fields which had to be given up for their keep. Therefore the ploughing which could be done in the autumn was limited; some of the cornfield stubble had to be left over winter, and their weeds continued to grow and shed their seeds until caught by the frost or ploughed in early the following year. Today ploughing with tractors is much faster; a few extra machines are easily stored, or sometimes hired, so that most cornfields are ploughed within a few days of the crop being harvested. Far too often the stubble is burnt first, but in any case it is ploughed in and the cornfield weeds which have been starved of light in the standing crop have no chance to ripen fruit. This, far more than the use of herbicides, is the cause of the spectacular decrease in our cornfield weeds. Changes in the crops have also had some influence: wheat has decreased from 20,254 acres in 1897 to 13,860 in 1971 (though this shows some improvement since 1936); oats have fallen from 23,490 acres in 1897 to 4,784 in 1971, but their place has been taken by barley which has increased over twenty-fold from 1,286 to 28,292 acres. Barley is a more valuable crop than the others and hence receives more thorough cultivation, so that fewer weeds are to be found in barley fields.

These changes explain the reduction in arable weeds generally, but some of the rarer species have been reduced to the point of extinction. This is due additionally to the near freedom of agricultural seed from impurities. Corn-cockle, *Agrostemma githago*, Pheasant's Eye, *Adonis annua*, Thorowwax, *Bupleurum rotundifolium*, and Corn Goosegrass, *Galium tricornutum*, are natives of the Mediterranean regularly introduced to this country in corn seed in the past. They are no longer replenished when they die out.

Surrey has suffered very much less than many counties from the grubbing up of hedges to make larger fields, but there have been some occurrences. On the other hand, pylons carrying the electric grid or other wires across arable fields provide a useful haven for wild plants. Tractor drivers avoid going too close to the metal struts and so leave a strip of varying width round the base which is used by arable weeds. The space inside the base is seldom disturbed, and thus protects woody plants, such as blackberries, hawthorns and roses, which arrive as pips voided by birds sitting on the crossbars.

Grazing

The botanical interest of considerable areas of Surrey is the result of regular grazing in the past and depends on grazing for its maintenance. Sheep are the most efficient grazing animals and have produced the short downland turf of our chalk grassland areas which we are now in danger of losing. In Tudor times sheep, which must have been kept in very large numbers to provide wool for the cloth industry of Guildford and Godalming, were grazed on the light Greensand pastures of west Surrey as well as on the chalk. More recently sheep have been fattened rather than bred in Surrey, bought in for the purpose in the autumn and sent to market in spring. On the chalk, Banstead Downs were noted for their sweet mutton and the excellent flavour was said to be due to the fineness of the 'pile' (turf) which was made up of 'English White Clover, Trefoil and Thyme' (James & Malcolm, 1749). The number of sheep in Surrey has fallen from 128,000 in 1868 to 69,648 in 1909 (Hall & Russell, 1911) to 24,550 in 1971 (*Agric. Statistics*). Few of these are pastured on the chalk; their place there, and that of cattle which graze less closely, was taken by the rabbit, but when the rabbit population was reduced by myxomatosis grave conservation problems arose (p. 68).

Similarly cessation of grazing has changed the character of many village greens and commons. The right to graze cattle, horses, geese and sometimes sheep, and to cut furze, was valued by Surrey commoners until they became less needy, and the exercise of their rights became difficult. More houses brought more dogs, and untrained dogs chased cattle and horses. More motors brought dangers to unfenced animals, and the cost of fencing-in commons was usually prohibitive, even when the necessary consents were obtainable. So during the last forty years most of our commons have become overgrown with coarse grass, and then scrub and even trees.

We have got so used to these changes that it is difficult to appreciate their extent, but it was brought home to a few of us in a very impressive way. The London Natural History Society had made a detailed ecological survey of Limpsfield Common from 1936 to 1941. In 1964 a party of members, including three of the original committee, went back there with the vegetation map they had completed in 1939, and line transects, to see what changes had taken place. Three areas showed little change: the golf course and cricket ground, where mowing had controlled the vegetation, and Ridlands Wood, where changes would necessarily be slow. On the rest of the common the vegetation has completely altered. Ling, which covered considerable areas in 1939 and was up to 2ft tall, had almost disappeared. Gorse, then locally dominant, was drawn up under trees where it survived. Bracken had replaced some of the grass heath and ling; scrub and young trees, with many oaks about 10ft tall, the rest of the open ground (Lousley, 1965). The rate and nature of the changes are typical of what is happening to commons all over Surrey in the absence of grazing, although at Limpsfield there are conservators who exercise what control they can.

The losses on village greens are especially serious. Round the ponds the droppings from geese and ducks produced an area of very rich mud and, when the water level dropped, this was the habitat that suited Red Goosefoot, *Chenopodium rubrum*, Mudwort, *Limosella aquatica*, and a rare sedge, *Cyperus fuscus*. The well-cropped grassy area above this was the home of sheets of Camomile, *Chamaemelum nobile;* also Small Fleabane, *Pulicaria vulgaris*, and Pennyroyal, *Mentha pulegium.* Forty years ago Surrey was famous for rare mints on its village greens—Shalford, Ripley Green, Dawes Green near Leigh, Nutfield Green, Broadham Green near Oxted, are a few that spring to mind—but it seems that they have nearly all gone. Village greens no longer play the same part in the economy of those who live round them and the loss of botanical interest is immense. This is shown by the use of so many ponds for the dumping of metal or plastic refuse. On the other hand, some are converted into ornamental ponds with planted water-lilies. Aquarist's waterweeds, like *Lagarosiphon major*, are also planted. This is the sadder as so few field ponds of botanical interest remain (p. 64).

Industry

Whatever may be the case in other counties, in Surrey industry has been a good friend of the botanist in opening up important new habitats. The benefits started with the Tudor iron industry, which dug out its Hammer Ponds to become, in some cases, useful habitats for aquatics. The winning of chalk was of far greater importance for it left many pits, large and small, which became some of our best localities for calcicoles. The gradual building up of soil over bare chalk produces for a time a fascinating series of habitats often rich in orchids, though in the absence of management this is eventually lost by the growth of scrub, and ultimately of trees. Seale Chalkpit was the first reserve acquired by the Surrey Naturalists' Trust.

Sand and gravel pits are recolonised by plants much more quickly because soil, of a sort, is already there. Most of the early arrivals have windborne seeds—like willowherbs, cudweeds, willows and ragworts—or are bird-carried or brought on tyres of lorries. We have few sand-pits, the largest I recall being near Waverley Abbey; there are others near Farnham and scattered round the county. In the Thames valley there are the Ham gravel pits, near Richmond, and many large ones near Thorpe, while the Wey valley has another group of recent excavations near Send. In most cases our pits fill with water. Brick pits, excavations on clay, are sometimes interesting—such as the one near Brook, Witley, where G. M. Ash demonstrated to so many botanists a fascinating colony of willowherbs with numerous hybrids.

People

Public pressure has affected many species, especially the primrose and cowslip. In my youth these were still plentiful quite near London, but as roads and transport improved, and people had more leisure, they disappeared over an ever-widening circle. They were picked and dug up in vast quantities, so that to see a good display of primroses it is necessary to visit private ground or the remoter parts of the county. Bluebells have suffered from trampling more than picking. A wood at Norbury was blue with them until it was thrown open to the public—in three years they had disappeared. They still persist in places quite near London where trampling is not too severe. Uprooting and picking for sale by gipsies and vagrants had a considerable influence in the past. Ferns were raided over a wide area to supply the carts which I can remember in 1914 rumbling along surburban streets to sell roots to housewives. There has been a considerable recovery of ferns in Surrey over the last forty years, and several of the rarer species are less rare. Primrose roots are still offered by gipsies and street traders, and wild daffodils from the Weald are occasionally sold in street markets in the south of the county. The main threat to our ornamental wildflowers today is from the sheer pressure of private motorists at weekends.

Pressure on the rare species comes from people knowledgeable about wildflowers. The vasculum is fortunately extinct—or so nearly so that, on the rare occasions when one appears, it is used to carry the lunch rather than specimens. University students still make collections of dried plants, but few amateurs have herbaria. The main danger to rare plants comes from the enormously increased interest in them, and the large numbers of people who visit the rarities. Photographers are the worst offenders. Even with the greatest care it is difficult to photograph rare orchids, for example, without damage. In Kent the habitat of the Monkey Orchid has had to be closed to photographers and so has the Military Orchid in Buckinghamshire, while in Suffolk visitors are restricted to walking on duckboards on two days a year. In Surrey the main damage observed has been to bog plants, which are plunged below the surface by visitor's boots, and to woodland rarities not easily seen in the dim light. Trampling is one of the most difficult conservation problems because, whereas committees are very reluctant to turn visitors away, the alternative may be to lose the species. Sadly, those who love them most do the greatest damage to our rarest species—often without being aware of it.

Chapter Five

Conservation of the Flora

Although Surrey is so fortunate in its large areas of unspoilt country, there can be a few counties where the pressure for land is so great. The insatiable demand for housing for Londoners, for new roads converging on London, and for industry ensures that the pressure is growing every year. A single new estate may destroy an important habitat for ever; yet, as with the recent destruction of the historic site at Molesey Hurst for the Autumnal Squill and other fine plants, the economic value of the land may be so great that it is hopeless to resist. For this reason it is essential that naturalists do all they can to safeguard sites before they are threatened.

It is sites—that is, habitats—rather than individual rare species that need to be safeguarded. Our records show that about fifty-one native species have become extinct in Surrey since 1550 (Table 2). At first sight this seems a frightening total, but it is spread over 400 years and some of the species listed come and go in most of their localities. Others hardly belong to the county, since they are here outside their main geographical range; and nine require maritime or brackish conditions which disappeared with changes in the tidal Thames and the construction of concrete embankments. Taking into account species like the Fir Clubmoss, *Huperzia selago*, and probably the horsetail, *Equisetum hyemale*, which are likely to be lost owing to climatic changes, and the list of those whose loss is really important becomes quite short. A more serious aspect is that the rate of loss is getting faster (Table 3). Only seventeen species were lost in the first 350 years and twice as many in the last seventy years. The heaviest losses were nine in the decade 1921–30, and seven in 1951–60, but the only conclusion to be drawn from this is that our environment is changing more rapidly with the increasing power of modern tools.

The loss of habitats is obvious to all. It arises mainly from the spread of commuter housing in wide circles round railway stations and, with universal ownership of cars, over areas away from the railways. Sometimes the building is in unexpected places, as when the London County Council Sheerwater Estate in 1950 destroyed Sheerwater Bog. This was a bog with a fine flora, including such plants as Lesser Wintergreen, *Pyrola minor*, Lesser Bladderwort,

Table 2 Species which have become extinct in Surrey since 1550

	Last reported
Huperzia selago	1905
Equisetum hyemale	1855
Asplenium viride	1889?
A. septentrionale	1917
Ranunculus baudotii	1900
Fumaria martinii	1959
Viola lactea	1930
Polygala austriaca	1888
Elatine hydropiper	1935
Agrostemma githago	1958
Chenopodium vulvaria	1954
Atriplex littoralis	c1875
Althaea hirsuta	1922
Linum bienne	c1915
Erodium maritimum	c1861
Trifolium ochroleucon	1880
T. glomeratum	1953
Vicia tenuissima	c1950
Bupleurum tenuissimum	c1915
Carum verticillatum	1967
Oenanthe fluviatilis	c1920
Glaux maritima	1910
Samolus valerandi	c1849
Scrophularia umbrosa	1934
Limosella aquatica	1934
Calamintha nepeta	c1875
Valerianella rimosa	1923
Senecio integrifolius	c1930
Antennaria dioica	c1836
Aster tripolium	1956
Sonchus palustris	c1830
Triglochin maritimum	1910
Potamogeton compressus	1898
Fritillaria meleagris	1876
Gagea lutea	c1746
Juncus gerardii	1931
Hammarbya paludosa	1925
Ophrys fuciflora	1833
O. sphegodes	1959?
Himantoglossum hircinum	1927
Orchis purpurea	c1960
Wolffia arrhiza	1882
Sparganium angustifolium	1888
S. minimum	c1915
Scirpus triquetrus	c1946
S. cernuus	1922

	Last reported
Blysmus compressus	c1923
Cyperus fuscus	c1950
Carex depauperata	1972
C. diandra	1923
C. dioica	1933

Table 3 Number of Species which became extinct in each decade

Before 1900	17
1900–10	3
1911–20	5
1921–30	9
1931–40	5
1941–50	3
1951–60	7
1961–70	1
1971 onwards	1
	51

Utricularia minor, and Hare's-tail Cotton-grass, *Eriophorum vaginatum* (Lousley, *BSBI Year Book 1951*, 60–65, 1951). With such pressures it is remarkable that so much of Surrey has escaped the builder, leaving us so many rich botanical habitats. This is mainly due to the high proportion of commonland, the large areas used for army exercises, the National Trust, the Green Belt, and numerous golf courses.

Surrey has 26,871 acres of commonland, which was about 5.8 per cent of the acreage of the administrative county in 1956–8 (*Report Ryl. Commission on Common Land, 1955–1958*, 1958). This had fallen from 42,936 acres in 1873. Rights of common do not give complete protection against building and other exploitation, but they make it very much more difficult, especially in the face of commoners aware of their ancient rights. Many of the rich botanical sites which we owe to their protection as commons have been much reduced in size as a result of inroads by the lords of the manor and others, but considerable areas are left. Barnes, Wimbledon and Mitcham Commons are still valuable habitats; Bookham Common is the subject of one of the most exhaustive ecological surveys ever carried out in Britain; Merrow Downs, Ranmore Common, Banstead Downs, Farthing Downs and Riddlesdown have good chalk floras, while Runnymede has Thames-side and aquatic flowers. Epsom Common, Oxshott Heath, Esher Common, Horsell Common, Ockham and Wisley Commons, Whitmoor, Tilburstow, Holmwood, Hurtwood, Hankley, Hindheath, Witley and Frensham Commons crop up frequently in these pages. Some of these are managed by the local authority or the County Council, a few by Conservators (usually handicapped by lack of funds), one by a Preservation Society and several by the National Trust.

In addition there are fourteen commons totalling 8,200 acres in the Bagshot, Bisley, Chobham and Pirbright areas which were acquired by the War Office. They took steps to extinguish the common rights between 1854 and 1890, and sold Normandy Common (69 acres) to the Parish Council as an open space, and 250 acres of Old Dean Common are being released to the Urban District Council for housing. Nevertheless the rest of these commons, and especially the parts used for army exercises with live ammunition, have been conserved with great efficiency. They contain some of our finest bogs, with a group of species—including *Schoenus nigricans*, *Carex dioica* and *Eleocharis quinqueflora*—which are more characteristic of base-rich flushes, in addition to acid-bog plants mostly to be found elsewhere in the county. There has been some damage from fires and tracked vehicles, but superb bogs remain to confirm that the most efficient warden of all is a notice saying: 'Keep out—unexploded missiles'.

The National Trust is one of the largest landowners in Surrey. It owns 12,520 acres and has arrangements for the protection of a further 495 acres. Its property includes some of our finest habitats, such as Box Hill (841 acres owned and 285 protected); Mickleham Downs (73 acres); Reigate, Colley and Juniper Hills (149.5 acres); Ranmore Common and Denbies (715 acres); Norbury Park (74 acres); South Hawke, Woldingham (9 acres and covenants over 44); Selsdon Wood (198.5 acres) and Netley Park. Common rights have existed over part or the whole of some of their sites: Bookham and Banks Commons (447 acres); part of Frensham Common (905 acres); Hackhurst Downs (13 acres); Headley Heath (482 acres); Hindhead (1,076 acres with neighbouring commons and copses); Holmwood Common (632 acres); Leith Hill and large areas nearby; Runnymede (183 acres) and part of Witley Common, and many others. The National Trust provides a tremendous service to naturalists, but it must be remembered that its main purpose in holding land is for amenity—for the benefit of the general public.

The Green Belt policy, initiated in the 1930s, kept the builder off large areas and provided considerable amenity advantages, but did less to protect the flora than might have been expected. Builders transferred their operations to areas beyond the Green Belt, while within the protected area farming became difficult owing to the activities of the public, their children and dogs, so that fields went out of cultivation, becoming derelict and overgrown, and many plants suffered accordingly. Several Acts amended the areas covered, and in 1972 most of rural Surrey became part of the Metropolitan Green Belt round London, when 110,000 acres in the county were added. Two growth areas were excepted: the first takes in Frimley and Camberley, Bagshot, Windlesham, Lightwater and Bisley and is related to the growth area of Aldershot-Reading-Basingstoke; the second covers the parishes of Horley and Charlwood, linked with the proposed growth areas in Sussex. Building developments are to be expected within these areas, which include important botanical sites.

Fortunately Surrey County Council have also shown great foresight over the last forty years in acquiring commons and other land for public open spaces. Land they have bought, or which they manage under access agreements, now exceeds 7,000 acres. Those of botanical interest include Chobham Common, part of Hackhurst Downs, Norbury Park, Ockham and Wisley Commons, Sheepleas and Whitmoor, Rickford and Stringer's Commons. Local Nature Reserves have been declared for part of Chobham Common, Hackhurst Downs and Staffhurst Wood, Limpsfield. The conservation work of Surrey County Council has been strengthened by the designation in 1958 of 160 sq miles of the North Downs, the Leith Hill range and the hills around Hindhead as the Surrey Hills Area of Outstanding Natural Beauty. Although the significance is secondary to that of a National Park, the designation assures the council of support in certain planning decisions, and makes government grants available towards wardening costs and the acquisition of land.

The 'Coulsdon Commons', purchased by the Corporation of London in 1883, are a remarkable example of foresight and generosity since they are about 18 miles outside the Corporation's boundary. It has since added to the original purchase, and Riddlesdown and Farthing Downs have considerable botanical interest. The total area to which the public has access in Surrey is about 36,000 acres, and this includes land owned by district councils, as well as by the National Trust, Surrey County Council and the Corporation of London.

Most of the land mentioned so far is of interest to botanists, and

has contributed very heavily to this *Flora*, but the primary purpose
for which it was acquired was for amenity. Far too little Surrey land
is held for scientific reasons—for research, conservation and educational use. The Nature Conservancy Council, which holds so much
land as reserves in other counties, including Kent and Sussex, holds
no Surrey land at all. The Surrey Naturalists' Trust was incorporated
in 1959 for the study and protection of natural history interests, and
with power to acquire and hold land. The reserves it owns or leases
include Godstone Bay Pond; Nower Wood, Headley; Seale Chalkpit,
and Bagmoor near Milford. It manages other areas under agreements, and advises and does conservation work on many more. With
the help of the Surrey Flora Committee and other bodies, it has
surveyed and recommended a large number of areas to the Nature
Conservancy as Sites of Scientific Interest (SSI's)—the early warning
system which gives notice of proposed developments requiring
planning consents.

Golf clubs have also played an important role in resisting the
builder and managing the land in a way favourable to naturalists.
Limpsfield Common provided a very striking example of this. The
golf course was one of the few areas of the Common which had
changed very little in the twenty-five years since the London Natural
History Society made their survey in 1939 (see p 51). Mitcham
Common is another good illustration since the area occupied by the
golf course is almost the only part where a flora characteristic of
Taplow Gravels can be seen; much of the remainder is buried under
municipal refuse or levelled and seeded for playing fields (Lousley,
1971). The New Zealand Golf Course, Woking; the Wentworth Golf
Course, Virginia Water; the Tyrrell's Wood at Leatherhead and
courses at West Byfleet, Walton Heath, Purley Downs, Wimbledon
Common, Banstead Downs, Merrow Downs, Reigate Heath,
Ditton Common and Puttenham Heath are others that have protected
an interesting flora. Golfers require the scrub to be kept under
control and some of the grass to be kept short, and this maintains a
good range of flowers.

Botanists also owe something to horse racing. The best example was
Hurst Park Racecourse. This was on the site of a gravel pit at Molesey
Hurst by the Thames from which H. C. Watson, in the early part of
the nineteenth century, recorded many lovely plants. Most of these,
including Autumn Squill, *Scilla autumnalis*, were still plentiful until
about 1965 when the site was sold for building. The use of Epsom
Downs for racing has preserved an interesting chalk flora and introduced the Cypress Spurge, *Euphorbia cyparissias*. Although blasted

to the ground by the crowds at the Derby and the Oaks, and by the preparations for these meetings, the plants soon recover.

The Forestry Commission—a more recent and very welcome landowner in Surrey—manages its woodlands and those of other owners with considerable goodwill to conservation interests.

Important Plant Habitats

Thus much land has been preserved up to the present time in Surrey on account of rights of common, use by army or for sports, and general amenity purposes. It happens that this includes a magnificent series of botanic habitats of great interest. The following selection also includes a few sites in private ownership.

Chalk Grassland, Scrub and Woodland

The south-facing chalk escarpment of the North Downs stretches across the county from the Kent border almost to Farnham. South Hawke, near Woldingham, NT*, has a small area of chalk grassland, which was restored by the Council for Nature Conservation Corps. It has the only locality for *Aristolochia rotunda*, which is now swamped in scrub; and there is an excellent example of beechwood near the road. Reigate Hill and Colley Hill, NT, and the Buckland Hills form a magnificent stretch of chalk grassland on a steep south-facing slope with much short grass and many orchids, and some rarities such as *Salvia pratensis* and *Galium pumilum*. Next comes Dawcombe, Pebblecombe, P, SNT, to which there is no public access, and then Box Hill and Mickleham Downs, NT, with a wide range of chalk habitats and clay-with-flints woodland (Lousley, 1950, 1969; Bridges & Sankey, 1969). Across the River Mole there is Ranmore Common and the Denbies Estate, NT, with some chalk grassland and much woodland, leading on to White Downs, with an important flora of bare chalk and short grassland, and then Hackhurst Downs, NT, SCC, SNT, with our finest colony of juniper, and much scrub which is being cleared and managed (plate 2). Newlands Corner, P, managed by SCC, and Merrow Downs, of which part is managed as a golf course, have some useful chalk grassland. Most of the Hog's Back vegetation has been destroyed

---

*NT: National Trust ownership; SCC: Surrey County Council; CL: Corporation of London; LA: Local authority; P: Private ownership; SNT: owned or managed by Surrey Naturalists' Trust; WD: War Department lands.

by road widening, but Seale Chalkpit, SCC, SNT, conserves an interesting sample with some orchids.

North of the escarpment there are some important chalk exposures. Riddlesdown, CL, has an excellent flora, while Farthing Downs, Coulsdon, CL, is of botanical importance mainly as leading to the Happy Valley and Devil's Den Wood areas of the London Borough of Croydon. The same authority manages Selsdon Wood, east of Croydon, NT. Park Downs, NT, Banstead Woods and Fames Rough, LA, offer a wide contrast in chalk habitats, including our best localities for Cut-leaved Germander, *Teucrium botrys* (plate 21) and Ground Pine, *Ajuga chamaepitys* (plate 22). Banstead Downs, LA, has our most reliable station for *Gentianella anglica*, while Epsom Downs and Walton Downs both have uncommon plants (Lousley, 1969, 63–4). The Headley Warren Reserve of the Surrey Naturalists' Trust is in private ownership; it is our finest area for education and research on the chalk flora, but its scientific value has been decreased by the introduction of plants from elsewhere. The Sheepleas near Horsley, SCC, is of special interest for orchids.

Heathland

Surrey is just as fortunate with its acid soil habitats for calcifuges as it is with those for calcicoles, and Headley Heath, NT, just across the valley from Headley Warren, makes it easy to compare the two floras. On Lower Greensand, Leith Hill and Wotton Common have fine stretches of Ling, *Calluna vulgaris*, and Whortleberry, *Vaccinium myrtillus*, heathland. Holmwood Common, NT, south of Dorking, is more wooded. In the south-west of the county is Frensham Common and Devil's Jumps, NT, designated as a Country Park and managed by Hambledon RDC. Apart from the large areas of heathland, the Common is important to botanists for the sandy areas by Frensham Great Pond with so many plants usually found on the coast: Sea Mouse-ear, *Cerastium diffusum*, Sand Sedge, *Carex arenaria*, and three grasses—*Poa bulbosa* (plate 32), *Vulpia ambigua*, *V. membranacea*. Then Thursley and Ockham Commons, and Bagmoor, SNT, the finest bog area in south-east England (with the possible exception of some in military occupation), has lovely displays of colourful bog plants, as well as rare ones, such as *Deschampsia setacea*, *Rhynchospora fusca*, and *Utricularia minor*, and also some good dry heath areas. Witley and Milford Commons, NT, are mainly dry, and have suffered more interference owing to use for army hutments during the war. There, and on Rodborough Common on the other side of the Portsmouth Road, are some interesting established plants which have been introduced: *Coch-learia danica* and Moonwort, *Botrychium lunaria*. On the Hythe Beds there is Hindhead Common, NT, including the Devil's Punch Bowl, a wild area with several types of heathland and the subject of early ecological work by Fritsch and Salisbury, and others.

On the Bagshot Beds there are commons with first-class boggy areas of which pride of place must be given to the West End Common, Bisley Common, Lightwater Bog, and Colony Bog series. These are used heavily by the army and live ammunition is fired, so that we have only been able to explore the fringe as a taste of the treasures within. In addition to masses of plants characteristic of acid bogs, there are some which favour less acid or fen conditions, such as Lesser Butterfly-orchid, *Platanthera bifolia*, *Schoenus nigricans*, *Carex pulicaris* and *Eleocharis quinqueflora*. There are two fine colonies of Marsh Fern, *Thelypteris palustris*, and several of Hare's-tail Cotton-grass, *Eriophorum vaginatum*. Olddean Common, WD, and the adjoining heathland near Camberley have sheets of the lovely grass *Agrostis setacea* and our only station for Crested Buckler Fern, *Dryopteris cristata*. Pirbright and adjoining commons, WD, mostly out-of-bounds for civilians, have probably the best colony in Britain of the lovely but fugacious Slender Cotton-grass, *Eriophorum gracile* (plate 3).

Whitmoor, Richford and Stringer's Commons, north of Guildford, SCC, have been important sites in the past, but their value has fallen sharply in recent years. Unfortunately much the same is true of Chobham Common, SCC. Here the motorway cut a broad swathe of destruction right across the Common but left colonies of the beautiful Marsh Gentian, *Gentiana pneumonanthe*, and the elusive *Kalmia polifolia*, which has been established for at least sixty-five years. Horsell Common, north of Woking, managed by conservators, is rather overgrown, but still has at least two wet acid areas; in one there is a very rare Water Crowfoot, *Ranunculus lutarius*, and in the other a grass *Deschampsia setacea*. The latter also grows on Ockham and Wisley Commons, SCC, which have boggy areas and a lake, Boldermere, with interesting margins (plate 4) as well as good areas of dry heath. Nearer London, Esher, Arbrook, Oxshott and West End Commons, Esher UDC and Oxshott Heath Conservators, offer a good series of heathland much studied by ecologists. Banstead Heath, LA, near Kingswood, verges on Walton Heath and together they form a good continuous stretch of heathland of several types, including grass-heath. It has one exceptionally fine colony of Stagshorn Clubmoss, *Lycopodium clavatum*.

Woodland

Beechwoods are well represented in the county though none of them is very large. Of the 'hanger' type on steep chalk slopes there are good examples in Headley Lane behind Box Hill, NT (plate 5); in Norbury Park near Leatherhead, SCC; in Marden Park near Caterham, and in Netley Park, NT. These have a characteristic flora of Bird's-nest Orchid, *Neottia nidus-avis*, Yellow Bird's-nest, *Monotropa hypopitys*, and in the lighter parts other orchids, and in a few places the Green Hound's-tongue, *Cynoglossum germanicum* (plate 18). Examples of the 'plateau' type can be seen in the woods at the top of Box Hill, NT, and Ranmore Common, NT, but there are many other examples of both types. There is a superb boxwood on the rivercliff of Box Hill, NT; and the Druid's Grove, in Norbury Park, NT, is a fine grove of ancient yews.

Pure oakwoods of any size are scarce owing to the extensive use made of oak in the past, but we have many mixed woods in which the oak is the most common tree, and especially on the Weald Clay in the south of the county. Probably the best example of this ancient woodland is Staffhurst Wood, south of Limpsfield, SCC, which is made up of oak, beech and hornbeam, with a shrub layer of holly and hazel and was formerly coppiced. Glover's Wood, Charlwood, on calcareous clay has oak standards over mixed hornbeam, hazel and ash coppice and there is a ravine with Paludina limestone. Woods on Osbrook's and Oakdale Farms, near Ockley, SNT, have fine coppiced woodland on Weald Clay. Frillinghurst Wood, Grayswood, Forestry Commission, and the woods round Vann Lake, Ockley, P, SNT, are other examples of woodland on the Wealden Clay series. These woods are of Common Oak, *Quercus robur*. The best examples of Sessile Oak, *Q. petraea*, woodland in the county are relics of the Great North Wood on the slopes below the Crystal Palace. On the Dulwich College Estate these extend continuously for just on a mile through Low Cross, Peckarman's, Ambrook Hill and Lapse Woods.

Nower Wood, Headley, SNT, on Reading Beds is very mixed woodland useful for educational and amenity purposes, but of no great botanical interest. Alder occurs, a few trees at a time, along several river valleys, but there is a good alder holt at the bottom of Colyers Hanger, Albury (there is oak woodland at the top); a deep alder swamp at Moor Park, near Farnham, and another by Godstone Bay Pond, SNT.

Meadows

Old pastures hardly exist in Surrey now—they have nearly all been ploughed, drained and reseeded, and perennials such as the Green-winged Orchid, *Orchis morio*, which formerly grew in them in plenty, are scarce. Of water-meadows we still have an interesting series round the Eden Brook near Lingfield, where *Oenanthe silaifolia* and *Carex vulpina* are to be found. The Thames-side meads, slightly calcareous with chalk brought down by the river, are sadly reduced, but Chertsey Mead can still produce most of the species for which it is botanically famous, while Runnymede, NT, has a grand list of water-loving plants in spite of being overrun with people.

Habitats for Aquatics

Our outstanding habitat for water plants is the Basingstoke Canal which the Surrey and Hampshire County Councils hope to preserve for amenity purposes. Their action will need to be swift as, since through-water flow ceased and it became broken up into a series of nearly stagnant ponds, deterioration has been rapid. The canal was at its best for plants when there was still some barge traffic, and the pounds were cleaned and dredged in turn, but even in its present state the stretch near Ash Vale Station and Mytchett Lake (plate 6) has much to offer the botanist. The all too abundant Water Soldier, *Stratiotes aloides*, is introduced (plate 25), but sheets of *Eleocharis acicularis*, *Carex serotina*, several pondweeds, *Potamogeton* spp, Lesser Water Plantain, *Baldellia ranunculoides*, Narrow-leaved Water Plantain, *Alisma lanceolatum*, and Shoreweed, *Littorella uniflora*, are a few of the species to be seen here to perfection. The Wey is our best river for aquatics, especially the River Wey Navigation of which part is owned by the National Trust.

Most of our ponds have either disappeared or lost their botanical interest. Britten's Pond on Stringer's Common, north of Guildford, is one of the latest to lose its botanical interest—here the construction of concrete embankments for fishermen destroyed the habitat for Starwort, *Damasonium alisma*, and other interesting plants. However, we still have a few good ponds. Boldermere (=the Hut Pond) on Ockham Common, SCC, has a grand flora on and near its margins—two Sundews, *Drosera rotundifolia and D. intermedia;* Pillwort, *Pilularia globulifera*, and the long-established *Calla palustris* are among the plants found there. Godstone Bay Pond, SNT, is an interesting small reserve where Golden Dock, *Rumex maritimus*, has one of its very few Surrey localities.

Plate 1   *Box Hill, showing chalk grassland and yews*

Plate 2   *Hackhurst Downs, the invasion of chalk grassland by scrub*

Plate 3   *Acid pond, Pirbright Common, with a Cottongrass,* Eriophorum gracile *(1973)*

Plate 4   *Exposure of mud with an interesting flora at Boldermere, Wisley*

Commons

Examples have already been given of commons with mainly acid soils and bogs, but a few others need to be mentioned. Barnes Common, LA, the home of many historic records, still has an interesting flora, including several patches of Burnet Rose, *Rosa pimpinellifolia*, known here since 1666. Wimbledon Common and Putney Heath, LA and Conservators, had two bogs which are now overgrown, but it has an excellent flora with a large colony of Early Marsh Orchid, *Dactylorhiza praetermissa*, and abundant Creeping Willow, *Salix repens* (Pearson, 1918; Castell MS). Mitcham Common, LA, Conservators, has been disgracefully treated and much of it used for municipal refuse dumps, but the golf course saved a representative sample of the former flora of the Taplow Gravels, and there are other small areas which have escaped destruction. *Salix repens* is also on this London common, and a Dropwort. *Oenanthe lachenalii* persists in two places in its only locality in the county (Lousley, 1971).

Bookham Common, and Banks Common, NT, are on London Clay, and have been the subject of an exhaustive ecological survey by the London Natural History Society since 1941, which still continues. This has revealed a great deal of information on the changes taking place in the vegetation (*Lond. Nat.* passim). Blindley Heath, Godstone, PC, an interesting example of a common on Weald Clay, is very wet and has some interesting sedges. Shalford and Wonersh Commons, P, LA, are partly on River Gravels and partly on Atherfield Clay. They have produced a lot of uncommon plants in the past, and still have a good list of aquatics from round the ponds, Fiddle Dock, *Rumex pulcher*, and several established aliens of interest.

The Management of Reserves

Suitable and regular management is essential for the conservation of interest for scientific purposes. Very often this is provided by the present management for agricultural, military or sporting use, but the site will deteriorate unless it is continued, and this is the problem which so often has to be faced. If the site is left without management it will probably soon become overgrown and useless. Two Surrey cemeteries are dramatic examples of this: one, on Barnes Common, has the new graves buried under sycamores; the other, at Nunhead, has graves distorted by the roots and shaded by tall dense elms. Fencing-in the area in which some rarity grows is a temptation, but can often accelerate its destruction. Keeping out cows, sheep or rabbits, or whatever animals are doing the grazing, only encourages the growth of coarse grasses, and permits the rapid establishment of shrubs and trees.

That is what has been happening on our chalk grassland. Formerly it was grazed by sheep, probably heavily in view of the woollen industry in the west of the county, and these four-legged lawn mowers eventually produced the short springy turf characteristic of the chalk. Cattle graze less closely, but keep down the growth of grasses and trample or eat off seedling shrubs. It is many years since I last saw sheep or cattle in large numbers on chalk grassland, but when regular pasturing ceased their place was taken by rabbits. Grazing in circles of decreasing pressure round their warrens, rabbits were more selective; they ate many attractive chalk plants, including orchids, and left others such as Ragwort. When myxomatosis struck in 1954 the effect was immediate—an increase of colourful flowers which were then swamped by the growth of coarse grass and herbs, and then of scrub.

The Conservation Corps of the Council for Nature, of which I was then Honorary Secretary, was set up to help with problems such as this, and one of its first tasks was to attack the growth of scrub, mainly Dogwood, on Juniper Top, at the back of Box Hill. The plan was to clear as much as the group of young people could manage the first year, and go over it again the next—which took less time— and then clear a bit more, leaving occasional trees. The experience revealed the enormous effort which would be required to make any substantial impression on the growth of scrub in Surrey, but in one respect the results exceeded the hopes of most of us. The return of grassland species was remarkably quick; evidently many of them including the orchids, had persisted, without flowers, underneath the bushes. By the third year there was a fine show of plants in flower. In 1965 the Surrey Naturalists' Trust set up its own Conservation Corps which is doing excellent work. Most parties number twenty to forty volunteers for Saturday or Sunday work, and occasionally there have been more than a hundred. In 1971–2 (the most recent figures available) there were thirty-six such parties working on various types of habitat. Work is also done in the county by the National Conservation Corps, run by a nationally orientated trust.

Scrub is equally a problem on many of our commons with acid or basic soils, but here fire and its consequences are also a conservation worry. The worst fires occur after a spell of dry weather and are caused by careless smokers, picnickers and barbecues, broken glass and vandalism. Large recently burnt areas can always be seen on the heaths of south-west Surrey, especially round Frensham Ponds. The shallow-rooted ling is replaced after a few fires by bracken, with its rhizomes too deep to be harmed. The top humus is destroyed with its varied flora and associated fauna (often of great interest), leaving a dreary sea of uniform green bracken. The public, by using the beaters provided, can do much to stop fires in their very early stages; once out of control they are likely to devastate many acres.

In the wet areas, water levels need to be maintained at the height which provides vegetation of maximum interest. Thursley Common is an excellent example of this. Water is impounded by a sluice (plate 10) which is adjusted by the SNT warden to keep a large area of bog reasonably wet. Readings from water gauges provide a record of the levels maintained, so that the increase and decrease of the species can be related to the control of water. In some other places ditches have to be dug—or cleaned out—to maintain drainage. Ponds give rise to special problems. Nearly all are artificial; those on village greens were dug out to water domestic animals and used for geese and ducks. Most of these ponds are now neglected rubbish dumps or perhaps planted out with ornamental aquatic plants. The field ponds were mostly dug out after the fields were enclosed less than 200 years ago, and filled in during the last twenty-five years. Modern farm animals need more hygienic water supplies brought in pipes or by tankers. The Surrey Naturalists' Trust is making a special survey of our ponds. Starting in 1965 they had reports of 128 ponds by the end of 1972, but many of these are in private grounds. Silting, water diversion, erosion of banks, litter and the use of herbicides are the main problems with which they have to contend.

It is the policy of most of the landowners, including the Naturalists' Trust, to give the public free access to as much of their land as possible, but there are times when it is essential to fence part of it off. This can arise from the sheer volume of public pressure. A footpath on a slope will be gradually deepened as it is worn down by the walkers' shoes, then water is channelled into the depression after storms, and eventually the turf is destroyed and the surface rutted over a band spreading from the original path. This can be seen in many places on the chalk escarpment and there is a well-known example on the path up Box Hill from the car park by the Burford Bridge Hotel. Another form of erosion on Box Hill is caused by children using the slopes by the Zig-zag for dry sledging. At some risk to their limbs they descend these slopes on wooden and cardboard boxes so that the fine turf, with *Viola hirta* subsp. *calcarea*, is being destroyed. Horse-riding is another threat. It is exceedingly popular in Surrey and especially in the Banstead, Walton, Ashtead, Headley, Box Hill and Bookham areas (*Surrey Nat.* 1968, 1970). As the horses diverge more and more from the track they have made impassable, the vegetation destroyed is considerable.

The sheer pressure of large numbers of people can break up turf and expose the soil as it does the loose sand by Frensham Ponds and the chalk at Box Hill and Reigate Hill. Public pressure is not always harmful; below the Salomon's Memorial at Box Hill it has preserved the open turf with such plants as the Autumn Lady's Tresses, but too many people can do an awful lot of damage simply by trampling the vegetation into a desert. Cars too have to be controlled by the provision of ample parking facilities and obstacles to prevent them driving off the road. Controls such as these are to the benefit of everyone in the long run.

It is the sites rather than individual rare species which need to be conserved. This is true because scientific knowledge is at last beginning to reveal some of the facts necessary for the control of habitats, but extremely little is known about the precise requirements of particular rarities. Many of these live on an ecological knife-edge, a brief stage in the succession of a wood, or heath, or bog. We can acquire and manage an area of reasonable size, such as Thursley and Ockham Commons, with a range of micro-habitats, and expect success, but a lot more research is necessary before we can hope to keep a single rarity like *Eriophorum gracile* (plate 3) for more than a fairly brief period.

In a few cases we have had spectacular success in recent years. Starfruit, *Damasonium alisma*, had not been seen in Surrey for several years when, in 1973, a bulldozer was put through the end of a pond where it was formerly known on Headley Heath and in July it reappeared in quantity (plate 24). Cut-leaved Germander, *Teucrium botrys*, requires an open habitat (plate 21) and I have seen it come and go for fifty years in the Chipstead Valley. There, in recent years, it has been reduced to a few plants, or even failed to appear at all. Early in 1973 a strip of the field was ploughed and by the autumn the plant was back in great quantity but too late for flowering that year. As a bonus there was also a mass of Ground Pine, *Ajuga chamaepitys*, and other nice plants. These are two examples of the success which followed advice given by the SNT. Probably other rare species could be brought back by putting part of one of their old habitats in a suitable condition. This is a very different matter from transplanting rarities from other places, which destroys the scientific value of the habitat and usually involves the loss of the plants. Our experience with two rare orchids transplanted from Kent, *Orchis purpurea* and *Ophrys sphegodes*, is an example of this (pp 359–60).

The central body for the organisation of conservation in Surrey is the Surrey Naturalists' Trust, Juniper Hall, Dorking. It deserves the support of every botanist in the county.

Chapter Six

Distribution of Flowering Plants and Ferns

The main purpose of a county *Flora* is to record distribution. By plotting this on maps based on a suitable grid, and on a sufficiently large scale, patterns may be revealed which are easily related to rivers, ponds and railways, to soils or to climate. This is a routine exercise, but the critical interpretation of the maps against a knowledge of the background and the history of the species can be of far greater interest.

The distribution maps (pp 405–67) raise many interesting problems. In most cases the general pattern revealed is very much that expected by experienced workers—a broad correspondence with calcareous soils, or with areas where acid soils prevail, for example. The detail, however, often reveals unexpected features, showing that soil preferences are much more complicated than the text books suggest, or quite unexpected features, like the water plants left behind as relics along the course of a canal which has long since been filled in except for occasional pond-like stretches.

The maps record distribution and this must not be confused with frequency. Some of the most abundant Surrey species are restricted to limited areas, while much scarcer plants occur in very small numbers over wider areas. The dots on the map indicate that the species has been found in the tetrad marked, and the dot is the same whether there is one plant or thousands. For example, Scaly Male-fern, *Dryopteris pseudomas*, M13*, has been found in very small numbers, sometimes only a single plant, in many of the 101 squares from which it is recorded. Tutsan, *Hypericum androsaemum*, M67, though locally more plentiful, has often been recorded from isolated plants, though shown from seventy-four squares. Another aspect to bear in mind is that the flora is constantly changing whereas, for practical reasons, distribution maps usually show records collected over a number of years. The present maps exclude records made before 1950; nevertheless some species have become much rarer during the period the records have been collected, so that the maps overstate the frequency, while other species have increased so rapidly

---

*Map number 13

that frequency is understated. Decreases have been most marked in the case of the agricultural weeds, such as Corn Buttercup, *Ranunculus arvensis*, M25, Venus' Looking-glass, *Legousia hybrida*, M311, and Red Hemp-nettle, *Galeopsis angustifolia*, M301. The increases have been most marked in the case of aliens, such as Juneberry, *Amelanchier lamarckii*, M158. Changes are recorded in the text, but it must be remembered that the distribution of plants today is only a stage in their history, and changes are taking place much faster than ever before.

Climate is the main factor which controls distribution over continents, but in a lowland area as small as Surrey the variations in temperature, rainfall, etc are not sufficient for the effect on plants to be very marked. Such changes as can be distinguished take the form of gradients across the county, so that there is a tendency for moister-loving species to be more common in the west with the heavier rainfall. For example, Ivy-leaved Bellflower, *Wahlenbergia hederacea*, is a delicate little plant restricted to a very few places with a humid atmosphere in the west. Soils have a much more marked effect on distribution patterns than climate.

Calcareous Soils

Some of the most impressive distribution patterns are those of the calcicoles—the species which affect the more important types of calcareous soils and are rare or absent on acid soils. The soils rich

Table 4 Species restricted to Chalk in Surrey

Hairy Rockcress, *Arabis hirsuta*. M56
Chalk Milkwort, *Polygala calcarea*. M66
Bastard Toadflax, *Thesium humifusum*. M186
Autumn Gentian, *Gentianella amarella*. M243
Early Gentian, *G. anglica*. M244
Common Gromwell, *Lithospermum officinale*. M256
White Mullein, *Verbascum lychnitis*. M265
Tall Broomrape, *Orobanche elatior*. M286
Red Hempnettle, *Galeopsis angustifolia*. M301
Ground Pine, *Ajuga chamaepitys*. M304
Round-headed Rampion, *Phyteuma tenerum*. M312
Squinancywort, *Asperula cynanchica*. M314
Musk Orchid, *Herminium monorchis*. M404
Sweet-scented Orchid, *Gymnadenia conopsea*. M405
Fly Orchid, *Ophrys insectifera*. M408
Man Orchid, *Aceras anthropophorum*. M415
Meadow Oat-grass, *Helictotrichon pratense*. M488
(Species of very restricted distribution not included)

in free calcium carbonate on which they grow are best known within the area where chalk is exposed, and especially on the south-facing escarpment of the North Downs where the soil is often very shallow over the chalk. Table 4 shows the main species which are known only from the chalk and in addition there are many very rare plants, such as *Fumaria densiflora*, *F. vaillantii*, *F. parviflora*, *Minuartia hybrida*, *Teucrium botrys*, *Cirsium eriophorum* and *Nardurus maritimus*, to mention only a few. Marsh Helleborine, *Epipactis palustris*, elsewhere usually a plant of fens and depressions on coastal dunes, in Surrey grows only in chalk grassland.

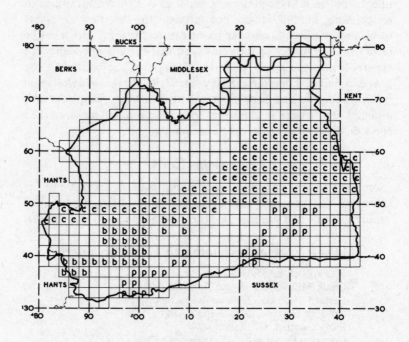

Fig 6 Calcareous Soils: Tetrads containing the main calcareous soils. c=Chalk; b=Bargate Beds; p=Paludina Limestone

The number of calcicoles which also occur off the chalk is far greater. Many of them are associated with outcrops of Bargate Stone about Godalming. Various geraniums are good examples of this—the Long-stalked, *Geranium columbinum*, M100; and the Shining Cranesbill, *G. lucidum*, M103, for example. All the recent stations for *Carex depauperata* have been on Bargate Stone, though formerly it was also on the chalk, and Narrow-leaved Bitter-cress, *Cardamine impatiens*, M54, is mainly in wet places on this formation, while Pale St John's-wort, *Hypericum montanum*, M71, is clearly influenced by it.

The alluvial meadows by the Thames are rich in calcium carbonate and have some fine calcicoles, of which characteristic examples include: Dropwort, *Filipendula vulgaris*, M144, Salad Burnet, *Poterium sanguisorba*, M151, Clustered Bellflower, *Campanula glomerata*, M309, Small Scabious, *Scabiosa columbaria*, M326, and Upright Brome, *Bromus erectus*, M480. Other calcicoles occur on small limestone outcrops in the Weald Clay and other formations, but even more commonly on the calcicolous habitats created by human activity. These include old walls of stone or brick held together by mortar; the sites of cottages and houses, especially those built with lime-mortar; and railway embankments, and roads and paths for which chalk has been used as a foundation. Traveller's Joy, *Clematis vitalba*, M24, a strict calcicole, is very efficient at reaching such places, which explains the scattered localities. On Chobham Common some of the roads are built on chalk which overspills on each side so that calcicoles grow in an area otherwise acid. An even more remarkable instance of this occurs on Hankley Common where round the half-demolished remains of a wall and building (84E1) Mr Raymond Fry found twelve fine Bee Orchids, *Ophrys apifera*, Spotted Orchid, *Dactylorhiza fuchsii*, Twayblade, *Listera ovata*, Ploughman's Spikenard, *Inula conyza*, and other calcicoles in the middle of a large area of acid vegetation.

A few species generally regarded as calcicoles grow equally well in Surrey in sandy or gravelly places where drainage is good but the soil is hardly basic. Vervain, *Verbena officinalis*, M288, and *Koeleria arenaria*, M487, are examples. Other species grow on apparently well-drained places on the basic chalk and elsewhere in very wet places with the soil water neutral or even acid. Thus Water Figwort, *Scrophularia auriculata*, usually a plant of very wet places, grows here and there on Surrey chalk. *Epipactis palustris* has already been mentioned, while Common Valerian, *Valeriana officinalis*, M322, may have different chromosome races for the swamps and chalk. There is food for thought also in the association of certain parasites and saprophytes with calcareous habitats. It is understandable that Tall Broomrape, *Orobanche elatior*, M286, which is a parasite on *Centaurea scabiosa*, is only known on chalk because its host rarely grows on other soils. We do not know enough about the hosts of *Euphrasia pseudo-kerneri* to ascertain whether these offer an explanation, but why should Toothwort, *Lathraea squamaria*, M285, be almost restricted to chalk when it grows as a parasite on hazel and elms which are so plentiful elsewhere? For saprophytes the basic conditions of calcareous soils produce mild humus and the maps of Bird's-nest Orchid, *Neottia nidus-avis*, M403, and Yellow Bird's-nest, *Monotropa hypopitys*, M232, relate to this.

Herb Paris, *Paris quadrifolia*, M385, has a restricted distribution in Surrey, and is mainly a plant of the Gault Clay and no doubt basic. Finally the maps will show that many species generally regarded as characteristic of chalk and limestone soils are much more widespread, for example: Mignonette, *Reseda lutea*, M61, Whitebeam, *Sorbus aria*, M159, Spurge Laurel, *Daphne laureola*, M174, Cowslip, *Primula veris*, M233, and Deadly Nightshade, *Atropa belladonna*, M263. Most of these show a strong preference for the chalk and occur as scattered plants elsewhere.

Acid Soils

Acid soils are widely distributed over Surrey. The most extreme examples with the richest flora are on the great commons in the west of the county, on Bagshot Beds and Lower Greensand. Here we have the fine bogs of Thursley Common and less well-known bogs near Bagshot. These and other commons explain the western type of distribution of the species listed in Table 5.

Table 5 Species restricted (or almost so) to acid commons in the west

Marsh Clubmoss, *Lycopodiella inundata*. M1
Shepherd's Cress, *Teesdalia nudicaulis*. M51
Marsh St John's-wort, *Hypericum elodes*. M72
Common Sundew, *Drosera rotundifolia*. M170
Long-leaved Sundew, *Drosera intermedia*, M171
Bog Myrtle, *Myrica gale*. M217
Bog Pimpernel, *Anagallis tenella*. M236
Marsh Lousewort, *Pedicularis palustris*. M279
Bog Asphodel, *Narthecium ossifragum*. M380
Early Marsh Orchid, *Dactylorhiza incarnata*. M413
Common Cottongrass, *Eriophorum angustifolium*. M424
Harestail Cottongrass, *Eriophorum vaginatum*. M425
Deer Grass, *Scirpus cespitosus*. M426.
Many-stalked Spikerush, *Eleocharis multicaulis*. M432
White Beak Sedge, *Rhynchospora alba*. M433
Sand Sedge, *Carex arenaria*. M453
Star Sedge, *Carex echinata*. M458
White Sedge, *Carex curta*. M459
Bristle Bent, *Agrostis setacea*. M494

Acid soils prevail over large areas of the higher ground on the Lower Greensand in the Leith Hill range, and large areas on the Bagshot Sands, as at Esher and Oxshott Commons, while many smaller scattered areas occur on the London Clay, Weald Clay and Hastings Beds and extensively over the Clay-with-Flints on the higher areas of the Chalk. This explains why the distribution of the commoner calcifuges is so widespread—examples are given in Table 6. The map of *Rhododendron ponticum*, M228, covers the drier acid heathland and that of Purple Moor-grass, *Molinia caerulea*, M462, most of the wetter acid areas of the county.

Table 6 Some widespread calcifuges

Stagshorn Clubmoss, *Lycopodium clavatum*. M2
Wood Horsetail, *Equisetum sylvaticum*. M3
Hard Fern, *Blechnum spicant*. M6
Moor Fern, *Oreopteris limbosperma*. M17
Round-leaved Crowfoot, *Ranunculus omiophyllus*. M31
Climbing Corydalis, *Corydalis claviculata*. M43
Marsh Violet, *Viola palustris*. M64
Ragged Robin, *Lychnis flos-cuculi*. M74
Annual Knawel, *Scleranthus annuus*. M87
Spring Beauty, *Montia perfoliata*. M90
Allseed, *Radiola linoides*. M97
Alder Buckthorn, *Frangula alnus*. M115
Spear-leaved Willowherb, *Epilobium lanceolatum*. M176
Narrow-leaved Sheep's Sorrel, *Rumex tenuifolius*. M210
Rhododendron, *Rhododendron ponticum*. M228
Partridge Berry, *Gaultheria shallon*. M229
Bilberry, *Vaccinium myrtillus*. M231
Lousewort, *Pedicularis sylvatica*. M280
English Eyebright, *Euphrasia anglica*. M283
Lesser Skullcap, *Scutellaria minor*. M303
Hard Rush, *Juncus squarrosus*. M386
Compact Rush, *Juncus subuliflorus*. M389
Bulbous Rush, *Juncus bulbosus* (sensu lato). M390
Great Woodrush, *Luzula sylvatica*. M393
Heath Woodrush, *Luzula multiflora*. M394
Heath Spotted Orchid, *Dactylorhiza ericetorum*. M412
Floating Clubrush, *Scirpus fluitans*. M430
Green-ribbed Sedge, *Carex binervis*. M435
Bottle Sedge, *Carex rostrata*. M438
Pill Sedge, *Carex pilulifera*. M446
Purple Moor-grass, *Molinia caerulea*. M462
Mat-grass, *Nardus stricta*. M504

A further complication in interpreting distribution maps of calcifuges is that some of them can grow on the chalk in the absence of competition. Thus in 1938 Sheep's Sorrel, *Rumex acetosella*, was plentiful and exceptionally large on exposed chalk at the side of the new Leatherhead By-pass, and in July 1941 it was plentiful in a chalky field in the Chipstead Valley. Patches of 'chalk heath' with calcifuges form on the chalk owing to leaching, but, apart from this, occurrences of calcifuges are not rare in places where they are not in direct competition with the better-suited calcicoles. Devil's-bit Scabious, *Succisa pratensis*, and Heath-grass, *Sieglingia decumbens*, are sometimes found on chalk in Surrey in the Chipstead Valley and elsewhere.

Sands, Gravels and Clays

Sands and gravels occur mainly in the west and north of the county with smaller patches elsewhere, and this is reflected in the distribution maps. Many of the plants of these soils are spring ephemerals and disappear in early summer because the drainage between the

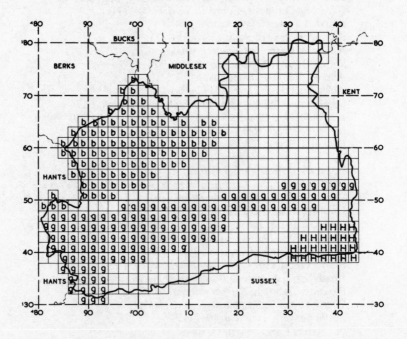

Fig 7 Sands: Tetrads in which these occur—b=Bagshot Beds *sensu lato* (Barton Sand, Bracklesham Beds, Bagshot Beds *sensu stricto*); g=Lower Greensand (Folkestone Sands, Sandgate Beds, Bargate Beds, and Hythe Beds); H=Hastings Beds (Tunbridge Wells Sands and Ashdown Sands)

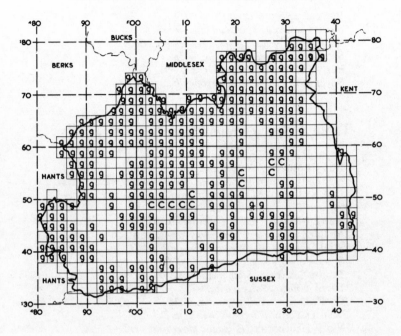

Fig 8 Gravels and Drifts: Tetrads in which these occur—g=Plateau and River Gravels; C=Calabrian (Early Pleistocene) gravels, such as Netley Heath deposits

Table 7 Species of sands and gravels

Shepherd's Cress, *Teesdalia nudicaulis*. M51
Little Mouse-ear, *Cerastium semidecandrum*. M77
Ciliate Pearlwort, *Sagina ciliata*. M83
Common Sandspurrey, *Spergularia rubra*. M86
Common Storksbill, *Erodium cicutarium*. M104
Haresfoot Clover, *Trifolium arvense*. M127
Knotted Clover, *Trifolium striatum*. M128
Rough Clover, *Trifolium scabrum*. M129
Subterranean Clover, *Trifolium subterraneum*. M130
Birdsfoot, *Ornithopus perpusillus*. M137
Spring Vetch, *Vicia lathyroides*. M139
Hoary Cinquefoil, *Potentilla argentea*. M147
Slender Parsley-piert, *Aphanes microcarpa*. M150
Meadow Saxifrage, *Saxifraga granulata*. M166
Spear-leaved Willowherb, *Epilobium lanceolatum*. M176
Fiddle Dock, *Rumex pulcher*. M212

Bugloss, *Lycopsis arvensis*. M251
Small Cudweed, *Filago minima*. M338
Smooth Cat's-ear, *Hypochoeris glabra*. M357
Sand Sedge, *Carex arenaria*. M453
Early Hair-grass, *Aira praecox*. M491
Silver Hair-grass, *Aira caryophyllea*. M492

Table 8 Species found mainly on clay soils

Great Horsetail, *Equisetum telmateia*. M4
Hairy Buttercup, *Ranunculus sardous*. M26
Medium-flowered Wintercress, *Barbarea intermedia*. M55
Dyer's Greenweed, *Genista tinctoria*. M116
Zig-zag Clover, *Trifolium medium*. M126
Narrow-leaved Trefoil, *Lotus tenuis*. M133
Midland Hawthorn, *Crataegus laevigata*. M156
Wild Service-Tree, *Sorbus torminalis*. M160
Stone Parsley, *Sison amomum*. M193
Pepper Saxifrage, *Silaum silaus*. M198
Eared Willow, *Salix aurita*. M226
Chaffweed, *Anagallis minima*. M238
Wild Daffodil, *Narcissus pseudonarcissus*. M397
Violet Helleborine, *Epipactis purpurata*. M401
Smooth-stalked Sedge, *Carex laevigata*. M434
Bladder Sedge, *Carex vesicaria*. M439
Pendulous Sedge, *Carex pendula*. M442
Thin-spiked Wood-Sedge, *Carex strigosa*. M443
Pale Sedge, *Carex pallescens*. M444
Carex x *pseudoaxillaris*. M451
Oval Sedge, *Carex ovalis*. M460
Meadow Barley, *Hordeum secalinum*. M486

large soil particles makes them too dry for shallow-rooted plants.
The same characteristic makes them warm soils so they attract some
species with a Mediterranean type of distribution, such as Fiddle
Dock, *Rumex pulcher*, M212. The sandy area round Frensham
Ponds, south of Farnham, has attracted a number of species characteristic of sandy coasts (p 61). Sand Sedge, *Carex arenaria*, M453,
Cerastium diffusum, *Poa bulbosa* and *Vulpia membranacea* are
examples, though from their situation it is possible that the grasses
were introduced by motorists' children who last used their buckets
and spades on the coast.

Clay soils are most common on the Weald Clay and London Clay,
as is shown on the map of Stone Parsley, *Sison amomum*, M193. They
also occur on the Clay-with-Flints in the Chalk area as is shown on

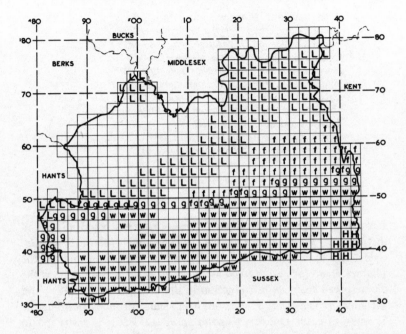

Fig 9 Clays: Tetrads in which the following occur—L=London Clay; f=Clay-with-Flints; g=Gault; w=Wea.d Clay; H=Hastings Beds (Grinstead Clay & Wadhurst Clay)

the map of Violet Helleborine, *Epipactis purpurata*, M401, for example. Nearly all the species listed in Table 8 are characteristic of the squelchy clay woodlands of the Weald. Mention has already been made of Herb Paris, *Paris quadrifolia*, M385, as a calcicole growing on Gault; this clay also has many of the species associated with this soil in the Weald.

Aquatic Species

The pattern of rivers, streams and canals is reflected in many distribution patterns, though often obscured because the species concerned also grow in ponds. Those in Table 9 have been selected

Table 9 Rivers and Canals

Water Crowfoot, *Ranunculus fluitans*. M33
Water Crowfoot, *Ranunculus penicillatus*. M33
Meadow-rue, *Thalictrum flavum*. M35

Black Mustard, *Brassica nigra.* M44
*Great Yellowcress, *Rorippa amphibia.* M60
*Water Chickweed, *Myosoton aquaticum.* M78
Orange Balsam, *Impatiens capensis.* M108
Garden Angelica, *Angelica archangelica.* M199
Water Dock, *Rumex hydrolapathum.* M211
Greater Dodder, *Cuscuta europaea.* M261
Hemp Agrimony, *Eupatorium cannabinum.* M342
Tansy, *Tanacetum vulgare.* M348
Arrowhead, *Sagittaria sagittifolia.* M370
Flowering Rush, *Butomus umbellatus.* M371

*Especially the Wey and Arun Canal

as the best examples. The distribution by rivers is based mainly on the Thames, Wey, Mole and Blackwater, though the other smaller rivers also play a part, and far fewer plants grow in and by the Thames than formerly. *Impatiens capensis*, M108, shows how the plant spread from its introduction in the Tillingbourne Valley, and *Angelica archangelica*, M199, from its ancient introduction along the Thames. Greater Dodder, *Cuscuta europaea*, M261, is restricted to the Mole, Thames and Wey, though it is difficult to see why this should be so when it grows on the Common Nettle, which is plentiful in other habitats, and also grows on a very wide range of hosts abroad.

The Basingstoke Canal (p 44) provided a highway for plants while it was active and is a haven for several rare species now that it is disused. Cut-grass, *Leersia oryzoides*, now grows only on its banks, though much reduced; *Eleocharis acicularis*, M431, has its best stations by the canal, and many commoner species produce the characteristic right-angle pattern that represents the Basingstoke Canal in squares 85 and 95 on the map—Narrow-leaved Water-plantain, *Alisma lanceolatum*, M369, is an example. The influence of the Wey and Arun Canal on distribution is still evident on some of the maps, though it has been disused since 1871 (p 45). Great Yellow-cress, *Rorippa amphibia*, M60, Water Chickweed, *Myosoton aquaticum*, M78, and Reed Sweetgrass, *Glyceria maxima*, M467, are among the plants which show patterns derived from the old course of the canal, where they survive in small sections still holding water. Ponds scattered over the county, though going out of use, influence the pattern of many aquatics—perhaps the two Watercresses, *Nasturtium officinale*, M57, and *N. microphyllum*, M58, are as good examples as any.

Aliens

It is the plotting of the alien species which has produced the most unexpected results. Most botanists would have expected Oxford Ragwort, *Senecio squalidus*, M330, and Sticky Groundsel, *Senecio viscosus*, M332, to show a much closer relationship to the railways than they do (see Fig 5). There is no question that in Surrey the railways have played a part in the spread of these plants, but in recent years other vectors, including the wind, have been more active so that now outside London they occur in as many tetrads without railways as with them. Broad-leaved Everlasting Pea, *Lathyrus latifolius*, M142, and Rat's-tail Fescue, *Vulpia myuros*, M474, are more closely associated with railways; the first because it thrives on slopes and in many cases has reached the railway from adjoining gardens, and the second because it requires well-drained soils which abound on railways. Surrey has produced much less botanical interest on its railways than some other counties, but it must be remembered that workers have been inhibited by the electrified third-rail. Railway brickwork has provided a habitat for ferns, some of them uncommon, of which the map of Black Spleenwort, *Asplenium adiantum-nigrum*, M8, is an example.

Another unexpected feature is the large number of aliens which show a distribution pattern consistent with a spread from London— ie a concentration of records in the north-east corner of the map thinning out to scattered dots as one gets deeper into the county. This is to be expected, perhaps, with species known to have started from the London area, such as Perennial Wall-rocket, *Diplotaxis tenuifolia*, M45, *Galinsoga parviflora* and *G. ciliata*, M328, M329, and Chinese Mugwort, *Artemisia verlotorum*, M349. It was quite unexpected with Prickly Lettuce, *Lactuca serriola*, M360, Annual Mercury, *Mercurialis annua*, M200, and others. Hoary Cress, *Cardaria draba*, M50, seems to have spread from Kent, while *Bromus carinatus*, M483, is spreading rapidly but still mainly in the Kew-Richmond area. Alien trees and shrubs are usually under-recorded, but the map of Turkey Oak, *Quercus cerris*, M220, which is based on self-sown trees only, shows how widely and well established this is in the county. Juneberry, *Amelanchier lamarckii*, M158, is shown to be established over wide areas, although we know that the records for this are still incomplete.

It has been my object in this chapter to draw attention to the interest which can be derived from the distribution maps. Those mentioned must be regarded as examples only and selected to help the reader to interpret the maps and the text for himself.

Part Two

The Flora

The purpose of this *Flora* is to describe the established native and alien flowering plants and ferns to be found currently in the botanical county of Surrey, Watsonian vice-county 17 (p 14). Records have been collected from 1950 onwards to cover the same period as the Botanical Society of the British Isles Mapping Scheme. As in that Scheme, the work is based on field records specially collected by a team of workers (the Surrey Flora Committee) and, as in that Scheme, earlier records are included for very rare or extinct or critical native species. Casuals (aliens which do not persist) are omitted unless very common, or for some other special reason.

Plan

SEQUENCE AND NOMENCLATURE

These are based on the *List of British Vascular Plants* (ed. J. E. Dandy, 1958) with some changes to conform with current practice. Entries for species no longer to be found in the county are placed in square brackets. Synonyms are given only where the new name is still unfamiliar, or to relate the entry to Salmon's 1931 *Flora*.

ENGLISH NAMES

These are based on Dony, Perring and Rob's *English Names of Wild Flowers* (1974), but recommended names which would be unsuitable for Surrey plants have been replaced by others. In a few cases it has been possible to use names special to Surrey.

STATUS

The status (Native, Denizen, Colonist, Naturalised Alien, Established Alien, Casual) refers only to Surrey, and is a judgment based on the evidence which may differ from the status elsewhere.

FIRST RECORD OR FIRST EVIDENCE

The dates given are usually the first published record and for most the place of publication is cited in Salmon, 1931. In other cases the full references are given. These dates are of special interest in the cases of alien plants which have spread since their introduction.

FREQUENCY

This is described by the usual terms: abundant, very common, common, frequent, occasional, rare and very rare.

DISTRIBUTION

The number which follows the statement of frequency indicates in how many tetrads—out of a total of 555 (460 wholly within the county)—the species has been reported by the Surrey Flora Committee workers. This number is a very good indication of how widespread the plant is in the county. The map number cited refers to the Distribution Maps on pages 405–67.

HABITATS

The habitats described are those in which the species has been found in Surrey; in some cases they differ from those in which it has been found elsewhere.

RECORDS

These are listed in the sequence of the numbered tetrads. For example, Godstone Pond, an SNT Reserve, is in tetrad 35C1 because:
1 The 10km Grid Squares (centrads) are numbered from their coordinates on the National Grid. Hence Godstone is on centrad 35.
2 The tetrads, which are 2km squares, are cited as shown on the following diagram. It will be noted that, by using the diagram, tetrads can be instantly converted to conventional four-figure grid references. These will only be approximate because tetrads are based on co-ordinates with even numbers, and those with odd numbers convert to the even number next below. They are, however, sufficiently accurate to locate within a mile any locality on any map bearing the National Grid. Thus Godstone Pond in 35C1 would be read off as 3450—the full reference is 353516.

A5	B5	C5	D5	E5
A4	B5	C5	D5	E5
A3	B4	C4	D4	E4
A2	B2	C2	D2	E2
A1	B1	C1	D1	E1

0 ___ 2 ___ 4 ___ 6 ___ 8 ___ 0 (top) 8, 6, 4, 2, 0 (left)

An '!' after a place name in the cited localities indicates that the author has seen it there.

Periodicals Cited

Rep. B.E.C.: *Report of the Botanical Exchange Club of the British Isles* (later Botanical Society and Exchange Club of the British Isles) 1879–1948

Watsonia: *Watsonia*, the Journal of the Botanical Society of the British Isles, 1949

Proc. B.S.B.I.: *Proceedings of the Botanical Society of the British Isles*, 1954–1969

Rep. Watson B.E.C.: *Report of the Watson Botanical Exchange Club*, 1884–1934

Herbaria Cited

Surrey specimens are to be found in most plant collections in Britain. The following are those most frequently cited:

Herb. Mus. Brit.: Herbarium of the British Museum (Natural History), Cromwell Road, London SW7. Includes the collections of Arthur Bennett, C. E. Salmon, A. L. Still, H. W. Pugsley, D. P. Young, and many others. (BM).

Herb. Croydon Nat. Hist. & Sci. Soc.: Herbarium of the Croydon Natural History & Scientific Society, 96a Brighton Road, South Croydon, Surrey, CR2 6AD. (CYN). Includes specimens from H. F. Parsons and D. P. Young.

Herb. Holmesdale: The Herbarium of the Holmesdale Natural History Club, The Museum, 14 Croydon Road, Reigate, Surrey. (RTE). Includes the collections of J. A. Brewer and J. A. Power.

Herb. Kew.: Herbarium of the Royal Botanic Gardens, Kew, Richmond, Surrey. (K). Includes the collections of A. Bennett (Potamogeton), C. E. Britton, John Stuart Mill (part) and H. C. Watson.

Herb. Lousley: Private collection of J. E. Lousley, 7 Penistone Road, London, SW16 5LU. Surrey specimens collected from 1919 onwards.

Herb. SLBI: Herbarium of the South London Botanical Institute, 323 Norwood Road, London SE24 9AQ. (SLBI). Includes specimens from Surrey from many collectors, voucher specimens for this *Flora*, and the herbarium of W. H. Beeby.

Herb. Wallace: The private collection of E. C. Wallace, 2 Strathearn Road, Sutton, Surrey.

List of Recorders

In addition to those whose names are given in full in the text, records have been included from the following. The names of those who have accepted the initial responsibility for one or more tetrads are prefixed by an asterisk. A few earlier workers whose names occur frequently have also been included.

Aberdeen, I. B.	IBA
*Adams, Miss A. W.	AWA
Allen, Mrs Betty	BA
Allen, D. E.	DEA
Alston, A. H. G. (1902–58)	AHGA
Andrews, C. E. A.	CEAA
Arcus, C. B.	CBA
Ash, G. M. (1900–59)	GMA
Ashbery, K. F.	KFA
Baldcock, David W.	DWB
Bangerter, E. B.	EBBa
Beadell, Arthur (1872?–1957)	ABead

Becher, Mrs M. B.	MBB
Beckett, K. A.	KAB
Bedell, Dr B. J.	BJB
Beeby, W. H. (1849–1910)	WHB
Bellamy, Mrs L. J.	LJB
Bennett, Arthur (1843–1929)	ABe
Benoit, P.	PB
Bishop, E. B. (1864–1947)	EBB
Blaker, G. B.	GBB
Boniface, R. A.	RAB
*Bonner, I. R.	IRB
Bostock, Miss M. W.	MWB
Brenan, J. P. M.	JPMB
Brewis, Lady Anne	AB
Briggs, Mrs E. A.	EAB
Briggs, Mrs Mary	MB
Britten, Harry	HB
Britton, C. E. (1872–1944)	CEB
*Brookes, B. S.	BSB
Browning, F. R.	FRB
Brummitt, Dr R. K.	RKB
Bull, K. E.	KEB
Bullock, Joy	JB
Burton, R. M.	RMB
Butcher, Dr R. W. (1897–1971)	RWB
Cadman, Michael L.	MLC
Carter, A. H. (1867–1939)	AHC
Cartwright, H. M.	HMC
Chapman, G. N.	GNC
Clarke, F. G.	FGC
*Clarke, R. A. R. = Ray	RARC
*Clement, E. J.	EJC
*Cole, J. A.	JAC
Collenette, C. L. (1888–1959)	CLC
*Conolly, Miss A. P.	AC
Cook, Dr C. D. K.	CDKC
Cooke, Rev. P. H. (1859–1950)	PHC
Cosser, R. F.	RFC
*Crawford, G. I.	GIC
Crow, Mrs G. L. A.	GC
Crow, H. A.	HAC

Crisp, Mrs L.	LC
Croydon Nat. Hist. & Sci. Society	CNHSS
Crundall, R. J.	RJC
Dallas, J. E. S. (d.1952)	JESD
Dandy, J. E.	JED
*Darter, Miss C. E.	CED
David, R. W.	RWD
Davies, Jack	JD
Denston, T. C.	TCD
Denyer, E. R.	ERD
Diver, Capt. Cyril	CD
Easton, Dr A. M.	AME
*Elliott, G. S.	GSE
Ellis, A. E.	AEE
Ettlinger, D. M. Turner	DMTE
Fagg, C. C. (1883–1965)	CCF
Farenden, W. E.	WEF
*Fawdry, Miss D. W. (1933–72)	DWF
Ferguson, Miss M. G. R.	MGRF
Fitter, R. S. R.	RSRF
Fowell, R. R.	RRF
*France, Miss Doris W.	DF
Fry, Raymond	RF
*Gardiner, J. C.	JCG
Gerrans, Miss M. B.	MBG
Gibson, I. J.	IJG
Gilbert, J. C.	JCGi
Gilbert, J. L.	JLG
Gilmour, J. S. L.	JSLG
Graham, R. A. H. (1915–58)	RAG
Groom, R. E.	REG
Grose, J. D. (1901–73)	JDG
Groves, E. W.	EWG
Gurteen, F. M.	FMG
*Gush, G. H.	GHG
Halligay, Peter	PHa
Hambler, D. J.	DJH

Hart, Miss W.	WH
*Heard, A. G.	AGH
Hepper, N.	NH
Hill, Derek	DH
Hills, Miss Mollie	MH
Hitch, Miss P. Ann	PAH
Hodgson, John	JH
Holland, Peter	PH
Holmesdale Nat. Hist. Society	HNHS
Home, D.	DH
Hooper, Miss Sheila S.	SSH
*Howard, Miss E. Maud	EMH
Howitt, R. C. L.	RCLH
Hunt, Peter F.	PFH
Hunt, Richard G.	RGH
Hurst, Miss B.	BH
*Inglis, Miss H.	HI
*Isherwood, Miss E. M. C.	EMCI
Jackson, G. W.	GWJ
Jackson, Miss R. Hartas	RHJ
Jackson, Major J. P. A.	JPAJ
Jermy, A. Clive	CJ
Jerrard, L. J.	LJJ
Jewell, Arthur L.	ALJ
Jones, A. W.	AWJ
*Kelly, D. C.	DK
Kelly, John	JK
Kent, D. H.	DHK
Kimmins, D. E.	DEK
*Kneller, Miss B. A.	BAK
Kusel, Mrs D. M.	DMK
Lang, Arthur	ALa
Lang, David C.	DCL
Laundon, J. R.	JRL
Lawrence, G. F.	GFL
Leather, Miss V. M.	VML
Le Brocq, P. F.	PFLeB
Leslie, Alan	AL

Leslie, Mrs Julia F.	JFL
*Little, G. J. S.	GJSL
Lodge, Edward	EL
Lousley, J. E. (1907–76)	JEL
Low, L. W.	LWL
Lunnon, Mrs. K.	KL
McClintock, D.	DMcC
Mackworth-Praed, H. W.	HM-P
*Marks, Miss Kathleen M.	KMM
Marshall, J. B.	JBM
Mason, John	JM
May, Miss Margaret R.	MRM
Meikle, R. D.	RDM
Melville, Dr R.	RM
Melville, Simon	SM
Milne-Redhead, E.	EM-R
Minnion, Miss E.	EM
*Morgan, Miss B. M. C.	BMCM
Mountford, J. O.	JOM
Newton, Alan	AN
Nichols, J. D.	JDN
Norman, Mrs E.	EN
Norman, Mrs Margot	MN
Norman, P. R.	PRN
Orton, P. D.	PDO
Page, K.	KP
Palmer, J. R.	JRP
Palmer, Richard	RP
Parker, R. N.	RNP
Parr, Don	DP
Pennington, T. D.	TDP
*Petch, Dr C. P.	CPP
Petch, Michael	MP
Philcox, D.	DPh
Phillips, H. W.	HP
Pigott, Prof. C. D.	CDP
Polunin, O. V.	OVP or OP

Ponsonby, Hon. Laura	LP
Powell, Miss Doris	DP
Pratt, H. M. (d. 1971)	HMP
*Prime, Dr C. T.	CTP
Proctor, M. C. P.	MP
Radcliffe, B. R.	BRR
Raven, Mrs J.	JR
Raven, P. H.	PHR
Rhys, David V.	DVR
*Richardson, F. D. S. (d. 1966)	FDSR
Riddick, Mrs L.	LR
Robbins, R. W. (1871–1941)	RWR
*Robertson, Dr J. P. S.	JPSR
Roffey, Miss Mabel	MR
Rose, Dr Francis	FR
Russell, Mrs B. H. S.	BHSR
Ryves, T. B.	TBR
Salmon, C. E. (1872–1930)	CES
Sandwith, N. Y. (1901–65)	NYS
Sankey, J. H. P.	JHPS
Savidge, J. P.	JPS
Scoulter, C. E. K.	CEKS
Sell, P.	PS
Sennitt, B.	BS
Sheldrick, Mrs J.	JS
*Sherwood, J. M.	JMS
Simpson, Geo. A.	GAS
Sims, Miss P. A.	PAS
Small, Mrs L. M. P.	LMPS
*Smith, Mrs J. E.	JES
Smith, S. B.	SBS
Spalding, I. E.	IES
Spreadbury, W. H.	WHS
Stace, Clive A.	CAS
Stearn, Dr W. T.	WTS
Stern, R. C.	RS
*Stevens, Dr A. James	AJS
Still, A. L. (1872–1944)	ALS
Summerhayes, V. S.	VSS
Surrey Flora Committee	SFC

Tahourdin, C. B. (d. 1942)	CBT
Thomas, P. L.	PLT
Thompson, R. P. H.	RPHT
Timson, Dr J.	JT
Townsend, C. C.	CCT
Tremayne, L. J.	LJT
Turner, Don	DT
Turrill, Dr W. B. (1890–1961)	WBT
Verdcourt, B.	BV
Wade, A. E.	AEW
Wallace, E. C.	ECW
Walters, Dr S. M.	SMW
Warburg, Dr E. F. (1908–66)	EFW
*Ward, C. W.	CWW
*Warren, W. E.	WEW
Webber, Miss T.	TW
Webster, Miss M. McCallum	MW
*Welch, Mrs B.	BW
Westrup, A. W. (d. 1964)	AWW
Whitehouse, Mrs M.	MW
Wilkins, Lt. Col. James S.	JSW
Wilkinson, Denby	DW
Williams, Iolo A. (1890–1962)	IAW
*Wilson, Mrs D. A.	DAW
*Wingfield, R. C.	RCW
Woodhead, J. E. (1883–1967)	JEW
Wurzell, B.	BWu
Yeo, Dr P. F.	PFY
*Young, Dr D. P. (1917–72)	DPY
Yule, Mrs P.	PY

Pteridophyta

Lycopsida

LYCOPODIACEAE

[*Huperizia* Bernh.
  H. selago L. Bernh. (*Lycopodium selago* L.) Native, now extinct. Last records: Bisley, 1887; Thursley, 1890, 1903, 1905; Swanton 1915.]

Lycopodiella Holub
  L. inundata (L.) Holub (*Lycopodium inundatum* L.)
  Marsh Clubmoss. Native. 1666. Rare, 15.
  Wet heathland in the west, usually on bare black peaty soil by tracks. Map 1. In three areas—round Thursley and Hankley Commons in 83, 84, 94; Smarts Heath and Whitmoor Common, 95; Chobham Common, Bagshot Heath, Bisley Common, 95, 96.

Lycopodium L.
  L. clavatum L. Stagshorn Clubmoss. Native. 1746.
  Rare, 13.
  Dry heathland, often in gravelly or sandy pits, heathy woods. Most frequent in the Winterfold, Peaslake, Leith Hill area, 04, 14. Map 2. (Plate 12). Seldom persists for long. Records increasing but this reflects closer search by more mobile botanists, as well as appearances along woodland rides following Forestry Commission activities.

Sphenopsida

EQUISETACEAE

Equisetum L.
  [*E. hyemale* L. Recorded as extending over 2–3 acres in Wanborough Woods (94), 1847–55, and in plenty at Deepdene, near Dorking (14), 1726–45.]

E. x moorei Newm. Established Alien? Very rare, 1. 1912.
Rampant in a garden; elsewhere in the British Isles only on the

coast of Wicklow and Wexford. (Plate 13). This interesting plant was first recorded from the garden of Lady Victoria Russell at the Ridgway, Shere (04C4) by G. C. Druce as *E. hyemale* (*J. Bot.*, **50,** 288, 1912). He corrected the name to *E. occidentale* (Hy) Coste (*Proc. Zinn. Soc.*, **1929,** 23) which is known from France, and in a longer paper stated that it has increased greatly since it was first found (*Rep. BEC*, **9,** 42–44, 1930). Since I first saw it in 1937 it has increased still more. Only a few abortive cones are produced, and cytologists regard *E. x moorei* as of hybrid origin, though views about the parentage differ.

E. fluviatile L.   (*E. limosum* L.)   Water Horsetail.
Native.   1838.
Frequent by ditches, canals, streams, ponds, marshy meadows and wet woods throughout the county.

E. sylvaticum L.   Wood Horsetail.   Native.   1838.
Local, 15, and confined to the SW of the county: most frequent about Haslemere, Grayswood and Chiddingfold. Wet woods. Map 3. In shady localities often represented by the form with long slender emerald green branches (var. *capillare* Milde).

E. arvense L.   Common Horsetail.   Native.   1666.
Abundant throughout the county. Roadsides, railways, fields, gardens, waste ground. A noxious weed in town as well as country gardens.
x *fluviatile* (*E. x litorale* Kuehlew.)   Rare, 9.
This hybrid was first recorded for Britain from Bisley Common (Colony Bog) 95B5, by W. H. Beeby (*J. Bot.*, 1886, 54). It is still there. In 1920 it was found by Lady Davy by the Basingstoke Canal near Byfleet, 06B1, and it was here that most British botanists saw it for the next 20 years. Recently it has been found in 83, 94, 95, 96, 05, 15 & 26, and is no doubt elsewhere. The hybrid resembles *E. fluviatile* in general appearance but is most easily distinguished by taking a section of the stem, when the central hollow will be found to be intermediate between that of the parents, and more deeply grooved than that of *E. fluviatile*. Cones are sparingly developed and the spores abortive.

E. telmateia Ehrh.   (*E. maximum* auct.)   Great Horsetail.
Native.   1842.   Locally Common, 100.
Wet shady places in woods and on railway banks. Most common

on clay soils south of the chalk and also in the London suburbs, but almost absent from large areas of the NE and central parts. Map 4.

Pteropsida
OSMUNDACEAE

Osmunda L.
O. regalis L.   Royal Fern.   Native.   1746.
Rare, 14.   Bogs and boggy woods.   Map 5.
Formerly locally plentiful but almost eradicated during the Victorian fern craze. Has reappeared in some old stations, and been found in new ones, but there is often difficulty in distinguishing truly native plants from those which have arisen from spores from gardens. Undoubted native localities include Thursley Common (94), Pirbright and Worplesdon, and perhaps Ockham and Wisley Commons (05).

DENNSTAEDTIACEAE

Pteridium Scop.
P. aquilinum (L.) Kuhn   Bracken.   Native.   1725.
Widespread and locally abundant, especially on gravels and sands; rare on chalk. Bracken normally reproduces by extensively spreading deeply buried rhizomes. These are undamaged by heath fires and it therefore tends to replace ling and heathers on some commons. Reproduction by spores is rare and the delicate rounded pinnules of the sporelings are very unlike those of the mature plant. They were plentiful in a gravel pit on Limpsfield Common in 1937, JEL, and on burnt areas on Ockham Common in 1949, FR, and again in thousands in 1957, JEL. It was juvenile plants like these which were the subject of the first notice of bracken in the county in 1725 (Martyn, *Pl. Cant.* 1763).

ADIANTHACEAE

Adiantum L.
A. capillus-veneris L.   Maidenhair Fern.
Established alien.   1862.
Very rare, 4, damp brickwork, especially when north-facing and on railways. Arises from spores, but these are likely to come from cultivated plants rather than from the native ones which are

restricted to frost-free coastal cliffs in the W of England. A magnificent patch was discovered by a bridge near Merstham, 25, by D. C. Kelly in 1960, and still there but very scarce in 1973, JEL.

BLECHNACEAE

[*Onoclea sensibilis* L. A single plant on Shalford Common by the Bradstone Brook (04A4) found in 1967 is an obvious escape from the garden across the stream.]

Blechnum L.
B. spicant (L.) Roth   Hard Fern.   Native.   1829.
Common, 128, in heathy places in woods and commons over much of the west of the county and in the extreme SE. Map 6. Many old records we have been unable to confirm. A calcifuge restricted to very acid habitats, usually ditches and banks, and absent from the chalk and other basic soils.

ASPLENIACEAE

Asplenium L.
A scolopendrium L.   (*Phyllitis scolopendrium* (L.) Newman;
Scolopendrium vulgare Sm.)   Hart's-tongue.   Native.   1718.
Frequent throughout the county, 206, on railway and other brickwork, garden walls, and in street drains and wells. Rarely in natural habitats such as a sandy bank at Puttenham, 94B4; edge of copse near Baynards, 03D3, Chilworth in woods, 04B4, and woods in Norbury Park, 15D3. A great contrast to its behaviour in the wet woods of western England where it is abundant in natural habitats. Map 7.

A. adiantum-nigrum L.   Black Spleenwort.   Native.
1666.   Occasional, 81.
Brickwork of railways, canal bridges, cemetery and other walls. Map 8. Less common in Surrey now there are fewer old walls, and today very much a railway plant.

[*A. viride* Huds. This fern of limestone in mountains grew on a brickwall at Mickleham, 15, from 1854 to ?1889.]

A. trichomanes L.   Maidenhair Spleenwort.   Native.
1597.   Occasional, 64.
Garden and churchyard walls, and railway brickwork. Mainly

dependent on old stone walls which are disappearing rapidly. Map 9.

A. ruta-muraria L.   Wall-rue.   Native.   1650.
Rather frequent, 93.
Railway brickwork, churches, churchyard, garden and other walls. This fern has decreased sadly in the last 40 years, partly due to the destruction of suitable habitats but perhaps also on account of increased atmospheric pollution. Map 10.

[*A. septentrionale* (L.) Hoffm. Forked Spleenwort. Grew on an old flint wall near Box Hill from 1911 to 1917, where it was found by C. E. Salmon. This I believe to be the wall in Headley Lane near Juniper Hall.]

A. ceterach L.   (*Ceterach officinarum* DC.)   Rusty Back.
Native.   1763.   Rare, 18, but increasing.
Churchyard, cemetery and park walls, railway brickwork. (Plate 14). This is best known from railway habitats but only 5 out of our 18 current records are from railways. Mainly of western distribution in Britain but increasing in the east. Map 11.

ATHYRIACEAE

Athyrium Roth
A. filix-femina (L.) Bernh.   Lady Fern.   Native.   1830.
Common, 251.
Damp woods, hedgebanks and ditches. Sometimes so abundant locally in wet woods that it forms almost pure communities, as by Lythehill Lake (93B1). Map 12.

Cystopteris Bernh.
C. fragilis (L.) Bernh.   Brittle Bladder-fern.   Native.
The only record known to Salmon was John Stuart Mill's 1824 report from Albury (Mill, 1841). In 1956 several plants were found by Mrs B. Welch near W. Byfleet, 06B1, and in 1960 R. Clarke found it near Lingfield, 34D2. Both records were on old brickwork and the plants are likely to have arrived very recently.

ASPIDIACEAE

Dryopteris Adans.
D. filix-mas (L.) Schott   Common Male-fern.   Native.   1834.

Plate 5 *Beech Hanger, Headley Lane, Mickleham, showing scarcity of ground vegetation*

Plate 6 *The Basingstoke Canal: Mytchett Lake, a pound for the canal, near Ash*

Plate 7   *The Basingstoke Canal: A derelict lock near Pirbright, showing overgrown towpath and canal bed choked with vegetation of stagnant water*

Plate 8   *Elms killed by Dutch elm disease, Jacob's Well, with a magpie's nest (1973)*

Very common throughout the county. Woods, hedges, ditches, brickwork, etc. in a wide range of habitats.

D. pseudomas (Wollaston) Holub & Pouzar (*D. borreri* Newm.). Scaly Male-fern.   Native.   1931.
Salmon, *Flora*, 657 as *Lastrea filix-mas* var. *paleacea* T. Moore. Locally frequent, 101, but concentrated mainly on the Lower Greensand and Weald Clay, though absent from large areas of these. Damp woods and especially in ghylls sheltered from frost. Map 13, under-recorded.

D. cristata (L.) A. Gray   Crested Buckler-fern.   Native. 1939, *Lond. Nat.*, **1938**, 29–31.   Very rare, 1.
Acid heathland with *Narthecium ossifragum*, *Carduus dissectum*, *Myrica gale*, etc. The discovery of this very rare British fern in the vicinity of Bagshot (86) by L. G. Payne in June 1938 was the first occasion on which it had been found south of the Thames. The colony has from time to time been reported extinct, but it has in fact been kept under observation by friends of the finder and the number of crowns counted has fluctuated from very few in 1956 to about 25 in 1948. At present there are about 15.

D. carthusiana (Vill.) H. P. Fuchs (*D. lanceolatocristata* (Hoffm.) Alston; *Lastrea spinulosa* C. Presl).   Narrow Buckler-fern. Native.   Frequent, 162.   1836.
Wet, and usually acid, woods and marshes often with alders or birches, thinly distributed over the county. Map 14, perhaps under-recorded.

D. austriaca (*D. dilatata* (Hoffm.) A. Gray)
Broad Bucker-fern.   Native.   1836.
Common throughout the county. Woods, copses, hedges, wet heaths.

Polystichum Roth.
P. setiferum (Forsk.) Woynar (*P. angulare* (Willd.) C. Presl)
Soft Shield-fern.   Native.   Frequent, 82.
Woods and hedges. Map 15. Scattered thinly over much of the county.

P. aculeatum (L.) Roth (*P. lobatum* (Huds.) Chevall.)
Hard Shield-fern.   Native.   Rare, 38.

Woods and hedges. Map 16. Rarer than the last species but with a rather similar scatter of localities. Both species are grown in gardens, often occur in small numbers near houses, and some localities arise from dumping of garden refuse or from spores from garden plants.

THELYPTERIDACEAE

Thelypteris Schmidel
T. palustris Schott (*Lastrea thelypteris* (L.) Bory).
Marsh Fern.   Native.   1809.   Very rare, 3.
Wet alder carr, swamps by streams, forming large patches. 83D5, both sides of a stream near 'Pride of the Valley' Inn, Dunn 1893, MH 1964; 95B5, Bisley Common, ECW 1952, RAB 1961; 96B1, carr W of Donkey Town, RAB & FR 1953, SFC 1972.

Oreopteris Holub
O. limbosperma (All.) Holub (*Thelypteris oreopteris* (Ehrh.)
Slosson; *Lastrea montana* Newman).   Moor Fern.   Native.
1838.   Rare, 27.
Heaths and heathy woods. Map 17. Widely scattered colonies of very few plants, but most frequent about Haslemere, Windsor Great Park, the Leith Hill area, and Lingfield.

Phegopteris (C. Presl) Fée
P. connectilis (*Thelypteris phegopteris* (L.) Slosson;
Phegopteris polypodioides Fée)   Beech Fern.   Native.
1885.   Very rare, 1.
Wet mixed woodland, one patch about 7x3m with scattered outliers. 93A1, Lythe Hill Estate.

[*Gymnocarpium* Newman
G. robertianum (Hoffm.) Newm. Established in 17E4, platform wall of Kew Gardens Station, 1954, JPMB; 1973, BWu.]

POLYPODIACEAE

Polypodium L.
P. vulgare L. *sensu lato*   Polypody.   Native.   1763.
Very common in south, but absent from the N of the county, 196.
Map 18. Usually an epiphyte on trees where not exposed to midday sun, but also on mossy banks, walls, brickwork and occasionally on the ground in woods. Now divided into three species on cytological

and morphological grounds: *P. vulgare sensu stricto* common, especially in the SW, 78; *P. interjectum* Shivas mainly in the SW, less common, 55. Map 19. The third, *P. australe* Fée has not been found in Surrey.

MARSILIACEAE

Pilularia L.
P. globulifera L.   Pillwort.   Native.   1724.   Rare, 3.
Shallow water on margins of rather acid ponds. 05D5, Boldermere, seen in small quantity most years but in immense abundance when water level fell in dry autumn of 1972; 18E3, Holmwood Common; 17E2 near Richmond.

AZOLLACEAE

Azolla Lam.
A. filiculoides Lam.   Water Fern.   American alien.
1905.   Rare, 13.
Still water of a canal and ponds. Map 19. Water Fern often appears suddenly, increases rapidly, and disappears equally abruptly after a few years. Water fowl are believed to be responsible for spreading it from ornamental ponds in gardens, and some of the disappearances are due to this subtropical plant being unable to survive in severe winters. From a pond at Blindley Heath it was eradicated by the proprietor of a restaurant whose customers' young children repeatedly tried to walk on the soft green carpet on the pond. Along the Basingstoke Canal it has been found for about 8 miles from Brookwood to Byfleet in 95, 05, 06.

OPHIOGLOSSACEAE

Botrychium Sw.
B. lunaria (L.) Sw.   Moonwort.   Native.   1809.
Very rare, 4.
Heathy or grassy places, often under bracken. 85E1, Camberley Golf Course, 1966, WEW; 85E5, Tekels Park, I. Renson comm. HJR, 1965–6; 94B1 & B3, Witley Common in at least 3 places, 1964 onwards. Formerly in 84, Farnham, but destroyed by ploughing during war, WEF; and 14B4 pasture above Wotton Church, c.1933, BMCM. Erratic in appearance, and there are often gaps of many years between the production of sporophytes.

Ophioglossum L.

O. vulgatum L. subsp. *vulgatum*   Adderstongue.   Native.
1746.   Frequent, 77.

Grassy places, often hollows on clay or chalk, scrubland and
woods. Map 20. Scattered over the county in a wide range of
habitats and, in most, irregular in appearance.

Spermatophyta

Gymnospermae
PINACEAE

Pinus L.
P. sylvestris L.   Scots Pine.   Probably native.
It is generally assumed that the native pines of southern England
were destroyed and that the present populations are descended
from those planted about the time of James I. No doubt there has
been a great deal of planting and those we have now are of mixed
origins, but there appears to be no convincing evidence that the
early pines were eliminated. In many parts of Surrey the Scots
Pine is abundant and regenerates freely. This is especially the case
on the large sandy commons in the west where it is often an important feature in the landscape.

Many alien conifers are reported as regenerating from seed in
Surrey but, as in most other counties, no attempt has been made to
collect records.

CUPRESSACEAE

Juniperus L.
J. communis L. subsp. *communis*   Juniper.   Native.
1666.   Rare, 25.
Chalk grassland; very rarely in scrub on Greensand. Map 21.
(Plate 15). On the chalk much less plentiful than it was in 1920, but
still occurs, usually in small quantity, along the chalk escarpment
from Guildford to Riddlesdown. The finest stand is on Hackhurst
Downs, 04E5. There has been some increase in young bushes
noticed in recent years which may be associated with reduced
rabbit-grazing following myxomatosis. On the Greensand, the big
bushes 15–30ft tall described by Elwes & Henry, 1912, in Juniper
Valley, near Hascombe, 94E1, have been felled, but Sankey reported 7 seedlings in 1968. Also on the Greensand, Rose found an
old bush amongst pines near Leith Hill, 14B2, in 1952, and a bush

105

near Friday St, 14B3, about 1960. The former abundance of
Juniper is suggested by the many place-names in which it features,
such as Juniper Valley, 94E1; in 15, The Junipers, Fetcham Downs,
C3, Juniper Hall, Juniperhill, and Juniperhillwood, D2; Juniper
Top, Juniper Bottom and Junipers, E2; Juniper Hill near Colley
Hill, 25B2, etc.

TAXACEAE

Taxus L.

T. baccata L.   Yew.   Native.   1666.

Common throughout the county. Yew is a characteristic tree of
sheltered slopes on the chalk and occasionally forms pure stands
with scanty ground flora. The Druid's Grove in Norbury Park, 15C2,
is an impressive example. Elsewhere it is frequent in mixed woods
on well-drained soils and especially on the Greensand. It is also
widely planted and there are many historic old trees in churchyards,
of which those at Tandridge and Crowhurst are thought to be
certainly pre-Conquest—see E. W. Swanton, *The Yew Trees of
England* (1958). Yew is a constituent of many place-names in
the county.

Angiospermae

Dicotyledones
RANUNCULACEAE

Caltha L.
C. palustris L.   Marsh Marigold.   Native.   1798.
Common except on the chalk. Marshy meadows, stream sides, alder holts and pond sides.

Helleborus L.
H. foetidus L.   Stinking Hellebore.   Native.   1832.   Rare, 12.
Woods on the chalk. Map 22. Almost certainly native in several places above Headley Lane, 15D2; where it has been known on steep slopes since Brewer, 1863; and likely to be so at the head of Pebblecombe, 25A2, BMCM; at Woodmansterne, 26E1, where it has been known since 1888, and in a steep wood near Oxted chalkpit, 35E3, WH. Most of our remaining localities are near gardens or on the sites of gardens.

H. viridis L. subsp. *occidentalis* (Reut.) Schiffn.
Green Hellebore.   Native.   1832.   Rare, 9.
Calcareous woods, usually as large isolated patches under hazel or beech. Map 23. Known in Cizden's Copse, Godalming, 94E2, since before 1900; from near Ranmore Common, 15B1, since 1838; from Fridley Copse, 15D2, since 1837, and from a copse near Walton Downs, 25A3, since 1890. Miss Kneller has provided two records not on calcareous soils: in Threpps Wood, 35A5, and doubtfully native on an earthwork near Lodge Farm, 34B4. The pretty form with purple blotches at the base of the sepals was described by Salmon as forma *maculatus*, Salmon, 1931, 103, from Chelsham.

Eranthis Salisb.
E. hyemalis (L.) Salisb.   Winter Aconite.
Established alien.
Persists in parks and copses, as at Mitcham for at least 67 years in a

relic of Ravensbury Park, 26D4, and in Oaks Park, Carshalton, 26D1, and islands in the Wandle, 26E3, and in many other places.

Aconitum L.
A. napellus L. subsp. *napellus* (*A. anglicum* Stapf)
Monkshood.   Established alien.
By R. Wey near Somerset Bridge, Elstead, 94B2, for 50 years, and also recorded from A3, B3, C2, C3, D2 and E3 in the same square. Elsewhere in the county *A. napellus* or allied species are found occasionally on sites of old gardens or as garden outcasts.

[*Delphinium* L.
D. ambiguum L. Larkspur. Formerly a cornfield weed, but probably never established for long and dependent on reintroduction from impure seed corn. Now only as a casual.]

Anemone L.
A. nemorosa L.   Wood Anemone.   Native.   1725.
Common throughout the county. Deciduous woodland, often in large colonies. Plants with purplish flowers (var. *purpurea* DC.) are not rare and others with sky-blue flowers are reported from Reffolds Copse, 24A2, BMCM. C. E. Salmon had specimens of var. *robusta* E. J. Salisbury from copses near Woodhatch Farm, 1927, and Meadvale, Reigate, 24D5, Hb. BM.

A. apennina L.   Blue Anemone.   Established alien.   1724.
Naturalised in parks. This persistent garden plant has a long history in Surrey. From 1724 until about 1909 it grew in Wimbledon Woods, ie Wimbledon Park, 27C2, where it was probably introduced from Italy in the seventeenth century. In 1864 it was recorded from Ravensbury Park, on the right-hand side of the road from Mitcham to Sutton, 26D4, and here it persists in a public recreation ground which is a part of the old park, in a copse, and a rough field near Watermeads. Other recent records from 94C5, 05C1, 05D2 and 36A1.

Clematis L.
C. vitalba L.   Traveller's Joy.   Native.   1782.
Locally abundant, 170. Map 24. Hedgerows, edges of woods and copses, calcicole. Traveller's Joy is abundant throughout the chalk area. South of the chalk it is found on Bargate and Paludina limestones, and north of the chalk it is frequent on alluvial soils

by the Thames which are often rich in calcium carbonate brought down by the river. It also occurs on mortar rubble (which is calcareous) in gardens and towns. This probably explains the records from 37 around Dulwich and the Crystal Palace, and other scattered occurrences. A variant with long, rather narrow lanceolate leaves (var. *timbali* Drabble in *J. Bot.* **70**, 83–4, 1932) is plentiful with normal plants in a lane near Underhill Farm, Buckland, 25B2, where it was shown to me by Miss B. M. C. Morgan in 1950.

Ranunculus L.
R. acris L.   Meadow Buttercup.   Native.   1836.
Abundant throughout the county. Damp meadows and pastures, roadsides, avoiding acid soils. Attempts to divide this species into segregates have not proved satisfactory when applied to Surrey— see Salmon, 1931, 98–9.

R. repens L.   Creeping Buttercup.   Native.   1836.
Abundant throughout the county. Wet meadows, pastures and woods; extremely luxuriant in very wet woods and on river banks; a pest in suburban and other gardens, especially on clay soils.

R. bulbosus L.   Bulbous Buttercup.   Native.   1666.
Common.
Dry meadows and pastures on neutral or basic soils; especially characteristic of chalk grassland slopes.

R. arvensis L.   Corn Buttercup, Hedgehogs.   Native.
1832.   Rare, 54.   Map 25.
Cornfields, gardens and waste ground. Formerly quite a common cornfield weed which I have seen in masses in fields below the chalk escarpment and in the Weald. Now so rare that it can be found in few of the places marked on the map as seen since 1950, and we have received one record since 1966. Purer seed-corn, burning the stubble, immediate ploughing after harvesting have all contributed to bringing it near to extinction, apart from casual occurrences from introductions in corn for poultry.

R. sardous Crantz (*R. hirsutus* Curt.)   Hairy Buttercup.
Native.   1777.   Rare, 28.   Map 26.
Cornfields, mainly on Weald Clay. Formerly more widespread over the county but mapping revealed an unexpected concentration in

the SE. Here it grows mainly in places where the crop is thin and damaged by cart tracks, etc, reproducing annually from seed.

R. parviflorus L.   Small-flowered Buttercup.   Native.
1666.   Very rare, 5.   Map 27.
Dry places. Only records—94: Whitley Common, A1, 1952–3, BMCM; Cutt Mill, on roadside, 1926–8, 1954, WEW. 35: Lacey Green, A4, 1929, JEL; 1955, HB; Halliloo Wood, C4, 1950, JPSR; Marden Park, D3, AWJ. In Surrey the plants are usually small and very inconspicuous, and their appearance is erratic. The records given by Salmon covered a period of over 250 years and the effective decrease is very much less than a direct comparison with our 20-year period would suggest. Recent habitats have included sandy ground by a heathland track, a roadside sandy bank, a garden, and disturbed soil on chalk.

R. auricomus L.   Goldilocks Buttercup.   Native.   1836.
Frequent, 126.   Map 28.
Woodlands, copses, hedges. As Salmon observed, it is locally distributed with a preference for chalky and clayey soils. It is thus fairly common on the chalk, Bargate Beds, and Weald Clay, but avoids the Bagshot Beds and River Gravel, Greensand and Hastings Beds. A variable species.

R. lingua L.   Greater Spearwort.   Established alien.
Formerly native.   1732.   Rare.   Ponds.
The last record known to us of this species as a possible native was from near Chertsey, 06 by C. E. Britton (*J. Bot.*, **74**, 355, 1936), but it has not been confirmed. In the following ponds it was almost certainly planted: 95E4, Hook Heath, golf course, 1964, WEW; 13B5, Gatton Manor, SFC; 16A2, Burhill golf course, JES; 25A2, Brimmer Pond, Headley Heath, 1947, EFW, *Watsonia*, **2**, 191, still there 1972, JEL; 25B3, Mere Close Nursery, 1964, DCK; 34D1, Wire Mill Pond, 1943, FR.

R. flammula L.   Lesser Spearwort.   Native.   1746.
Common throughout the county. Wet places—by ponds and lakes, moist depressions on heathland, wet woods, roadside ditches, etc. Varying widely from plants with broad and sometimes serrate leaves in very wet lush places, to subprostrate plants, rooting at the nodes, by ponds. These are habitat forms, but failure to realise that the lowest leaves are often relatively broad increases the confusion.

R. sceleratus L.   Celery-leaved Buttercup.   Native.   1836.
Frequent, 164.   Map 29.
Pond-sides, ditches, wet corners of fields. Scattered throughout
the county but rare on the chalk.

R. hederaceus L.   Ivy-leaved Crowfoot.   Native.   1763.
Rare, 29.   Map 30.
On mud by ponds and ditches. A scattered distribution difficult to
explain. In some localities it has a very long history, such as
Shalford Common, 94E4 & 04A4; Richmond Park, 17E1, and
Mitcham Common, 26E4.

R. omiophyllus Ten. (*R. lenormandi* F. Schultz).
Round-leaved Crowfoot. Native. 1846. Rare, 28. Map 31.
Non-calcareous ponds and ditches. Most frequent in acid, often
peaty, places in the west and south.

R. tripartitus DC (*R. lutarius* (Revel) Bouvet).
Native.   1848.   Very rare, 1.
Recent records only from a spongy hollow which is almost dry in
summer on Horsell Common, 95E5, 1960–65, WEW, Herb. L.
1973; WEW. A species of SW distribution in Britain which just
extends to Surrey and is dependent on temporary habitats.

[*R. baudotii* Godron, a species of brackish pools round the coast
formerly occurred at Wimbledon Park, 27C2, but there are no recent
records.]

R. peltatus Schrank (*R. floribundus* Bab.)
Common Water-crowfoot.   Native.   1836.   Uncommon, 52.
Ponds, ditches and slow streams. The reduced frequency of Water-
crowfoot reflects the rapid destruction of ponds during our recording period. Where ponds are still numerous, as in 03 and 34, the
plant is still abundant, but over the county as a whole the once-
common sight of a wayside pool white with the flowers is now
seldom seen.

R. aquatilis L. (*R. heterophyllus* Weber, *R. radians* Revel).
Native.   1871.   Rare, 6.
Ponds. Probably under-recorded. 05E2, Horsley Place, SR-C &
JRS, Hb Kew; 05E3, nr Horsley Station, BW det. CDKC; 06C2,
SRC; 26E4, Mitcham Common, det RWB; 26D5, RWD; 34B1,
Rowlands Farm, RARC.

R. trichophyllus Chaix ex Villars (*R. drouetti* F. Schultz).
Thread-leaved Water-crowfoot.   Native.   1863.   Rare, 7.
Ponds. Under-recorded. 15A4, nr Bank's Common, JES; 16B2,
West End Common, MMcW; 16E3, Hook, AEE; 25C1, below
Colley Hill, JR; 25E2, Gatton Park Lake, RARC; 35E2, Chalkpit
Wood, CNHSS; 36A3.

R. circinatus Sibth.   Fan-leaved Water-crowfoot.   Native.
1850.   Rare, 8.
Ponds. 85E3, Mytchett Lake, BWu; 94A4, Hampton Park, CDP;
96D5, Virginia Water, WEW; 04A4, Shalford Village Green, EJC;
14C5, Bury Park Pond, AWW; 25D2, E2, Gatton Park Lake;
RARC, EJC; 26B2, Bluegates Pond, RCW.

R. fluitans Lam.   River Water-crowfoot.   Native.
1853.   Rare, 5.   Map 33.
Fast-flowing streams. In the Rivers Wey, Bourne, Ember and
Wandle: 94A2, R. Wey, Elstead Bridge, SFC; 94B2, R. Wey,
Somerset Bridge, SFC; 06B2, R. Bourne by Hall's Farm, DPY;
16C4, R. Ember nr Orchard Farm, LJT; 26D4, Watermeads, JEL.

R. penicillatus Dumort. var. *penicillatus* (Dumort.) Bab.
(*R. pseudofluitans* (Syme) Newbould.) var. *calcareus* (R. W. Butcher)
C. D. K. Cook (*R. calcareus* R. W. Butcher) Fast-flowing cal-
careous streams. Map 33. 84B3, R. Wey, Wrecclesham, DPY;
04C4, Stream crossing Blackheath Lane, JCG det. RWB; Stream
below Blackheath, perhaps on Bargate Beds, RARC; 06A1,
R. Bourne at Dunford Bridge, EJC. 15D4, nr Fetcham Millpond,
BMCM; 26E3, old cress-beds, Hackbridge, 1957. [var. *vertumnus*
C. D. K. Cook. 95B4, Basingstoke Canal nr Cowshot Lock,
Frimley, 1932. Extinct.]

We have had great difficulty in recording the last two species and
especially so as the national collections do not appear to have
material named by Prof. Cook for comparison.

R. ficaria L.   Lesser Celandine.   Native.   1836.
Common throughout the county in woods, copses, grassy banks
and commons. Some attempt has been made to map separately the
distribution of subsp. *ficaria* and subsp. *bulbifer* Lawalrée, but the
results reveal similar frequencies (28 and 27 tetrads), and almost
identical distributions. It seems that *bulbifer* is especially frequent

in 06 on the light soils NW of Byfleet, and *ficaria* in the northern parts of 05 and 15 on the heavier soils there; but in general the maps merely show the areas worked by people interested in distinguishing the two subspecies.

[*Adonis* L.
A. annua L. Pheasant's Eye. A cornfield weed never permanently established in Surrey. Our best locality was on chalk downs above Abinger Hall, 14TA5, in 1902 and 1912, and the most recent one, as a garden weed at Epsom College, 25B5, A. E. Ellis, 1945.]

Myosurus L.
M. minimus L.   Mousetail.   Native.   1762.   Very rare, 5.
Wet corners of cornfields where water has stood during the winter, field gateways, arable, garden weed. Probably always on clay, decreasing rapidly. 84B2, Frensham, 1970, 1971, AB; 95B2, cornfield nr Normandy, 1957, OVP; 15A5, Brickfield Copse, 1967–71, WJD; 25C1, Reigate, garden weed, 1970–72, BMCM; 25E1, near Redhill railway station, 1950, BMCM. In two cases the fields had previously been used for pigs and as, near Brockenhurst, it appears regularly on heavily manured ground, it probably requires habitats rich in nitrogen. Always erratic in appearance it may well reappear in stations, such as Claygate and Chertsey Mead, fairly recently lost.

Aquilegia L.
A. vulgaris L.   Columbine.   Native.   1763.
Rare, 17.   Map 34.
Woods and copses on chalk; also as a garden escape elsewhere. Native blue- or rarely white-flowered plants persist in a number of old localities on the chalk, usually in small quantity. With escapes from more gardens in the wilder parts, it is becoming more difficult to be sure which are native.

Thalictrum L.
T. flavum L.   Meadow-rue.   Native.   1597.
Rare, 15.   Map 35.
Ditch sides and swampy meadows near rivers, decreasing through drainage and development of sites. Near R. Blackwater, York Town, 85D5, JES & WEW, 1966; nr R. Wey and Wey Navigation from Eashing to Pyrford and Wisley; by or near R. Thames from Runnymede to Ham (and 1945, between Hammersmith & Mortlake, RAB). Probably garden escape at Downside Common, 15A5,

1964, JES; Malden Rushett, 16D1, 1958, JES, and West End Common, Esher, 16B2, 1950, LJJ.

[*T. minus* L. Alien, established Ham tips, 17D1, from 1954 until 1964 or later, BW.]

BERBERIDACEAE

Berberis L.
  B. vulgaris   Barberry.   Established alien.   1836.   Rare, 11. Hedges, usually near gardens. Recorded from the following tetrads: 95B5; 05D1; 06A4; 16C5 & E1; 17D2; 24E4; 25D3; 26C1; 34A2; 35D2. First planted for ornament and its edible fruits, then almost eradicated when it was found to be the host of a rust which destroyed wheat, Barberry is usually found as an isolated bush.

  B. glaucocarpa Stapf. Established in a hedge E of Rectory Pond, Godstone, 35D1, 1957 onwards, RARC.

Mahonia aquifolium (Pursh) Nutt.   Oregon Grape.
  Established alien. 1931, Salmon. Locally abundant, 33. Map 36. Open woods, heathland and hedges. Originally planted as game cover and probably spread by birds eating its edible fruits and voiding the seed. Now throroughly naturalised over considerable areas especially in 94 and 25.

NYMPHAEACEAE

Nymphaea L.
  N. alba L.   White Water-lily.   Native.   1836.
  Rare, 43.   Map 37.
  Rivers and streams. This is a decreasing species in flowing water. It is now seldom seen in the Thames where the leaves are cut to pieces by river traffic, but is fairly common in streams, canals, and ponds where it is difficult to separate native stations from those where it has been planted for ornament. In addition to the 43 stations noted without comment from places where it may be native, we have 9 from golf-club and village-green and other ponds where it is almost certainly introduced. Thus the increase in the number of records over those known to Salmon hides a decrease in the native populations.

Nuphar Sm.
N. lutea (L.) Sm.   Yellow Water-lily.   Native.   1775.
Common, 106.   Map 38.
Rivers, streams, canals, lakes and ponds. Still in the Thames, though decreasing; frequent in the Mole, Wey and Basingstoke Canal; elsewhere mainly in lakes and ponds. Much less frequently planted than *Nymphaea alba*.

CERATOPHYLLACEAE

Ceratophyllum L.
C. demersum L.   Spined Hornwort.   Native.
Before 1873, J. S. Mill.   Rare, 33.   Map 39.
Lakes, canals, ditches and ponds. Of special interest are the records from a large metal water-storage tank on Earlswood refuse tip, 24D4, 1955, AWJ, and pond on Wandsworth Common, 27D3, 1966, DPY, both probably introduced. We have far more records than Salmon, but this reflects a more thorough search rather than an increase in the plant, for which possible habitats have decreased.

PAPAVERACEAE

Papaver L.
P. rhoeas L.   Common Poppy.   Colonist.   1827.
Common throughout the county. Cornfields and disturbed ground by roadside, edges of fields, tracks and refuse tips. Still common in smaller numbers, but the great sheets of scarlet in cornfields which were a feature of the scenery 30 years ago are now rare, due to modern agricultural methods. *P. rhoeas* is variable and it is of interest that Shirley poppies, the prevailing form of gardeners, were selected from variations which arose in the garden of Shirley vicarage in 1880 and were distributed by the Rev W. Wilks (*The Garden*, **57**, 385).

P. dubium L.   Long-headed Poppy.   Colonist.   1762.
Common over much of the county. Cornfields and disturbed ground on roadsides, in gardens, etc., especially on light soils.
x *P. rhoeas*. Plants claimed to be of this parentage were described and illustrated by C. E. Salmon from Chilworth (*New Phytol.*, **18**, 111–17, 1919). In this paper he also accepted records from Pyrford and Ham, and from Banstead (C. E. Britton, *Rep. BEC*, **3**, 67–8, 1912 & **3**, 227–8, 1913). More recently MacNaughton & Harper's

work has shown that hybrids of this parentage occur naturally and can also be produced experimentally, but there has been much confusion over them (*New Phytol.*, **59,** 15–26, 1960). Koopmans has shown that the problem is complicated by the existence of cytological and morphological races of *P. dubium* (*Acta Bot. Neerland.*, **19,** 533–4, 1970 & *New Phytol.*, **69,** 1121–30, 1970).

P. lecoqii Lamotte   Babington's Poppy.   Colonist.   1863.
Very rare, 5.
Arable fields on chalk, also on sand. The following have been distinguished from *P. dubium* by the presence of deep yellow latex: 14B5, arable below White Downs, 1964, EJC & BWu.; 25A1, Sand-pit, Betchworth, 1962, BMCM; B5 on tipped soil, Great Burgh, 1961, DPY, D1, garden weed, Redhill, EMCI; 84E5, on chalk nr Hog's Back Hotel, 1963, AJS.

P. hybridum L.   Rough Poppy.   Colonist.   1805.
Very rare, 9.   Map 40.
Arable fields on chalk; otherwise usually a casual introduced with poultry or caged bird food. 94B5, cornfield above Wanborough, 1959, OVP; 94C5, cornfield N side Hog's Back, 1964, JEL & JES; 05C1, cornfield, West Clandon, 1961, BW; 06D1, Sanway, S of Byfleet, 1958, BW; 25B5, Epsom College field, 1952, AEE.

P. argemone L.   Pale Poppy.   Colonist.   c1786.
Locally frequent, 63.   Map 41.
Arable fields, roadsides, wall-tops, and disturbed soils, generally on sandy or chalky soils.

P. somniferum L.   Opium Poppy.
Established alien, also casual.   1835.
Uncommon, 44, as established.   Map 42.
Cornfields and other arable, mainly on the chalk; elsewhere as a frequent casual on refuse tips and waste places. A variation with scarlet flowers on refuse tips is found occasionally. The plant of chalky cornfields with pale lilac flowers and a few bristles on the peduncle (forma *hispidum* Wats.) is fairly uniform over a considerable area from Bookham and Box Hill to Epsom and Coulsdon. It probably originated from crops grown in the early nineteenth century for drugs. When marginal land was ploughed in 1941 and 1942 it came up in sheets in fields by Headley Lane and Fetcham Downs, and soon disappeared when cultivation ceased. It also

appears in the same area when copses which have grown up on former arable are felled.

P. lateritium C. Koch. On a specimen in Herb. E. B. Bishop collected by R. W. Robbins in 1931 he noted that he had known it as a garden weed in Limpsfield for many years and it grew on a wall by Limpsfield Church, 45A2. The record appeared in print shortly after when W. B. Turrill commented that the specimen did not agree over well with the wild material of *P. lateritium* from Armenia at Kew (*Rep. BEC*, **10**, 960, 1935). Whatever the correct name may be, the plant is now frequent on the walls of gardens and on roadsides in several villages and especially Godstone, 35C1.

Chelidonium L.
C. majus L.   Greater Celandine.   Established alien.
1724.   Common.
Hedgebanks, roadsides, waste places. Occurs in many Surrey villages but never far from houses. A favourite ancient herbal remedy, likely to be grown for this reason and to persist. I have known it on the site of the old workhouse on Mitcham Common, 26TE5, and near Box Hill Station, 15TD1, for over 50 years; doubtless it is equally persistent elsewhere.

FUMARIACEAE

Corydalis Medic.
C. claviculuta (L.) DC.   Climbing Corydalis.   Native.
1814.   Frequent, 81.   Map 43.
Heathy rather acid woodland, usually on sandy soils. Absent from the chalk and rare in the east. Still plentiful near Caesar's Camp, Wimbledon Common, 27A1, 17B1, whence it was recorded by Manning & Bray in 1814. In Birch Wood on poor soil with planted conifers, 35C4, 1965, S. Ferguson comm. RARC. Woods in Bethlem Hospital grounds, 36D4, 1968, DPY.

C. lutea (L.) DC.   Yellow Corydalis.   Frequent, 26.
On old walls in many villages, well spread over the county. Still about Godalming, 94D2, whence it was recorded by Rusticus, 1849.

[*C. bulbosa* (L.) DC. Reported as naturalised in Ravensbury Park, 26D4 in 1948, RAB, but DPY failed to find it after several searches in 1958–9.]

Fumaria L.

F. capreolata L. Lingfield, casual on tipped rubble, 34E2, 1958, DPY.—the only record.

F. martinii Clavaud.   Native.   1915.
Very rare and perhaps extinct. Found in 1938 near Kingswood, 25C4, by A. E. Ellis (*Rep. BEC*, **12**, 34, 1939). It persisted in an arable field in corn and other crops, and was seen by many observers until 1950, when I found the field planted with trees and grass. It was there for at least 20 years. There are no further reports from C. E. Salmon's 1912 locality on Reigate Hill.

F. muralis Sond. subsp. *boraei* (Jord.) Pugsl.
Doubtfully native.   1801.   Rare, 6.
Hedges, refuse tip, sandpit. 94C1, hedgebank, Mousehill Lane, 1969, DWB—cf Salmon, 1931, p. 116; 95C5, hedge of cottage garden, Bisley Common, 1963, WEW; 97E1, Egham Cemetery, on rubbish tip, 1966, DPY; 04B4, arable between Lockner's and Postford Farms, 1963, JCG; 16B2, West End Common, 1948, RAB; 25C1, Bletchingley, edge of sandpit, 1963, CPP.

F. densiflora DC. (*F. micrantha* Lag.)   Native.   1844.
Very rare, 4.
Chalky arable fields. A decreasing species. 04A5, Guildford, on building site, plentiful, 1974, ACL,!; 04A5, Pewley Downs, scarce, 1974, ACL; 15C3, Fetcham Downs, 1941, JEL; 16B4, Esher Sewage Works, 1960, JES & MW; 25D4, Chipstead Valley, potato field, 1943, JEL; 35D4, Warlingham, Warren Barn, 1943, JEL, *Rep. BEC*, **12**, 702; 36B1, Mitchley Avenue, 1961, JEL, 1962, RARC; 36B2, Sanderstead, garden weed, 1951, DPY.

F. officinalis L.   Common Fumitory.   Native.   1836.
Common throughout the county. Arable fields, gardens, waste ground, refuse tips. On the chalk most of our plants belong to subsp. *wirtgenii* (Koch) Arcangeli.

F. vaillantii Lois.   Native.   1853.   Very rare, 2.
Cornfields on the chalk. 94D5, Guildford, the Mount, two plants 1974, JFL & ACL,!; 25A1, Pebblecombe, 1950, BW; 25C3, Mugswell, cornfield, 1942, AEE, Hb.L.; 26B1, Mizen's Market Garden, Epsom, 1942, AEE & ECW, Hb.L.; 36B2, above Haling

chalkpit, S. Croydon, 1935, JEL, Hb.L. A decreasing species which was formerly much more widespread in the county.

F. parviflora Lam.   Native.   1798.   Very rare, 2.
Chalky arable fields. 94D5, Hog's Back, by new road above Compton, 1974, ACL,!; 05C1, Clandon Downs, on cultivated land, 1938, ECW; 25B5, Epsom College field, 1952, AEE. It is interesting that all these are refinds in old localities; the first record for the county by T. F. Forster in 1798 was from 'near Epsom'. This species has decreased, but it has always been very rare in Surrey.

CRUCIFERAE

Brassica L.
B. napus L.   Rape, Swede.   Casual.   1836.   Rare, 19.
Refuse tips, sewage farms, nurseries and waste places. The name covers a bunch of relics and rejects of cultivation and it has enjoyed an unwarranted status in the British flora for far too long. *B. napus* is now a rather rare casual and our recorders were able to produce records from only 19 tetrads, and some of these may belong to the next species.

B. rapa L.   Bargeman's Cabbage.   Native.   1805.
Locally Common.
Banks of R. Thames and adjacent ditches. 'The Thames-side Brassica' occurs freely along the river from the Berkshire border to London and was first recorded by W. Borrer from Kew in 1805, though it was not until 1869 that H. C. Watson gave it the publicity it deserved. He showed that it germinated in autumn, produced a rosette of rough lyrate-pinnatifid radical leaves from which in spring a flowering stem arises with smooth and glaucous leaves, followed by ascending fruits. This he suggested was the wild stock of the turnip (*J. Bot.* **7**, 346, 1869; **8**, 369, 1870; **9**, 180, 1871). This was challenged by Dyer (*J. Bot.* **9**, 193, 1871) on grounds which are not convincing. The difficulty is, of course, to decide whether it arose from offspring of cultivated plants which became established, or was wild before the cultivated races were evolved. All the evidence seems in favour of the latter and this is supported by L. H. Bailey after a special study of the genus (*Gentes Herbarum*, **1**, 86, 1922, & *J. Bot.* **61**, 104–5, 1923). Good places to see this plant include Runnymede, 07A2, and Chertsey Mead, 06C4.
    In addition to these wild plants, cultivated turnips occur as casuals on refuse-tips, waste ground and edges of cultivated fields.

B. nigra (L.) Koch   Black Mustard.   Native.   1838.
Locally common, 50.   Map 44.
River and streambanks, occasionally on field borders. Abundant
on the banks of parts of the Wey and Mole, less so by the Thames
and Eden Brook.

Sinapis L.
S. arvensis L.   Charlock.   Native.   1836.
Common on chalk and clay, less so on sandy soils. Field borders,
weed in crops, waste ground, roadsides. Plants which are completely
glabrous, or very nearly so, are not uncommon (var. *glabra* Lindem.)
and have puzzled our helpers because they are not covered by the
usual descriptions.

S. alba L.   White Mustard.   Colonist.   1786.   Frequent.
Mainly in cultivated fields on the chalk, but also on refuse tips,
roadsides and waste ground.

Hirschfeldia Moench
H. incana (L.) Lagr.-Foss.   Hoary Mustard.   Alien.
This Mediterranean species, which has increased its range in
England considerably in recent years, has been well established in
S. Essex and Middlesex for about 20 years and was found in W. Kent
in 1970. It was reported from 17E4, 'Kew—Chiswick area', by
Miss K. M. Marks in 1965-6, but details were not available before
she died. The first certain record came from Mrs J. E. Smith in
1972 when she found two plants in a cindery car-park at Arbrook
Common, 16C2. Six weeks later it was found by Miss E. M. I.
Isherwood in the centre of Redhill, 25D1, in quantity on the site of
houses pulled down a year or so earlier. In July 1972 I found it at
Surrey Commercial Docks, 37D5, and in the same year Alan Leslie
reported it from Guildford Refuse Tip, 05A1. It is surprising that
in one year it should have been reported in four widely scattered
places and may now be expected to spread more rapidly.

Diplotaxis DC.
D. muralis (L.) DC.   Annual Wall-rocket.
Established alien.   1835.   Common.
Roadsides and waste ground, especially in villages; canal tow-paths,
sandy fields, uncommon on chalk. Luxuriant plants with leafy
stems (var. *babingtonii* Syme) occur occasionally.

D. tenuifolia (L.)   DC.   Perennial Wall-rocket.
Established alien.   1817.   Locally common, 72.   Map 45.
Mainly on old walls (far more so than the last species), waste places
and roadsides. A calcicole, thriving under conditions when buildings decay and mortar is exposed. Hence a rapid increase during
the 1939–45 war. First recorded from Lambeth Palace in 1817 this
plant has been spreading into Surrey, but has still not extended
beyond the Weybridge—Cobham—Box Hill—Reigate line, except
for four records from the extreme west. An interesting record was
from 15E2, Juniper Top, on rabbit mounds, 1955, BW.

Raphanus L.
R. raphanistrum L.   Wild Radish.   Colonist.   1666.   Common.
Cornfield weed, also in other crops, roadsides and waste places.

[*Conringia* Adans.
C. orientalis (L.) Dumort. (*Erysimum orientale* Mill.; *E. perfoliatum*
Crantz). Probably always a casual, although records go back as far as
1778 in Surrey, this was still sufficiently common for Salmon, 1931, to
give a good many fairly recent records. Now much rarer in Britain and
no reports have been received.]

Lepidium L.
L. campestre (L.) R.Br.   Field Pepperwort.   Native.
1640.   Common, 117.
In well-drained places, fields, hedgebanks, walls, waste places.

L. heterophyllum Benth. (*L. smithii* Hook.).
Smith's Pepperwort.   Native.   1763.   Rare, 34.   Map 47.
Well-drained places, most frequent on gravel or sand, roadsides
and field edges. Persists on slopes around the Temperate House,
Kew Gardens, 17E4, 1956, BW, whence recorded by Nicholson
in 1906.

L. ruderale L.   Narrow-leaved Pepperwort.
Established alien.   1837.
Common on dry, often cindery, waste ground and on most refuse
tips.

L. latifolium L.   Dittander.
There are old records for this native of our coasts from Wimbledon
Park, 27C2, 1763, and Battersea, 27D4, 1863, from places which
may have been reached by brackish waters. In recent years this

plant has appeared in various inland stations in England where there is no excess of salt. This appears to be the case where it grew on waste ground at Ham Pits, 17D1, where it was found in 1963 by P. C. Holland.

Coronopus Zinn
C. squamatus (Forsk.) Aschers. (*C. coronopus* Gilib.).
Swinecress.   Native.   1836.   Common, 220.   Map 48.
Roadsides, gateways, sides of fields and waste ground. Mainly on clay and avoiding sand. Hence absent from the light soils around Chobham and Bagshot in the NW, and from Churt, Milford and Thursley to Witley in the SW, and from sandy soils elsewhere.

C. didymus (L.) Sm. (*Senebiera didyma* (L.) Pers.)
Lesser Swinecress.   Established alien.   1835.
Common, 214.   Map 49.
Roadsides and wasteground. Increasing and seldom overlooked owing to the pungent smell of cress.

Cardaria Desv.
C. draba (L.) Desv. (*Lepidium draba* L.)   Hoary Cress.
Established alien.   1852.   Locally common, 107.   Map 50.
Field borders, roadsides, waste places, railways. Mainly NE of a line passing through Epsom and Horley, scattered reports elsewhere. Many of the localities are associated with reconstruction of roads, refuse tips and railways. It seems likely that the seeds are carried in mud on the wheels of carts and lorries, and progress since Salmon's *Flora* has been slow.

Isatis L.
I. tinctoria L.   Woad.   Established alien.   c1683.
'On a chalky hill by ye Tenters at Guildford' (Newton MS). Very rare. On a chalk cliff. Woad is a Mediterranean species, unlikely to be native in NW Europe, which was formerly grown to provide a blue dye and a mordant for a black one. Early records were of plants which had spread from cultivation or, in the words of Gerard, 'The wilde kinde groweth where the tame kinde hath been sown' (*Herbal*, 394, 1597). It still grows on the face of a chalkpit south of Guildford, 04E5, where it is protected by the owners of adjoining property. A wool industry flourished in Guildford from the fourteenth to the seventeenth century, and 'ye Tenters', the wrenches for stretching the cloth (often beyond the legal limit), were

probably on level ground below the present locality which is 'on a chalky hill'. There are repeated records from these pits from 1800 onwards, and our Surrey colony of Woad has the longest history of any in Britain.

Iberis L.
I. amara L.   Candytuft.   Native.   1856.
Hanson in Brewer, *New Fl. Reigate*, 91. Disturbed soil on chalk. (Occurs also as casual escape from gardens). The wild chalk plant is reported only from the vicinity of Box Hill: on a steep slope above Headley Lane towards Mickleham, 15D2, and from here it has extended to at least two places in Juniper Bottom, 15E2, AWJ, FR, BWu. Also planted in Headley Warren Reserve, 15E3. The Headley Lane locality is probably the one reported by A. Choules in 1872 as found near Dorking 'about 38 years ago', *Gard. Chron.*, 466. It was found later by H. Goss 'In a wood between Mickleham and Headley' and then after a long interval by me in 1935 (Lousley, *J. Bot.* **74,** 197, 1936). Its reappearance followed tree clearance opening up the wooded slope, and the size of the population still fluctuates from year to year.

Thlaspi L.
T. arvense L.   Field Pennycress.   Colonist.   1597.   Frequent.
Weed in arable fields, waste ground.

Teesdalia R.Br.
T. nudicaulis (L.) R.Br.   Shepherd's-cress.   Native.
1666.   Rare, 31.   Map 51.
Dry banks on heathland, usually on sand. Now almost confined to the west of the county where it is occasionally rather plentiful on the sandy commons. Of the numerous old localities round Richmond, Barnes and Wimbledon, 27, it has been seen only near Caesar's Camp, Wimbledon Common, 27B1, 1950, AWJ, & 1956, BW. A decreasing species.

Capsella Medic.
C. bursa-pastoris (L.) Medik.   Shepherd's Purse.   Native.
Abundant.
Weed in gardens and other cultivated ground, roadsides, waste places. The microspecies into which this very variable aggregate species has been divided received considerable attention in Surrey

from C. E. Britton and others some 50 years ago, Salmon, 1931, 143–4. I am not aware of any recent work in the county.

Cochlearia L.
C. anglica L.   English Scurvy-grass.
Plentiful in salt-marshes at the mouth of the Thames, the only previous record was from near Putney in 1910. A single plant occurred near the foot of the river embankment by Battersea Park, 27E4, in 1965, RARC. Mr Clarke was unable to approach near enough to examine it critically.

C. danica L.   Danish Scurvy-grass.
A large colony over about 50 yards on the W side of the Portsmouth Road, 94B1, 1968, DWB, 1973, 1974, MBB and others.

Bunias L.
B. orientalis L.   Warty Cabbage.   Established alien.
1870.   Rare, 11.
Chalk downs, gravel pits, railway banks, towpath, banks of reservoirs and fallow arable. 03C3, Clandon Downs, 1945, JEL; Clandon Cross Road, 1973, AL; 14A5, White Down, 1961, JCG; 16A5, many plants on banks of Chelsea & Lambeth Reservoirs, 1964, EJC—this is very near the Sunbury Lock locality which dates back over 50 years and where I saw it in 1930 and 1942; 17D2, Ham Pits, 1950, 1959, BW; 25A4, Epsom Downs, 1941, JEL, 1973, BRR; 25C3, Hogden Bottom, Kingswood, 1966, BWu; 27E1, railway bank, Lower Streatham, 1958, RCW; 35A4, New Hill, fallow field, 1958, AWJ; 35C5, Whyteleafe refuse tip, 1950, RARC; 36B1, Riddlesdown, large colony on ploughed downland, 1961, JEL and 600 yards W of this in a car park near Kenley Police Station, 1960, KCB; 36D1, Selsdon, between Firth & Court Woods, 1941, JEL. This is a remarkably persistent alien which has been about Clandon Downs for 80 years (S. T. Dunn, *Rep. BEC*, **1**, 436, 1894).

Lunaria L.
L. annua L., a garden escape, was collected by D. P. Young from an old chalkpit opposite Warren Farm, in Headley Lane, 15E2, in 1938, and reported as still there in great quantity and apparently permanent by L. J. Jerrard in 1948.

[*Alyssum* L.
A. alyssoides (L.) L. (*A. calycinum* L.), an alien found in clover and other arable fields and sometimes persistent, has not been reported in recent years.]

Berteroa DC.
B. incana (L.) DC. (*Alyssum incanum* L.)   Hoary Alison.
Alien.   Rare.
25A2, fallow field on Pebble Combe Hill, 1952, JEL & AEE; 26E4, waste ground N of railway, long established, Mitcham Common, 1957, DPY. Also found in refuse tips.

Erophila DC.
E. verna (L.) Chevall.   Common Whitlow-grass.
Native.   1837.   Locally common, 97.   Map 52.
Sandy commons and arable, dry banks, walls. Common on the sandy soils of W Surrey, and frequent in the SE of the county, but otherwise a rare and decreasing species. In the NE it persists in at least two places on the towpath between Kingston and Richmond, 17D2, but the destruction of old walls has reduced the number of available habitats in built-up areas. Critical work on the genus in Surrey has failed to yield any useful results, but subsp. *spathulata* (Lang) Walters was reported from a sandy footpath on Reigate Heath, 25B1, 1957, CNHSS, det. DPY.

Armoracia Gilib.
A. rusticana Gaertn., Mey. & Scherb. (*Cochlearia armoracia* L.).
Horse Radish.
This aggressive vegetable is thoroughly established in many places from roots thrown out by gardeners. It is especially common on railway banks in the London suburbs, but is also frequent on commons and roadsides throughout the county. First recorded from by the river at Battersea in 1811 and illustrated from there in Sowerby's *English Botany*, there is no evidence that it spreads by seed and mature fruit is rare.

Cardamine L.
C. pratensis L.   Cuckoo Flower.   Native.   1629.
Common throughout the county, but especially on heavy soils. Moist meadows, by ditches and streams, hedgebanks, woods and copses. The Cuckoo Flower is made up of a variable complex with a wide range of chromosome numbers, and attempts to divide

these into subspecies which stand up to practical application in the field have not been very successful. In Surrey *C. pratensis* s.s. ($2n = 56$) seems to be common, and I have specimens determined by D. E. Allen from near Box Hill Station, Mitcham Common, Reigate Heath, etc. *C. dentata* Schultes ($2n = 32$) he has named for me from by the road from Earlswood to Nutfield, Kew; by the Mole below Box Hill, laneside S of Capel, and near Leigh. There are also several putative hybrids. A CNHSS party at Buckland Alders, 25B1, in 1956, found material which Mr Allen named as *C. dentata*, *C. fragilis* (Lloyd) Bor., and the hybrid between them; and this hybrid he also named from an old meadow at Warlingham, 35C5, which he visited with Dr Young. Pretty double-flowered forms occur occasionally. They were abundant over three fields, in only one of which any single-flowered plants were found, near Mayes Court, Warlingham, 35D5, 1955, JEL.

C. amara L.   Large Bittercress.   Native.   1660.
Uncommon, 58.   Map 53.
Water meadows and copses, especially near the R. Wey (and former-ly the Thames), plentiful in alder woods. Still in many old localities, including a boggy hollow near the Royal (Paper) Mills at Esher, 16B3, where H. C. Watson found it over a century ago. By the Moat of Kew Gardens, 17D4, until an exceptionally high tide destroyed it in March 1949. An attractive colour form with lilac petals was found by the Wey Navigation Canal, Addlestone at Ham Haw, 06D3, in 1905, Hb Kew, and has been collected many times since (*Rep. BEC*, **8**, 562–3, 1928; *Rep. Watson BEC*, **4**, 58, 1931), but may have been destroyed by gravel extraction.

C. impatiens L.   Narrow-leaved Bittercress.   Native.
1847.   Rare, 7.   Map 54.
Banks of R. Wey, damp copses, sandpit and roadsides. Restricted to a small area round Godalming and mainly on Bargate Beds, slight outliers to the north at Shalford and Guildford, and in the south at Winkworth Arboretum—in 95C3, D3, E1, E3, E4, E5; and 05A1. In Surrey this usually grows in much wetter places than in the west of England where rainfall is higher. (In 1941 this was collected by A. E. Ellis in woods near Ockley, 13C5, well away from the known area, Hb.L.)

C. flexuosa With.   Wavy Bittercress.   Native.   1844.
Common.

Moist shady places, roadside, ditches and woodland rides. Our experience confirms Salmon's observation that it is more common in the county than *C. hirsuta*.

C. hirsuta L.   Hairy Bittercress.   Native.   1836.
Fairly common.
Bare and often sandy ground by arable fields and gardens, walls, banks and waste places.

C. bulbifera (L.) Crantz (*Dentaria bulbifera* L.)   Native.
1884.   Very rare.
By streams in wooded ravines. Limited as a native to a small area near the Sussex border, S of Capel and on the Weald Clay. Varies greatly from year to year with the time since the wood was last coppiced—as, for example, in a ravine in 13D4 where Dr Young found many plants in flower in 1949, but only one barren plant in 1966. On balance it is probably decreasing, helped on in one place by the ravages of pigs. It also occurs as a probable introduction near the Mole below Box Hill, 15D1, where it was found by Miss B. M. C. Morgan in 1966. Here it is in the company of garden plants planted by the head gardener of the estate across the river.

Barbarea R.Br.
B. vulgaris R.Br.   Common Wintercress.   Native.
1836.   Common.
Banks of rivers and streams, damp hedgebanks and roadsides.

B. stricta Andrz.   Small-flowered Wintercress.   Native.
1871.   Very rare, 3.
Ditches and swamps near the Thames. Formerly by and near the river from Putney to Ham, and perhaps higher; now greatly restricted in range and numbers. 17D2, Ham, wet meadow by the Thames, 1950, JEL, Hb.L.; 17D3, Old Deer Park, BW. Seen here by many observers, and varying from one plant in 1959 to many in 1960, and two in 1962; 17D4, Kew towpath, 1871 until 1946, LJJ, and perhaps later.

B. intermedia Bor.   Medium-flowered Watercress.
Colonist.   1863.   Rather rare, 35.   Map 55.
Roadsides and waste ground, arable fields. An alien which has made little progress in a century and indeed some of our records show remarkable persistence. For example, it is still a weed in

Reynolds Avenue, Chessington, 16E2, 1965, EJC, in gardens on the site of cornfields near the church; and at Buckland, opposite the 'Jolly Farmer', 25B1, 1961, BMCM. Our map shows a random scatter over the county mainly, but not exclusively, on heavy soils.

B. verna (Mill.) Aschers.   American Wintercress.
Colonist.   1805.   Rare, 6.
Roadsides, waste ground, old army sites. Grown as a cress for the table, this depends on current fashion and taste for the provision of seed. It has probably never established itself in the county for more than a few years and is now rarer than it was.

Arabis L.
A. caucasica Willd.
A garden plant occasionally established on walls.

A. hirsuta (L.) Scop.   Hairy Rockcress.   Native.   1778.
Rare and decreasing, 10.   Map 56.
Chalk grassland. Comparison of our records with those given in Salmon's *Flora* shows a very rapid and somewhat unexpected decrease in the last 40 years. It is still in several places in the Box Hill area in 15, with a rather isolated locality at Albury Downs, 04B5 to the west. To the east of the main area it is on the Buckland Hills, 25B2; Banstead Downs, 26C1; Quarry Hangers, 35B2, and near Coulsdon, 35A5. A great many old stations on walls and hedgebanks in the NE, and on sandy ground, Bargate stone and quarries in the SW and W, appear to be lost.

A. glabra (L.) Bernh. (*A. perfoliata* Lam; *Turritis glabra* L.)
Tower Mustard.   Native.   1746.   Very rare, 3.
Roadside banks. Houndon, Thursley, 84E1, 1959, VML, 1960, FR; near Frensham Great Pond, 84C1, JEL & WEW; Oxted Green near Milford, 94C1, 1971, DWB. A biennial which is unreliable in its best stations and reappears after very long intervals in others. Has disappeared from many stations in the NE and N of the county and is now restricted to a small area in the SW which continues over the Hampshire border.

Nasturtium R.Br.
N. officinale R. Br. (*Radicula nasturtium-aquaticum* (L.) Druce, *Rorippa nasturtium-aquaticum* (L.) Hayek)   Watercress.

Native.   1724 (for aggregate species).   Common, 65.
Map 57.
Ponds, ditches, streams.

N. microphyllum (Boenn.) Reichb. (*Rorippa microphylla* (Boenn.)
Hyland.)   Native.
First evidence: 1857 Gatton Park Pond, J. S. Mill (Herb. Kew) ex
Watsonia, **1**, 231, 1950. Common, 58. Map 58. Ponds, ditches,
streams.
x *N. officinale*.
First evidence: 1872, Barnes Common, J. H. Morgan (Herb. BM)
ex *Watsonia* **1**, 232, 1950. Uncommon in 18 tetrads. Map 59.

Our records show that *N. officinale* is slightly more common than
N. microphyllum (65 and 58 tetrads). They are both scattered all
over the county with a concentration in the SE which is probably
explained by the enthusiasm of the recorders. Their distributions
do not follow the rivers, but they do avoid chalk and light sandy
soils, and show some preference for clay. Records for the hybrid
are equally random and, although this is often the watercress grown
commercially in other counties, we have no reports of it from the
Surrey watercress beds.

Rorippa Scop.
R. sylvestris (L.) Bess. (*Nasturtium sylvestre* (L.) R.Br)
Creeping Yellowcress.   Native.   1805.   Frequent.
Banks of the Thames and other rivers, ditches and ponds, and also
as a weed in arable land away from water.

R. islandica (Oeder) Borbás (*Nasturtium palustre*).
Marsh Yellowcress.   Native.   1827.   Common.
Round ponds, ditches and riverbanks. Grows by the Thames, Wey
and Mole on mud or gravel, but is more characteristic of pondsides
and other places where the waterlevel falls in summer.

R. amphibia (L.) Bess. (*Nasturtium amphibium* (L.) R. Br.)
Great Yellowcress.   Native.   1827.   Frequent, 84.   Map 60.
Banks of rivers and canals, by ditches and ponds. The map shows a
most interesting distribution. It follows the Thames from Runny-
mede to Barnes, the Wey to above Shalford and the Basingstoke
Canal to Ash Vale, and the old course of the long-disused Wey and
Arun Canal to the county boundary. The Mole is followed to some

distance above Dorking, and it grows by the Blackwater, in New Pond, Merstham, and Gatton Park Lake and a few other ponds. We have not found it by the Tillingbourne.

x *R. islandica* = *R. x erythrocaulis* Borbás was discovered new to Britain by C. E. Britton, growing with the parents at several spots on the Thames embankment between Putney and Richmond, and especially abundant on the river wall near Hammersmith Bridge (*J. Bot.*, **47**, 430, 1909). I found it plentiful still by Hammersmith Bridge in 1932 and between Mortlake and Kew in 1942. Since then the embankment has been rebuilt and the flora greatly reduced, but this hybrid which occurred in such abundance is probably still by the Thames in 27A4 or 27B4.

R. austriaca (Crantz) Bess.   Austrian Yellowcress.
Established alien.
First record: 1953, Young in *BSBI Year Book*, **1953**, 104. 04C5, Weston Wood, Albury, 1963, JES; 17D2, Ham Pits, 1960, JCGi, 1964, LNHS; 26D5, Bunce's Meadow, Merton, 1956, RCW. 36B2, railway bank, Sussex Road, S. Croydon, known to DPY since 1937 or before, usually cut down before it flowers, but a good show in 1937, 1952 and 1967.

Matthiola R. Br.
M. incana (L.) R. Br. Hoary Stock. Established as a garden escape on a chalk railway cutting at 26C1, Banstead, 1956, RCW.

Hesperis L.
H. matronalis L.   Dame's-violet.   Naturalised alien.
1805.   Frequent, 31.
Hedgebanks, riverbanks and woods, also a casual on refuse tips. Well distributed over the county with the exception of the NW, most frequent in the Godalming-Milford area. Well established in quantity along R. Wey above Tilford for $\frac{3}{4}$ mile, 84D2, 1961, SFC, and frequent along Dunsfold River for at least 600 yards downstream from bridge E of White Beech Copse, 93E4, JCG, JES, BW, 1961.

Erysimum L.
E. cheiranthoides L.   Treacle Mustard.   Colonist.   1760.
Common.
Open gravelly places, arable fields especially on sand. Particularly frequent near the Thames and the Mole, where it could be native.

Cheiranthus L.
C. cheiri L.   Wallflower.   Established alien.   1838.
Old walls, chalkpits, railway cuttings. Common as an obvious garden escape but sometimes of a higher status, such as: 84E5, Seale chalkpit, 1959, VML; 06B5, Thorpe, walls near the church, 1965, EJC (recorded from here by Whale c.1875); 26E3, walls N of Beddington Orphanage, 1957, BW (recorded from Beddington by Irvine, 1838).

Alliaria Scop.
A. petiolata (Bieb.) Cavara & Grande   Garlic Mustard.
Native.   1836.
Abundant throughout the county. Wood margins, hedgebanks, wall bases and roadsides.

Sisymbrium L.
S. officinale (L.) Scop.   Hedge Mustard.   1836.
Abundant throughout the county. Roadsides, waste places, gardens, and fields. Equally abundant in built-up areas.

S. irio L.   London Rocket.   Alien.
First recorded in 1763. 24D4, refuse tip near Earlswood, and at another tip in the neighbourhood, 1957, BMCM, Hb Lousley.

S. loeselii L.   Alien.
First record: 1951, Kent & Lousley. Rare but increasing on waste ground. This species, often mistaken for London Rocket, seems to have been first noticed by D. G. Catcheside in 1925 on Mitcham Common, 26E5, Herb. L. Our records are: 16C2, Arbrook Common, 1 plant in car park, 1972, JES; 16E3, Tolworth, 1964, EJC; 17D1 & 17D2, Ham Pits, 1944, JEL, 1945 onwards, BW; 37D5, Rotherhithe, Surrey Commercial Docks, 1972, JEL.

S. orientale L. (*S. columnae* Jacq.)   Eastern Rocket.
Established alien.
First record: 1919, Latter, *Rep. BEC.*, **5**, 368, 1919. Rather rare. Railway stations, waste ground, refuse tips. Usually in very man-made habitats, but a chalk cliff in Seale quarry, 84E5, 1955, OVP, is wilder than most.

S. altissimum L. (*S. pannonicum* Jacq.).   Tall Rocket.
Established alien.   1877.   Frequent.
Waste places, roadsides, refuse tips, quarries. Mainly in the NE.

S. strictissimum L.   Garden alien.
First record: 1951, Kent & Lousley. Noticed in the churchyard of
St Anne's Parish Church, Kew Green, 17E4, by D. H. Kent in 1940,
probably originally planted, but visited by large numbers of
botanists annually, and still there, 1972, JEL.

S. wolgense Bieb.
Found on the Green Lane Refuse Tip, Malden, 34C1, 1957 & 1958,
RCW, det. DPY.

Arabidopsis (DC.) Heynh.
A. thalianum (L.) Heynh. (*Sisymbrium thalianum* Gay).
Thalecress.   Native.   1836.
Common on light soils in the western half of the county, frequent
elsewhere. Dry arable fields and gardens, walltops.

Camelina Crantz
C. sativa (L.) Crantz   Gold of Pleasure.   1763.
Common on refuse tips as a bird-seed alien but now very rare as a
weed in arable fields. One plant on the edge of an arable field N of
Betchworth Station, 25A2, 1950, Mrs H. P. Waldy.

Descurainia Webb & Berth.
D. sophia (L.) Webb ex Prantl (*Sisymbrium sophia* L.)
Flixweed.   Casual.   1827.   Very rare, 4.
Disturbed ground. 05C5, on dung heap near a farm, a single plant,
1965, RARC & DK; 15C5, Leatherhead Golf Course, several
plants on bare soil on fairway where turf had been removed, 1965,
JES; 27A1, horse-exercising ground by Beverley Brook with other
aliens, probably introduced with fodder; 35D4, Nore Hill Tip,
1961, RARC. Flixweed is probably always a casual alien in Surrey.

RESEDACEAE

Reseda L.
R. luteola L.   Weld.   Native.   1836.
Frequent over most of the county. Disturbed ground on the edges
of fields, roadsides, waste places, especially on the chalk, sometimes
very fine in woodland on the chalk after felling.

R. lutea L.   Wild Mignonette.   Native.   1832.
Common, 154.   Map 61.
Disturbed ground on chalk grassland, edges of arable fields, waste

Plate 9 *Headley Warren Reserve of the Surrey Naturalists' Trust, used for education and research: botany students recording quadrats*

Plate 10 *Thursley Common, control of water level by a sluice to maintain interest of vegetation*

Plate 11 *Reed invasion at Black Pond, Esher*

Plate 12 *Stag's-horn Clubmoss,* Lyco-podium clavatum, *Banstead Heath*

Plate 13 *A rare Horsetail,* Equisetum × moorei, *known at Shere since 1912*

ground, roadsides. Most plentiful on the chalk but, as the map shows, widely spread on other well-drained soils.

R. alba L.
Occurs at 17D1, Ham Pits, 1954, LNHS, and 26D4, Willow Lane, Mitcham, an established patch, and odd plants in four other places, 1957, RCW; 26E4, Mitcham Common, on embankment of new bridge, 1959, RCW & DPY.

VIOLACEAE

Viola L.
V. odorata L.   Sweet Violet.
Native, but also a relic of cottage garden cultivation.   1832.
Fairly common but most frequent on the chalk.
Wood borders, hedgebanks. Flowers are usually white: both var. *dumetorum* (Jord.) Rouy & Fouc. and var. *imberbis* (Leight.) Henslow occur. Var. *sulfurea* (Car.) Rouy & Fouc. was found in a clearing in the woods on chalk, Great Burgh, Epsom Downs, 25B5, 1950, DPY, *BSBI Year Book* **1951,** 121—this has ivory flowers shading to creamy yellow.

V. hirta L.   Hairy Violet.   Native.   1805.
Locally abundant, 96.   Map 62.
Chalk grassland, hedgebanks and wood borders. The map shows a remarkably close association with the chalk outcrops. Two localities in 94 are in a square where there are outcrops of Bargate limestone. In 05E5 on Wisley Common it was on imported chalky waste; in 17E1 it is near the Thames which brings down calcium carbonate from the Goring Gap. The records from 24E4, 34B5 and 34D5 are less easy to explain, but the last was on a railway bank and in two cases other calcicoles were noted for the tetrad.
x *V. odorata* = *V.* x *permixta* Jord. Frequent when the two parents occur in the same area, always on chalk.

V. calcarea (Bab.)   Gregory.
The small late-flowering plant, with narrow petals arranged in the form of a St Andrew's cross and an almost imperceptible spur, is plentiful in most years on the slopes above the Zig-zag, Box Hill, 15D1. This is the plant illustrated by Butcher, 1961 (*New Ill. Brit. Flora.*, **1,** 358). It is plentiful for only a brief period, which varies from year to year according to weather, and may well be a transitional stage towards later full cleistogamy. I have known the

plant here since 1923; the earliest date I have found it is April 28, the latest June 1. Elsewhere I have seen a few similar plants occasionally, but have never found a sizeable population. We also have records from 15C3, Fetcham Downs; 15D2, Headley Lane; 25B2, Buckland Hills; 26C1, Banstead Downs; 35C4, Warlingham, Tippett's Piece and 35D4, Nore Hill, but these include plants with larger petals and spurs. *V. calcarea* is now usually treated as a subspecies of *V. hirta*, but field observations suggest that they may be late-flowering plants of *V. hirta*.

V. riviniana Reichb.   Common Wood Violet.   Native. 1827.   Very common.
Woods, hedgebanks, and subsp. *minor* (Gregory) Valentine on commons and heaths. Some interesting plants from 15D2, open scrub in Norbury Park and 15E2, Juniper Top, Mickleham, 1965, EJC, and hirsute peduncles and petioles and were named forma *villosa* Neuman by Prof. D. H. Valentine.

V. reichenbachiana Jord. ex Bor.   Pale Wood Violet. Native.   1861.   Less common.
Woods and hedgebanks usually on chalk.
x *V. riviniana* occurs occasionally, as at 36D2, Selsdon Wood, 1959, DPY, and is probably overlooked.

V. canina L. (*V. ericetorum* Schrad.)   Dog Violet.   1724. Native.   Rare, 40.   Map 63.
Nearly all our records are from heaths or heathy commons, more rarely on ant-hills on the chalk, on gravel over chalk, and on a sandy bank.
x *V. lactea*. P. M. Hall suggested this determination for specimens I gathered on Mitcham Common, 26D4 in 1930, before *V. lactea* was known from there (see below). Similar plants persisted until very recently and may still be there.
x *V. riviniana*. Arises rather freely on heathy commons, such as Mitcham Common, 26D4, when *V. canina* grows with the small form of *V. riviniana*.

[*V. lactea* Sm. Native. 1883. We have failed to refind this at any of Salmon's localities (Chobham Common, Bisley Common, Copthorne or Hedge Court), but specimens were collected by C. Avery from Mitcham Common, 26D4, about 1930. Mowing the golf course has greatly reduced the number of violets allowed to flower and its is probably extinct, but it has left its influence in hybrids with other species.]

V. palustris L.   Marsh Violet.   1696.   Native.
Locally frequent, 40.   Map 64.
Wet acid places, bogs, hollows on heathland, sallow swamps, alder holts. Most of the records are concentrated on the commons in the far west of the county and in the Leith Hill area. Also on Oxshott Heath and Esher and Arbrook Commons, about Reigate Heath, Hedgecourt and Wire Mill Pond, Limpsfield and near Addington. Distribution is limited by availability of suitable habitats.

V. tricolor L. subsp. *tricolor*   Wild Pansy.   Native.
1836.   Rare and decreasing.
Cultivated ground, both on sand and chalk. Still persists S of Gomshall, 04E4, 1963, JCG, in an area where it has been known since 1909.

V. arvensis Murr.   Field Pansy.   Native.   1836.
Common throughout the county. Cultivated fields and gardens.
x *V. tricolor*. To this parentage belong some of the segregates of the last two species which attracted so much interest following Dr E. Drabble's paper in 1909 (*J. Bot.* **47,** *suppl.*) and subsequent notes and papers. Others are variants which are no longer recognised, to the great relief of botanists. Mr R. D. Meikle has confirmed specimens of the hybrid from 24C3 and 34A5.

POLYGALACEAE

Polygala L.
P. vulgaris L.   Common Milkwort.   Native.   1827.
Common.
Chalk grassland, commons, heaths, grassy banks and pastures.
P. oxyptera Reichb. (*P. dubia* Bellynck) is no longer regarded as worthy of separation.

P. serpyllifolia Hose (*P. serpyllacea* Weihe).   Native.
1724.   Frequent, 91.   Map 65.
Heaths and other rather acid places, especially on the Bagshot Sands, Lower Greensand and Weald Clay, often associated with Ling.

P. calcarea F. W. Schultz   Chalk Milkwort.   Native.
1833.   Locally common, 45.   Map 66.
Chalk grassland. A lovely feature of our chalk escarpment in May and early June.

[*P. austriaca* Crantz Kentish Milkwort. The roadside bank near Caterham where W. Whitwell found it in 1888 (*J. Bot.* **26,** 249) is now part of a garden and the species has not been refound in Surrey in recent years. In Kent it grows within 1½ miles of our county boundary and has several stations not far away; yet, in spite of thorough search by R. A. R. Clarke and others for several years, a Surrey locality still eludes us.]

GUTTIFERAE

Hypericum L.
H. androsaemum L.   Tutsan.   Native.   1763.
Locally frequent, 74.   Map 67.
Hedges and borders of copses and woods, with a preference for heavy soils and somewhat damp conditions. Most frequent on the Weald Clay and Hastings Beds in the S of the county.

H. inodorum Mill. (*H. elatum* Ait.) is a garden plant occasionally found under somewhat wild conditions, as at 04D5, Netley Park, 1961, BMCM. Similarly *H. hircinum* L. is sometimes established, as at 06A5 at Stroude where it persists by an out-building of Whitehall Farm in spite of the farmer's efforts to eradicate it, 1965, EJC.

H. calycinum L.   Rose of Sharon.   Naturalised alien.
1835.   Locally common.
Woodland rides, railway banks, hedgebanks. Between Leatherhead and Dorking this is thoroughly established in many places, especially in the woods of Norbury Park and above the railway tunnel and on railway banks, 15C2 and 15D3. By the drive at Denbies, 15C1, is another old and well-known station. There are many other places, especially on the chalk, where it has proved persistent and spreading vegetatively once planted, but we have no evidence that it spreads without human assistance.

H. perforatum L.   Perforate St John's-wort.   Native.   1836.
Abundant on the chalk, frequent elsewhere. Chalk grassland, open woods, commons and hedgerows.

H. maculatum Crantz (*H. dubium* Leers; *H. quadrangulum* auct.).
Imperforate St John's-wort.   Native.   1884.
Locally frequent, 73.   Map 68.
Damp places on the edges of woods, by canals and streams and
roadsides. All our plants belong to subsp. *obtusiusculum* (Tourlet)
Hayek.

x H. perforatum = *H. x desetangsii* Lamotte. Formerly abundant
in the Chipstead Valley but destroyed by agricultural activities in
1941, this has proved to be more widespread. 96E5, Windsor
Great Park, 1965, EJC & WEW; 06C2, New Haw, 1964, EJC;
14E2, Kingsland, Newdigate, 1960, BMCM & BW; 15D1, Stepping Stones, Box Hill, 1964, EJC; 24E2, Langshott Wood, 1956,
BMCM; 26A4, nr Malden Church, 1956, RCW; 34A2, Langshott
Wood as above; 34C2, Frogit Heath, 1961, RARC—most of these
det. DPY.

H. tetrapterum Fr.   Square-stalked St John's-wort.
Native.   1781.
Common throughout the county. Marshes, ditches, ponds, canals,
and rivers.

H. humifusum L.   Trailing St John's-wort.   Native.
1763.   Common, 185.   Map 69.
Heaths, roadside banks, rides in woods. In the chalk areas only on
clay-with-flint, and scarce in the NE of the county.

H. pulchrum L.   Slender St John's-wort.   Native.
1763.   Common, 212.   Map 70.
Dry heaths, banks and tracksides in woodland on somewhat
acid soils.

H. montanum L.   Pale St John's-wort.   Native.
1814.   Rather rare, 28.   Map 71.
Round scrub and wood borders on chalk but also in sandpits and
on roadsides. This species clearly prefers chalk, and the localities
mostly fall into three groups: round Godalming, influenced by the
Bargate stone; the chalk escarpment round Box Hill, and Croydon
southwards on chalk—but even in these areas there are habitats on
neutral soils.

H. elodes L.   Marsh St John's-wort.   Native.   1666.
Rare, 30.   Map 72.

Spongy bogs and on the margins of acid ponds on heaths. This
is still to be seen in quantity in the W of the county on our larger
commons and elsewhere, but we have no records E of the Black
Pond, Esher, 16B2. Drainage no doubt accounts for its disappear-
ance from some of the old stations, but it seems surprising if it has
really gone from the Leith Hill area, 14, and Hedgecourt, 34C1.

CISTACEAE

Helianthemum Mill.
H. nummularium (L.) Mill. (*H. chamaecistus* Mill.)
Common Rockrose.   1666.   Native.
Locally common, 81.   Map 73.

Banks and hill-slopes on chalk and abundant along most of the
chalk escarpment; also in 94 along the Bargate escarpment, in old
Bargate Stone quarries and on sand. *Cistus surrejanus* was a form of
this species described by Linnaeus (*Sp. pl.* **1,** 527, 1753) and said to
grow 'in Angliae comitatu Surrejano prope Croydon'. It was first
recorded in Dillenius' edition of Ray's *Synopsis*, ed. 3, 341, 1753,
described and illustrated in several other books, and rediscovered
in Croham Hurst Wood, 36B4, early in the nineteenth century,
but has not been seen recently. In any case it would now be regarded
as a mere form with laciniate petals—see Salmon, 1931, 154–5,
for a more detailed account of our Surrey Rockrose.

ELATINACEAE

Elatine L.
E. hexandra (Lapierre) DC.   Waterwort.   Native.
1824.   Very rare, 4.

Submerged in large ponds or canals, and flowering when exposed
on the muddy margins in dry summers. 86D1, pond in Staff
College Grounds, Camberley, 1952, RAB; 95C4, canal, Brookwood, 1957, RWD & WEW; 96D5, bed of drained Virginia Water,
1946, CPP; 34C1, Hedgecourt Pond, 1943, JAY, Hb.Y. All four
localities have a long history—Virginia Water dating back to the
first record for the county. Two records given are before 1950 but in
places where they may persist undetected.

[*E. hydropiper* L. Native. 1844. Probably extinct. Many old records
from Cut Mill Pond and Frensham Great Pond. In 1935 I collected

barren material, with the long petiole characteristic of this species, in water 2ft deep at Cut Mill Pond, 94A3, and it was collected again by H. S. Redgrove a few days later, Hb. Mus. Brit.]

CARYOPHYLLACEAE

Silene L.

S. vulgaris (Moench) Garcke (*S. inflata* Sm.; *S. cucubalus* Wibel) Bladder Campion.   Native.   1836.
Roadsides, hedgebanks, field borders, and chalk grassland. Common.

[*S. conica* L. Formerly an established alien but no recent records.]

S. dichotoma Ehrh.   Forked Catchfly.   Alien.
In several arable fields in the Chipstead Valley, 25D4, 1954, BMCM, Hb L. Seen there in decreasing numbers by JEL and others for a year or two. 25A2, Pebblecombe Hill, in fallow field, 1952, PDO.

S. nutans L. (*S. dubia* Herbich)   Nottingham Catchfly.
Native.   1908.
17D2, on a gravelly bank near the Thames at Ham, 1946–57; could not find in 1958, LJJ. This is var. *salmoniana* Hepper in *Watsonia* **1**, 80–90, 1951, where plants of this locality are discussed. It has been known here since before 1915 (*J. Bot.* **53**, 177), and the gravels here are likely to be calcareous. 15B2, Norbury Park, 1 plant on hillside, 1955, PDO; 2 plants, 1956, AEE.

S. gallica L. (*S. anglica* L.)   Colonist.   1760.   Very rare, 5.
Sandy fields. 95C3, arable field, Pirbright, 1950, DPY; 03C3, cornfield by Greenlane, 1957, JCG; 16B1, 1 plant by roadside, Fairmile Park, 1961, JEL, JCG, JES; 25B1, Reigate Heath near Clifton Lane, 1951–5, LJJ; 26B3, 1 plant, Barnes Common, 1951, LMPS. In addition this is recorded from a poultry farm in 94C1, 1960–1.

S. noctiflora L. (*Melandrium noctiflorum* (L.) Fr.)
Night-flowering Campion.   Colonist.   1756.   Very rare, 10.
Cornfields and other arable ground, also as a casual on refuse tips. 05B1, stubble field between Merrow Church and Fairway, 1962, KMM; 15E4 & 25A4, arable field along Headley Road, Ashtead, 1958, DPY; 26B1 & 26B2, cornfields, Ewell, abundant in places,

1962, DPY; 26E2, spontaneous and permanent in a garden at
Wallington, 1929, still there 1963, RG. Also on refuse tips at
Mitcham Common, 26E5; Caterham, 35C2, etc.

S. dioica (L.) Clairv. (*Lychnis dioica* L.; *Melandrium dioicum* (L.)
Coss. & Germ.)   Red Campion.   Native.   1836.
Very common.
Woods, copses, hedgebanks and damp thickets, avoids acid
habitats.

S. alba (Mill.) E. H. L. Krause (*Lychnis alba* Mill., *Melandrium
album* (Mill.) Garcke).   White Campion.   Native.
1836.   Very common.
Roadsides, borders of cultivated fields, waste places and refuse tips,
x *S. dioica*. This hybrid is not rare in the county, but there is a
tendency to over-record by ignoring the variations in flower colour
of the parents and assuming that all pink-flowered plants are
hybrid. Salmon (p 174) has a useful note and H. G. Baker showed
that the length of the calyx teeth provides one of the most convenient characters (*Genetica*, **25**, 125–56).

Lychnis L.
L. flos-cuculi L.   Ragged Robin.   Native.   1827.
Very common, 250.   Map 74.
Wet meadows, ditches and wet woods, thriving under rather acid
conditions. As the map shows, this is absent from most of the chalk
and rare in the NE where there are few undrained wet fields.

Agrostemma L.
A. githago L. (*Lychnis githago* (L.) Scop.)   Corn Cockle.
Colonist.   1775.   Very rare, 14.
Cornfields, especially on the chalk, also a casual on poultry farms.
This handsome plant was formerly frequent owing to its repeated
reintroduction from the Continent as an impurity in seed corn.
Now, with clean seed and modern cultivation, it has become
exceedingly rare as a cornfield weed. 14E2, abundant in a cornfield
N of Newdigate, 1945, FR; 14B5, near Landbarn Farm, 1950,
AWW; 25B5, arable field by Reigate Road, Epsom Downs,
formerly regularly, 1950, DPY; 25D4, field below Banstead Wood,
Chipstead Valley, 1955, JMS; 26E1, in barley crop, Grove Lane,
Carshalton, 1954, EWG; 35C5, in corn near Hamsey Green,
1946–57, LJJ, 1948, DPY; 35D3, field by bridleway, Marden Park,

1958, JPSR; 35E3, field W of Titsey Plantation, 1956, JMS. 44B4, abundant in barley field ½ mile N of Upper Barn, 1956, DPY. At the present rate these may well be almost the last records of Corn Cockle in corn in Surrey, but it still comes in with cheap grain for poultry and occurred, for example, on poultry farms in 94C1 and 15C3.

Dianthus L.
D. armeria L.   Deptford Pink.   Native.   1746.   Very rare, 3.
Well-drained sandy or chalky banks. 96D3, Chobham Common, about 50 plants on made-up soil by gates to army depot, 1970, WEW; a few plants after disturbance of habitat, 1972, JEL & JES; 15D4, in two places by Leatherhead by-pass, 1961, 1963, M. L. Cadman; 84D1, Rushmoor, Frensham Common, pathside in sandy woodland, 1965, DCK. Deptford Pink was formerly an undoubted native in the county, as, for example, near Addington where I knew it for many years, but all the above records are subject to some doubt about status. It was certainly casual in 27A2, Richmond Park, 1952, LMPS.

[*D. carthusianorum* L. grew from 1899 to 1910 or after in a sandy field near Byfleet (I believe in 05C5), but has not been seen in recent years.]

[*D. plumarius* L. was abundant on an old wall in Shalford-street, 04A4, from 1849 until at least 1935, G. Watts, but we have no recent reports.]

D. deltoides L.   Maiden Pink.   Native.   1778.
Very rare, more frequent as garden escape.
Dry banks and fields on sandy or gravelly soil. 96D4, Wentworth Golf Course on Bagshot Beds gravel, 1963, JES; 96E5, in turf, Virginia Water churchyard, 1965, EJC; 17D2, Ham, in gravelly pasture, 1971, FNH (recorded from here in 1835); 25B1, Reigate Heath, just below Golf Club House, 1963, M. Beattie comm. BMCM. These are our only records as possibly native, the remainder are all by gardens, or garden flowers. Exceedingly abundant at 94A2, Elstead Cemetery, 1974, DWB, RF, JEL.

Saponaria L.
S. officinalis L.   Soapwort.   1765.   Naturalised alien.
Frequent.
Roadsides, hedgebanks, waste places usually near houses, and often an obvious garden throw-out which has spread vegetatively. It is

possible that this is native in the W of England and Wales by streams and in damp woods, but we have not found it in these habitats in Surrey.

Cerastium L.
C. arvense L.   Field Mouse-ear.   Native.   1762.
Rare, 29.   Map 75.
Chalk grassland, roadside banks and fallow fields. Locally plentiful on the chalk in an area bounded by Leatherhead, Dorking, Reigate, Caterham and Sutton, also on sandy soils in the SE and NE, and on gravel or sand by the Thames. Still on Wimbledon Common, 27B2, where it was recorded by Pugsley in 1910, and on Hurst Park Racecourse, 16B5 & C5, where H. C. Watson recorded it in 1836.

C. tomentosum L. (including *C. biebersteinii* DC.)
Snow-in-Summer.   Alien.
This pretty but aggressive plant thrives in gardens on the chalk and sometimes on the sand, and mats of it are often established on banks by houses. The name covers at least two taxa which have not been worked out for Surrey, and it is recorded for 59 tetrads. Map 76.

C. fontanum Baumg. subsp. *triviale* (Murb.) Jalas (*C. holosteoides* Fr., *C. vulgatum* auct.)   Common Mouse-ear.   Native.
1724.   Common.
Fields, chalk grassland, heaths, waste places and wall-tops. Subsp. *triviale* is exceedingly variable in Surrey. Plants which are much more glabrous than usual, and with the stem leaves without hairs, occur by the tidal Thames and have been found from Battersea, whence they were recorded by Doody in 1724 nearly to Richmond, 17D3, JEL, Hb. L. They are probably allied to the plant collected by W. H. Beeby (no. 404) from wet meadows by the Thames at Putney in 1887 discussed by Jalas & Seel as *C. holosteoides* Fries subsp. *pseudoholosteoides* Moschl. (*Watsonia*, **6**, 294, 1967).

C. glomeratum Thuill. (*C. viscosum* auct.)
Broad-leaved Mouse-ear.   Native.   1836.   Frequent.
Arable fields, heaths, tracksides, walls, most common on sand. Apetalous forms are quite common.

C. diffusum Pers. (*C. atrovirens* Bab.; *C. tetrandrum* Curt.)
Sea Mouse-ear.   Established alien.   Very rare, 4.
Gravel track and railway tracks. 84C1, on sand by Frensham
Great Pond, plentiful, 1972, BWu; 04C5, Newlands Corner on
edge of old gravel track, 1964, DPY, conf. EM-R; 05E3, between
tracks, Horsley Railway Station, 1957, BW, Hb. L.; 34E2, Lingfield
on railway line, 1967, RARC; 44A4, railway line near Partridge
Farm, 1963, RARC. In the last 25 years *C. diffusum* has spread
extensively from its usual maritime habitats along railway lines
in Britain and abroad. It also grows on canal banks and sandy or
gravelly places as an introduction. In spite of the localities given
by Salmon (p 177) it is doubtful if it was ever native in Surrey.

C. pumilum Curt.   Dwarf Mouse-ear.   Native.
1794 (first British record).   Very rare, 1.
Chalk banks, with little competition from other species. 26C1,
Banstead Downs, many observers. A lovely little plant, glistening
with glands and flushed pink as it goes over. We have not refound
it on Epsom Downs or Walton Downs where there are still suitable
habitats.

C. semidecandrum L.   Little Mouse-ear.   Native.   1806.
Locally frequent, 115.   Map 77.
Sandy and gravelly places on commons, fields and in pits; rarely
on chalk. The distribution conforms closely to the Bagshot Beds
and Lower Greensand, with a few localities on the chalk, etc.

Myosoton Moench
M. aquaticum (L.) Moench (*Stellaria aquatica* (L.) Scop.).
Water Chickweed.   Native.   1836.   Frequent, 116.
Riverbanks, ditches, water meadows. Map 78. This species follows
the Wey and Tillingbourne, the Mole and tributaries, the Hogsmill
and the Thames, and the course of the long-disused Wey and Arun
canal, also the Wandle and Blackwater, and various minor streams.

Stellaria L.
S. media (L.) Vill.   Common Chickweed.   Native.   1836.
Abundant throughout the county in built-up as well as rural areas.
Gardens, arable fields, roadsides, waste ground.

S. pallida (Dumort.) Piré. (*S. boraeana* Jord.; *S. apetala* auct.).
Lesser Chickweed.   Native.   1850.   Rare, 15.
Loose sandy ground on commons and round cultivated fields.
Heavily under-recorded. We have numerous records from Thursley,
Frensham Great and Little Ponds, Puttenham, etc. in 83, 84 and 94
and scattered records from 85E1, 95A3, 04C4, 05C4, 16B5 and
17E2.

S. neglecta Weihe.   Greater Chickweed.   Native.
1882.   Rare, 29.   Map 79.
Damp borders of copses, shady ditches and hedgebanks. Mainly in
the W of the county, frequent about Chobham, and a group of
localities about Tilford, Thursley and Godalming, otherwise
rather scattered. Most of our material has been confirmed by
Dr F. N. Whitehead or I. J. Gibson.

S. holostea L.   Greater Stitchwort.   Native.   1836.
Common throughout the county. Wood borders, and hedgebanks.
Interesting variations of the flowers occur such as var. *laciniata*
Bromf., 15B4, Bookham Common, 1950, WHS, Hb. L; var.
lousleyi (Druce) Brenan & Lousley, 35D4, roadside near Wolding-
ham, 1927, JEL, Hb. L; 96D3, Valley End, Chobham, 1891,
J. D. Hooker, Hb. K; var. *apetala* (Rostrup) Asch. & Gr., 25E4,
Netherne House, 1902, CEB, Hb. K; 05D5, Wisley, 1928, Turner,
Hb. SLBI.

S. palustris Retz (*S. glauca* With).   Marsh Stitchwort.
Native.   1724.   Rare, 12.   Map 80.
Marshy meadows and by ditches. A decreasing species. Now very
rare in marshy meadows by the Thames and reported in only
two places, persisting in about seven localities by the R. Wey and
its branches, and still near the Blackwater.

S. graminea L.   Lesser Stitchwort.   Native.   1836.
Common over most of the county but avoiding the chalk. Heaths,
commons and roadsides, particularly abundant on commons on
sand or gravel.

S. alsine Grimm (*S. uliginosa* Murr.)   Bog Stitchwort.
Native.   1827.
Frequent and well distributed throughout the county. In shallow
water by wet tracks, streams, ponds and woodland rides.

[*Holosteum* L.
H. umbellatum L. This was found by Miss R. M. Cardew on and by the ruins of Newark Priory, 05C4, in April 1905 and was seen annually for at least 25 years. The last definite record known to me was that of Lt Col G. Watts who was shown the plant in fruit on 9 May 1930. Lady Davy took me to see it the following year but we failed to find it, and it is presumed extinct.]

Moenchia Ehrh.
M. erecta (L.) Gaertn., Mey. & Scherb.
Upright Chickweed.   Native.   1827.   Rare, 16.
Map 81.
In short turf on sandy or gravelly commons. A decreasing species but it often varies widely in quantity from year to year, and the white flowers rarely open, so that it is seldom safe to regard it as extinct in localities which still appear suitable. It occurs on at least four golf courses and a cricket pitch which provide the short grass it requires.

Sagina L.
Dr Young was keenly interested in Saginas and especially in *S. apetala*, *S. ciliata*, and *S. filicaulis*, which he regarded as distinct species. His views are followed here. Doubtless all three are somewhat under-recorded but the accounts are based on specimens checked by him.

S. apetala Ard.   Annual Pearlwort.   Native.   1832.
Common, 111.   Map 82.
On, and at the base of, walls, less often on gravelly or sandy tracks.

S. ciliata Fr.   Ciliate Pearlwort.   Native.   1863.
Locally frequent, 67.   Map 83.
Gravelly and sandy places on heaths and commons, rarely in churchyards and on railway sidings. Mainly on the commons in the W, and on the sand W of Reigate.

S. filicaulis Jord.   Native.   1908.
Rather rare, 38.   Map 84.
Garden paths and flagged steps in gardens, walls, cindery tracks, on graves. Much more a plant of man-made habitats than *S. ciliata*. In 06C1 at Byfleet on former heathland, 051616, Dr Young in 1966 found *S. filicaulis* abundant on the N side of the road, and *S. apetala* in the S side. His notes include only one case of two of

these species growing together—in 24D1, at Povey Cross, he found
S. filicaulis growing with *S. apetala* on the path of an abandoned
garden, 269419, but keeping absolutely distinct.

S. procumbens L.   Procumbent Pearlwort.   Native.   1836.
Abundant throughout the county, and especially so in built-up
areas. Walls, pathways, field borders, tracks in woods, in damp
and shady places.

S. subulata (Sw.) C. Presl   Heath Pearlwort.   Native.
1778.   Very rare, 2.
Moist places on sandy or gravelly heaths. 94D1, N of Hydon
Heath, 1960, GSE; 96D5, Virginia Water, 1955, 1963, WEW.
Always a rare plant in Surrey, this has decreased rapidly and is
approaching extinction. It is not easy to suggest a convincing reason,
but it may be that our commons are less wet than formerly.

S. nodosa (L.) Fenzl   Knotted Pearlwort.   Native.
1746.   Very rare, 3.
This is usually a plant of wet places which dry out in summer, but
our two most recent records are from broken chalky ground.
84C1, dry bed of Frensham Little Pond, 1947, ECW; 05D1,
E. Clandon Downs, 1969, WEW; 26C1, Banstead Downs, 1957,
EM & RCW, 1967, abundant, ECW (previous record from here
in 1882).

S. glabra (Willd.) Fenzl   Cemetery Pearlwort.
Established alien.   Rare.
On and between graves in cemeteries, garden lawns. 16B3, lawn in
Claremont Lane, Esher, 1960, JES; 17E3, East Sheen Cemetery,
1959, BW; 17E4, Hammersmith Cemetery, Mortlake, 1959, BW;
25C4, lawn in Outwood Lane, Kingswood, 1961, DPY; 37E3,
Nunhead Cemetery, 1965, RARC. This plant is receiving further
study and it will be necessary to change the name.

Minuartia L.
M. hybrida (Vill.) Schischk. (*Arenaria tenuifolia* L.).
Native.   1666.   Very rare, 2.
On chalky banks, formerly also on sand. 94C4, Puttenham Heath,
on chalk; 1957, OVP (first recorded from here in 1849); 26C1,
Banstead Downs, known here since 1837, it varies in size and
quantity from year to year. (I have also seen this species in 94D3,

Godalming railway goods-yard, 1934, JEL, where it was probably an introduction. It has spread along railways in Ireland).

Moehringia L.
M. trinervia (L.) Clairv. (*Arenaria trinervia* L.)
Three-nerved Sandwort.   Native.   1763.   Frequent.
Well-drained soils in woods, very common in beechwoods.

Arenaria L.
A. serpyllifolia L.   Thyme-leaved Sandwort.   Native.
1836.   Common.
Weed in arable fields and gardens, walls, bare soil on chalk, open sandy grassland.

A. leptoclados (Reichb.) Guss.   Native.   1863.
Common, 127.   Map 85.
Very dry places, walls, railway tracks, cornfields on chalk and sandy banks. Almost absent from the Weald Clay and scarce on the Bagshot Beds. Probably best treated as a subspecies of *A. serpyllifolia*.

A. balearica L.
A garden plant which is well established on walls and brickwork has been reported from: 04B3, wall near lodge at entrance to Barnet Hill House, and 04B5, wall of St Martha's churchyard, both 1957, JCG; 15E4, Ashtead Park, 1965, DPY; 35C1, Godstone, 1961, JEL.

Spergula L.
A. arvensis L. (including *S. vulgaris* Boenn. & *S. sativa* Boenn.)
Corn Spurrey.   Native.   1836.   Common.
A weed in arable fields, waste ground, and sandy tracksides.

Spergularia (Pers.) J. & C. Presl
S. rubra (L.) J. & C. Presl   Common Sandspurrey.
Native.   1827.   Frequent, 152.   Map 86.
Rather bare dry places on sand and gravel. The map shows it as recorded for a number of tetrads in 25 and 35 which are mainly chalk, but the occurrences are on clay-with-flint or even on gravel paths, as in Headley Churchyard.

Polycarpon L.
P. tetraphyllum (L.) L.
Found by Miss M. Whitelaw in 1944 in her garden at Longdown, Hindhead, 83E3. It is still there as a garden weed in quantity, 1974, DB, JEL. No doubt the plant was introduced, but it is not known how it arrived and this was probably when a former owner was in residence. Introduction with cage-bird seed has been suggested, but would have been unlikely in wartime.

ILLECEBRACEAE

Scleranthus L.
S. annuus L.   Annual Knawel.   Native.   1836.
Rather common, 99.   Map 87.
A weed in arable fields indicating poor, acid soils, also on sandy or gravelly heaths and commons, a calcifuge Especially common on Lower Greensand.

PORTULACACEAE

Montia L.
M. fontana L.   Blinks.   Native.   1763.
Frequent, 44.   Map 88.
Wet places, by streams, damp arable fields and gardens, especially on sandy soils and in nurseries, and by ponds. The following subspecies occur:

subsp. *chondrosperma* (Fenzl) Walters. Most of our material belongs to this and specimens have been checked from 15 localities as shown on Map 89.

subsp. *amoritana* Senn. (subsp. *intermedia* (Beeby) Walters; *M. lusitanica* Samp.) Very rare, 3. 05D5, Wisley, weed in flower-beds of RHS Gardens near R. Wey, 1963, BW; 45B2, Moorhouse, Limpsfield, marshy edge of stream, 1957, DPY; 26E4, Mitcham Common, 1930, JEL—all det. SMW.

subsp. *variabilis* Walters. Very rare, 1. 04C4, in stream, Postford Farm, Albury, 1966, RARC, det. SMW.

M. perfoliata (Willd.) Howell (*Claytonia perfoliata* Donn)
Spring Beauty.   Naturalised alien.   1853.
Locally common, 102.   Map 90.
Weed in cultivated ground, plantations, hedgerows, especially on sandy soils. Common on light soils in the W of the county, especially

on Bagshot Beds and Lower Greensand, and also on the sands about Reigate and Bletchingley.

M. sibirica (L.) Howell (*Claytonia alsinoides* Sims)
Pink Purslane.   Naturalised alien.   1915, roadside bank, Gomshall in plenty, W. H. Griffin, *Rep. BEC*, **4**, 63.
Rare, 16.   Map 91.
Tracks, ditches, by streams and in woods. Most of our localities are obvious garden escapes, but round Blackheath and Holmbury St Mary it is well established: 04B4, well established in lane from Blackheath to Chilworth, 1935, JEL; 04E4, track near Burrows Cross, 1963, JCG; 14A3, roadside nr Felday, G. E. Stephens in Salmon, 1930; roadside copse, Felday, 1936, JEL; Felday by track to Holmbury Hill, 1961, BW; plentifully naturalised for considerable distance along stream N of Holmbury St Mary School, 1960, FDSR; by stream in NW corner of Pasture Wood, 1962, JCG. Other tetrads where it may become established are 83D4, 86E1, 96D4, 16C2.

[*Portulaca* L.
P. oleracea L. seems to have been established for a time round Kew and Richmond, Salmon 1930, but we have no recent records.]

AMARANTHACEAE

Amaranthus L.
A. hybridus L. (*A. chlorostachys* Willd.).   Casual.
04A4, potato field at Clinthurst Farm, Wonersh, 1965, 1966, JFL conf. DPY; 15D2, field by R. Mole, Norbury Park, 1956, JEL, det. JPMB.

Several other species of *Amaranthus* appear as casuals on refuse tips.

CHENOPODIACEAE

Chenopodium L.
C. bonus-henricus L.   Good King Henry.
Naturalised alien.   1793.   Rare, 21.   Map 92.
Mainly in village streets or by farmhouses. Formerly much used as a vegetable like spinach, Good King Henry is very persistent in patches near old gardens. Our records which seem related to much older ones include: 84E5, Seale, 1959, VML; 15B4, between

Slyfield House and Barracks Farm, 1954, AWJ; 35B1, by wall of Brewer St farmhouse, 1956, BAK.

C. polyspermum L.   Many-seeded Goosefoot.   Native.
1793.   Frequent, 74.   Map 93.
Bare ground by rivers and ponds, cultivated fields and gardens, usually on rather rich soils. A variable species.

C. vulvaria L.   Stinking Goosefoot.
Formerly native, 1802, this is now extinct; but it occurred in 1954 in a garden at Morden, 26C4, where it was introduced with peas, RAG.

C. album L.   Fat Hen.   Native.   1836.
Abundant throughout the county.
Cultivated land, gardens, farmyards, waste places and roadsides. Extremely variable and yet difficult to divide into satisfactory segregates. Several allied species have been found on Surrey refuse tips as introductions.

C. opulifolium Schrad.   Alien.   1863.
Many plants, fruiting well, on formerly cultivated ground in edge of Walton Common, 06E3, 1965, RAB & JES. Also occasionally on refuse tips.

C. ficifolium Sm.   Fig-leaved Goosefoot.   Native.   1832.
Common, 122.   Map 94.
Cultivated fields and gardens, especially nurseries, waste ground. Most common on Weald Clay and London Clay, but also on sands and other soils.

C. murale L.   Nettle-leaved Goosefoot.   Casual.   1793.
Very rare, 5.
Cultivated ground, no longer persistent. 04A4, field near Clinthurst Farm. 1959, JCG; 25B5, Epsom College, 1951, AEE; 26B1, Downs Farm, Ewell, 1962, DPY; 26D3, waste ground, Carshalton 1958, RCW; 26E3, Female Orphanage, Beddington, plentiful 1942, JEL.

C. hybridum L.   Maple-leaved Goosefoot.   Casual.   1782.
Rare, 6.
Cultivated and waste ground. 84B4, car park, Farnham, 1961,

VML; 84C4, cultivated field near Farnham Cemetery, 1959, VML; 95E5, garden, Horsell, 1964, 1965, WEW; 06C1, tip, Byfleet, 1960, MW; 26D4, Morden, 1957, RCW; 27C3, Wandsworth Park, 1951, RAB. No longer persistent in Surrey.

C. rubrum L.   Red Goosefoot.   Native.   1793.
Frequent, 156.   Map 95.
This occurs as a small form (var. *pseudo-botryoides* Syme) on the exposed margins of ponds, and larger forms on manure heaps, cultivated ground and waste places. Formerly it was distributed by horses, passing through them and germinating in the farm yard or fields. The present distribution is patchy; it is scarce, as one might expect, in the areas of large commons in the NW and SW, and also in the Leith Hill area, Richmond and the outer London suburbs.

C. glaucum L.   Oak-leaved Goosefoot.   Casual.
1793.   Very rare, 5.
Gardens and refuse tips, probably distributed in horse-manure. 93A2, Spring Mill garden, 1964, EMH; 06C1, tip, Byfleet, 1960 and onwards, MW, AB; 06D1, tip near Sanway, 1963, EJC; 15B4, manure heap by Slyfield House, 1954, AWJ; 17E4, weed in Kew Gardens, 1960, BW; 26C2, garden, Cheam, 1964–72, CPP.

Atriplex L.
[*A. littoralis* L. This maritime species formerly ascended the tidal Thames as far as Kew, but has been extinct for at least 60 years. It occurred, however, as a casual on a refuse tip near Earlswood, 24D5, 1948 & 1949, BMCM.]

A. patula L.   Common Orache.   Native.   1836.
Very common.
Cultivated fields and gardens, waste ground.

A. hastata L. (including *A. deltoidea* Bab.)   Native.   1827.
Frequent.
Cultivated fields and gardens, waste ground.

PHYTOLACCACEAE

Phytolacca L.
P. americana L.   Poke-weed.
Garden plant which spreads outside by seed or persists when an outcast. 15D1, by Mole below Box Hill, established for several

years, 1948, RAB; 26B2, 1 plant spontaneous in garden of 'Priest's Hill', 1962, L. Garland; 17E4, passage between Leybourne Park & Cumberland Road, 3 seedlings in passage from parent in garden, 1958, further along in passage 1 fine plant, 1959, BW. It has not been possible to check these identifications.

TILIACEAE

Tilia L.

T. platyphyllos Scop.   Large-leaved Lime.   Alien.   1666.
Rare, probably always planted.
25B2, Pilgrim's Way path, 'naturally sown', 1972, RARC; 36E1, Fickleshole, planted in hedge on county boundary, 1965, RARC.

T. cordata Mill.   Small-leaved Lime.   Alien?   1856.
Occasionally planted in woods as at Birch Wood, Caterham, 35C4, where there is a group of old heavily coppiced trees, 1972, RARC. It may be native in Glover's Wood, Charlwood, 24B1, where there is a group and an isolated tree in a ghyll on Paludina limestone; here it grows under very similar conditions to those in Dallington Forest and other Sussex woods where it is regarded as native.

T. x europaea L. (*T. vulgaris* Hayne).   Common Lime.
Alien.   1838.
Commonly planted throughout the county and occasionally appearing naturalised. The Common Lime, which is generally regarded as a hybrid between the two preceding species, produces fertile seed and this may be responsible for such records as 'naturalised in the Caterham Valley', 35B4, 1957, DPY.

MALVACEAE

Malva L.

M. moschata L.   Musk Mallow.   Native.   1763.
Common, 193.   Map 96.
Dry places, rough fields, roadsides, hedgebanks, sandy, gravel and chalk soils. Most frequent in the S of the county, inexplicably not recorded from several areas.

M. sylvestris L.   Common Mallow.   Native.   1836.
Common.
Mainly a roadside and waste-ground plant, field borders. Thrives in

towns and villages where, as Salmon pointed out, it loves the shelter
of walls.

M. neglecta Wallr. (*M. rotundifolia* auct.)   Dwarf Mallow.
Native.   1827.   Common.
Dry places round buildings, farmyards, roadsides, waste places.
Thrives in the company of man and still extends right into London,
as at Festival Hall, 37A5, and Surrey Commercial Docks, Rother-
hithe, 37C5.

M. pusilla Sm.   Small Mallow.   Casual.
1878.   Rare, 5.
Laneside and waste ground. 04B4, Chilworth Railway Station,
1961, JCG; 04D1, Fowl's Farm, 1961, JCG; 16D3, waste ground,
Esher, 1958, JES; 16E5, refuse tip, Kingston, 1958, JES & BW;
25B1, laneside near Kemp's Farm, Buckland, 1950, BMCM & BW,
JEL, *BSBI Yearbook* **1950**, 101. It is perhaps only coincidence
that Salmon, 1931, records this from 'Ditch near Buckland Lane'—
there is no Buckland 'Lane' today.

M. parviflora L.   Alien.   1866.
16C2, Hare Lane, JES, det. JEL; 16C3, Littleworth Common,
1961, JES.

Althaea L.
A. officinalis L.   Marsh Mallow.
Native on the coast but probably a hortal introduction at 36A4,
waste ground nr Croydon 'B' Power Station, 1956, RCW. This is
in the same area as Beeby's 'waste ground by the roadside between
Croydon and Mitcham Common' which is at least 60 years old,
given by Salmon, 1931, p 203.

[*A. hirsuta* L. Hairy Mallow. Salmon claimed that this was native on
the chalk hills near Reigate 25B2? where it was found by W. B. Alexander
in 1902 (*J. Bot.*, **40**, 409) and continued until 1908 when the field was
taken into cultivation. It was rediscovered in some quantity in 1922 but
has not been reported recently and is probably extinct.]

Sidalcea A. Gray
S. malvaeflora A. Gray.   Naturalised alien.
94E4, by R. Wey in Artington parish, S of Guildford, 1955, EBB &
EWG, a clump 2m tall, seen by many observers and still there,
1967, DPY, 1974.

LINACEAE

Linum L.

[*L. bienne* Mill. (*L. angustifolium* Huds.) Native, presumed extinct. None of Salmon's records came from permanent stations and it has not been reported recently.]

L. usitatissimum L., Common Flax, is quite frequent as a casual from cage-bird seed.

L. catharticum L.   Purging Flax.   Native.   1664.
Common.
Chalk grassland, dry pastures, heaths and roadsides.

Radiola Hill

R. linoides Roth   All-seed.   Native.   1666.
Rare, 16.   Map 97.
Sandy or peaty depressions on heaths where water has stood during winter; damp depressions in tracks in woods. A decreasing species. The localities fall into three groups: the commons in the W of the county; Reigate Heath, 24B5; and Glovers Wood, 24B1, and rides in woods on Hastings Beds in the extreme SE.

GERANIACEAE

Geranium L.

G. pratense L.   Meadow Cranesbill.   Native by R. Thames and perhaps a few other places; elsewhere as a garden escape.
1696.   Frequent, 79.   Map 98.
River banks and wet meadows, hedgebanks, chalkpits, roadsides. Meadow Cranesbill is accepted as native from water meadows higher up the Thames, and also from grassland on chalk or limestone. In Surrey it is an unquestioned native by the Thames where it still occurs from Runnymede down to Kew, with a gap from Surbiton to about Ham due to the destruction of suitable habitats. For this reason it is decreasing. Over the rest of the county new localities are being found nearly every year, many being obvious garden escapes, some looking very 'wild'. Since it is impossible to decide on their status on historical or ecological grounds, it is safer to regard the species as an alien except in the 16 tetrads by or near the Thames.

G. endressii Gay.   French Cranesbill.   Established alien.
We have records from: 95C1, 95C3, 14B3, 15C5, 16D1, 35E2,

35D3, 36B1, 36B2. At Broadmoor, Wotton, 14B3, it was thoroughly established by a stream before 1947 and is still impressive; and it has persisted in some of the other stations for lengthy periods. The first records traced for Surrey are from Littleworth Common, 16C3, and Chelsham, 35D5?, by C. E. Britton (*J. Bot.*, **70**, 316, 1932). x *G. versicolor*. Established garden escape. 94E3, Tilthams Corner, Shalford, 1958, LJJ; 06A5, established near Hurst Farm, Thorpe Green, 1965, EJC.

G. versicolor L.   Pencilled Cranesbill.   Established alien.
1838.   Very rare, 2.
33B5, Horne in green lane on county boundary, 1963, RARC; 45B1, Limpsfield Chart, 1957, RARC.

G. phaeum L.   Dusky Cranesbill.   Established alien.
1836.   Rare, 6.
We have records from: 94A3, 95D3, 03A5, 05E1, 25A1 and 34C2, The plant from Worplesdon, 95D3, 1953, 1954, BMCM & BW, has blue flowers and when DPY collected it in 1960 he commented, 'Looks to me like a "wild" type plant such as occurs normally in the Pyrenees'.

G. sanguineum L.   Bloody Cranesbill.   Garden outcast.
35, Oxted on sand, 1957, RARC; 83E3, Hindhead Common, site of old radar station, 1964, DWF, 1970, BMCM; also on refuse tips and near houses.

G. x magnificum Hyl. (=*G. ibericum* auct.; *G. platyphyllum* Fisch. & Mey.)   Established alien.
17D1, river wall a little downstream from Teddington Lock, 1944–50, BW, 1967, DPY.

G. pyrenaicum Burm. f.   Hedgerow Cranesbill.
Established alien.   1780.   Locally common, 127.
Map 99.
Hedgebanks, railway banks, and roadsides. Often regarded as a native, the rapid spread since this was first recorded for Britain in 1762, and the nature of the habitats, strongly suggest an alien. The distribution in Surrey is clearly not associated with geology or natural features, but many of the localities are along railways. For example, the dots on the maps follow the Brighton line faithfully from Purley to the Sussex border, the line from Dorking through

Redhill to the Kent boundary, and the East Grinstead line through Oxted and Dormans. There are a good many occurrences away from railways, but if the initial spread was by roads and railways this will explain the large areas which the species has not yet reached.

G. columbinum L.   Long-stalked Cranesbill.   Native.
1832.   Rather rare, 35.   Map 100.
Grassland and open woods on chalk, sandy banks. The localities fall into two sharply defined groups. Those to the NE of Shere are on chalk, where the species is often found on rather bare ground. The group in SW Surrey is on sandy ground on Lower Greensand in the area of Bargate Stone but not closely associated with it.

G. rotundifolium L.   Round-leaved Cranesbill.   Native.
1762.   Rare, 17.   Map 101.
Sandy and gravelly banks on roadsides. This has been known from a roadside bank, Guildford Road, E of Farnham, 84D4, for at least 70 years, and it also occurs in Waverley Lane, 84C4, 1963, AJS. 'Very abundant at Farnham for at least 40 years—now decreasing', 1961, WEF. The 7 localities in the SW are all on Lower Greensand; the remainder on various rocks. It is still on Barnes Common, 27B4, where it was found by F. A. Paley in 1877–81. Probably native in some places, but the recent increase is due to introductions.

G. molle L.   Dovesfoot Cranesbill.   Native.   1827.
Very common.
Dry grassland on commons and downs, pastures, cultivated fields, roadsides.

G. pusillum L.   Small-flowered Cranesbill.   Native.
1827.   Frequent, 124.   Map 102.
Mainly in sandy soils, commons, arable ground, waste ground.

G. lucidum L.   Shining Cranesbill.   Native.   1633.
Rather rare, 52.   Map 103.
A calcicole, mainly on walls but also on dry banks. Especially characteristic of the Godalming district, where the map shows a concentration of localities. Here it is associated with the Bargate Stone used to build walls, but also occurs on banks. Elsewhere, further east, it is on chalk, and also on isolated limestone walls.

G. dissectum L.   Cut-leaved Cranesbill.   Native.
1666.   Very common.
Arable fields and gardens, roadsides and waste places. Extends right into London as a garden weed.

G. robertianum L.   Herb Robert.   Native.
1827.   Very common.
Wood borders, shady lanes and roadside banks, walls.

Erodium L'Hérit.
[*E. maritimum* (L.) L'Hérit. This grew on a sandy bank near Farnham from 1846 to 1861 or after, but has not been seen recently. It has been found in Hampshire as a wool alien recently, and may have been introduced at Farnham with shoddy used as manure for hops.]

[*E. moschatum* (L.) L'Hérit.   Musk Storksbill. No recent records. A common wool alien and in SE England occurrences inland are often connected with the use, or former use, of shoddy as manure on light soil. This probably explains some of the earlier Surrey records.]

E. cicutarium (L.) L'Hérit.   Common Storksbill.   Native.
1724.   Locally common, 161.   Map 104.
On sand and gravel on commons, by tracks, and in arable fields. Salmon, 1931 (pp 215–6), attempted to divide this species into *E. triviale* Jord. and *E. pimpinellaefolium* Sibth. (*E. praetermissum* Jord.) but although our plants are very variable in leaf-shape, presence or absence of black spots at the base of petals etc., it seems that we have only one taxon, subsp. *cicutarium*. The distribution map shows that it is very frequent on the light soils of the Bagshot Beds and Lower Greensand in the western half of the county and scattered on light soils elsewhere.

OXALIDACEAE

Oxalis L.
O. acetosella L.   Wood Sorrel.   Native.
1763.   Common.
Woods, and especially woods rich in humus, beech and oakwoods.

Dr D. P. Young made a special study of the species of *Oxalis* which occur as weeds in gardens, nurseries and elsewhere. His work revolutionised the treatment of these plants in British *Floras*, which is now usually based on his paper 'Oxalis in the British Isles'.

Watsonia **4,** 51–69, 1958. He continued working on this genus until his death, which prevented publication of his final views on several difficult taxa.

O. corniculata L. Naturalised alien. 1837, *Fl. Met. Suppl.* Frequent, 51. Map 105.
Paths, pavements and flower-beds in gardens, churchyards and cemeteries—no records contributed from outside such places.

O. exilis A. Cunn. (*O. corniculata* var. *microphylla* Hook. f.)
This forms a low compact mat of wiry creeping stems and has short fat capsules. It is frequently grown on rockeries and is occasionally established just outside gardens, such as 14A2, foot of wall, Holmbury Hill, 1932, ECW, Hb.L., in Young, 1958. Dr Young also named the following, but no doubt records were also included in those for *O. corniculata:* 96E1, Woking, garden weed, 1963, WEW; 16E3, Hook, garden weed, 1964, EJC, Hb. Clement; 36, Croydon, back of parish churchyard, 1958, RCW.

O. europaea Jord. (*O. stricta* auct.) Garden alien.
1838, John Stuart Mill, *London Fl.*, as *O. corniculata*.
Frequent, 74. Map 106.
A pestilent weed in flower-beds, nurseries, churchyards and shrubberies, occasionally in arable fields and plantations. var. *rufa* (Small) D. P. Young. Named by Dr Young from the following localities: 94C2, Milford, confined to one garden although *O. europaea* is frequent in other gardens in the district, 1952, DPY; 24A1, Horley, garden on Balcombe Road, 1953, DPY; and 24E1, Horley, 1953, FMG and DPY (*Proc. BSBI*, **1,** 463–4, 1955; *Watsonia,* **4,** 69, 1958); 25C5, Banstead, weed in churchyard, 1960, DPY.

O. corymbosa DC. Garden alien.
Locally common, 39. Map 107.
A pestilent weed in flower-beds, nurseries, churchyards, sometimes established in hedges both in and out of gardens. Probably first recorded by D. P. Young, 1955 (*Proc. BSBI*, **1,** 577). A common plant in the old gardens of London and London suburbs but, as the map shows, now known to be widespread over the county. Almost impossible to eradicate owing to the persistence of bulbils. One potential way in which this species may spread into wilder habitats is indicated by the observation of a great pile of plants dumped in the stream from a nearby garden at Bridge End, Ockham,

05D4, 1967, DPY. Cultivation by rotary cultivator produces a rapid increase.

O. debilis Kunth
Very close to the last species from which it differs in having smaller (2–3mm) bulbils and salmon-pink or brick-red flowers, and Dr Young hesitated in treating it as specifically distinct. He noted it at Kew Gardens, 17E4, as a weed in the Rose Garden and other flower-beds mixed with *O. corymbosa*, 1951, DPY (*Watsonia*, **4**, 63, 1958); also from 25B1, spontaneous weed in garden at Buckland, 1957, BMCM, and 25C1, derelict yard at Reigate, 1960, DPY; several gardens in the Reigate and Redhill district, more frequent than thought previously, and spreading, 1974, BMCM; 26E3, Wallington, The Grange, garden weed with *O. corymbosa*, 1957, DPY (*Proc. BSBI*, **3**, 294, 1959).

ZYGOPHYLLACEAE

[*Zygophyllum* L.
Z. fabago L. Syrian Bean Caper. A garden alien from N. Africa, was established in 26E4 on the edge of Mitcham Common, 1955, JEB, Hb. BM, and in an old gravel pit at Beddington Corner, 1955, JEP, 1956, DPY & JEL, where it continued for some years but had disappeared by 1963.]

BALSAMINACEAE

Impatiens L.
I. noli-tangere L.   Touch-me-not Balsam.
Naturalised alien.   1835.
This has been established for at least 40 years in 34C1, by the stream which runs into Hedgecourt Pond, 1933, JEL, Hb.L, 1955, DPY, 1974, JEL. It is also in the field on the S side of the road opposite the Pond, 1953, RWD, but has not persisted in the rather unsuitable localities cited in Salmon, 1931.

I. capensis Meerb. (*I. biflora* Walt.; *I. fulva* Nutt.)
Orange Balsam.   Naturalised alien.   1829.
Common, 91.   Map 108.
Riverbanks, streamsides, swamps and wet fields, and copses. This native of N. America was first noticed in Britain in 1822 by John Stuart Mill, who found it by the Tillingbourne at Albury, 04C4; it had been carried down from Albury Park gardens, where it was cultivated. Seven years later it was found by S. Palmer by the

R. Wey at Guildford—the first printed record for Britain. From here it was carried down to the Thames and had reached Barnes, 27A4(?), by 1843 and Nine Elms, 27E4, by 1859. The present distribution reflects this history. It is common along the Tillingbourne and the Wey, and on the banks of the Thames from the Berkshire border down to Mortlake Brewery, 27A4, after which continuous concrete embankments prevail. It is also along the Mole and the Basingstoke Canal. It moves upstream as well as down— probably enabled to do so by its seed propelling mechanism. It also occurs in wet places away from rivers. On Wimbledon Common, it occurs in abundance right down the Farm Ravine, 27A1, to the Beverley Brook, but is always cleistogamous, AWJ, JEL.

I. parviflora DC.   Small Balsam.   Naturalised alien.
1855.   Common, 127.   Map 109.
Woodland rides and borders, especially on the chalk, roadsides and waste places. The first record is from Nine Elms, Battersea, but it has been known for over a century from Norbury Park and Mickleham, where it is well established. Probably the earliest evidence from here is an undated specimen from a roadside at Mickleham collected by H. Trimen, Hb. Mus. Brit., likely to have been collected between 1869 and 1879. In Norbury Park Woods, 15D2, E. B. Bishop collected it in 1918, Hb. Mus. Brit., and in 1919 it was observed by BENA ramblers to have become abundant by the R. Mole and near Mickleham (*Countryside Leaflet*, **5**, 109.) It is still plentiful in 15C2, 15D2 and 15D3, in and around Norbury Park. Around Titsey Park, in 35E3 and 45A3, it has been known under very similar conditions since 1924.

I. glandulifera Royle.   Indian Balsam.   Naturalised alien.
1900, Dunn ex Britten, *J. Bot.* **38**, 51.   Locally common, 84.
Map 110.
Established and spreading on the banks of rivers and streams, alder holts, gardens and waste places. Along the whole length of the Mole, and by the Thames from the junction with the Mole down to Barnes, in several places by the Wey, and Wey Navigation; scattered, and mainly less permanent, records elsewhere. Dunn first recorded this species as *Impatiens noli-me-tangere* in 1893 (*Flora SW Surrey*, 17) from Sampleoak Lane, Chilworth, 04B4, and corrected it as above in 1900, when he said it had established itself and was spreading freely. Whitwell soon followed this up with a record from Weydown Common, 93A3, where he said it was

established from self-sown seed from plants in a cottage garden.
There is no evidence that it spread from these places, but it seems
that of the few localities in Salmon, 1931, the only one where it
persists is by the Beverley Brook, Barnes, 27B4. The spread in
Surrey is thus very recent and rapid. I found it on river gravel by
the Thames below Richmond, 17D3, in 1932. It is still near there
and, if it can be accepted that the propelling mechanism of the
seeds enables it to spread upstream, it may be that the colonies
up the Thames and Mole originated from here.

SIMAROUBACEAE

Ailanthus Desf.
A. altissima (Mill.) Swingle.   Tree of Heaven.   Alien.   1936.
A young plant or two on the river-wall between Kew and Mortlake,
RNP & JPMB, *Rep. BEC*, **11**, 26. This fine tree reproduces freely
from seed, and seedlings and saplings have been noted in 05B5,
EJC; 17E4, BW; and 27C3, BW. They are probably widespread.

ACERACEAE

Acer L.
A. pseudoplatanus L.   Sycamore.   Naturalised alien.
1838, Luxford, *Fl. Reigate*, 33.   Far too common.
A tree weed which spread rapidly through the county in the first
half of the nineteenth century, so that by 1863 Brewer described
it as 'frequent in woods and plantations throughout the county'.
In suburban Surrey it has changed the character of much of the
surviving woodland; and cemeteries, unless carefully maintained,
are submerged under sycamore woodland even while interments are
still going on.

A. platanoides L.   Norway Maple.   Naturalised alien.
1927, G. C. Druce, *Rep. BEC*, **8**, 110.   Locally common, 69.
Map 111.
Edges of woods, plantations, hedgerows. Druce's first record was
from Box Hill, and Norway Maple first attracted my attention
in Headley Lane, 15E2, in 1927, Hb.L. It is now common and there
has been a very rapid increase in the last 30 years. In late April
the yellow flowers make the plants conspicuous and in autumn
the foliage turns yellow or deep red. It it more common than our
records suggest.

A. campestre L.   Field Maple.   Native.   1832.   Common.
Woods, hedges, grassy fields, especially on the chalk.

HIPPOCASTANACEAE

Aesculus L.
A. hippocastanum L.   Horse Chestnut.
Seedlings are common from 'conkers' in the vicinity of old trees,
but I have no evidence that it appears in new localities unless
planted.

AQUIFOLIACEAE

Ilex L.
I. aquifolium L.   Holly.   Native.   1657.   Very common.
Woods, hedgerows, commons. Grows on all soils except those which
are very wet.

CELASTRACEAE

Euonymus L.
E. europaeus L.   Spindle-tree.   Native.   1763.
Locally common, 194.   Map 112.
Woods, hedgerows, downs, especially on the chalk. An interesting
distribution, apparently almost absent from large areas in the N
of the county, except by or near streams or the Thames. The very
rare form with white capsules (var. *leucocarpus* Druce) was found
in 35C4 near Caterham, 1967, R. Latham.

E. latifolius Scop.
Was first recorded by C. E. Britton from a field hedge at Chaldon,
35A4?, in 1933 (*Rep. BEC*, **10**, 752, 1934), when he said it was also
elsewhere in Surrey. Also 04D5, Netley Park, under yews, regenerating, 1961, BMCM, 1974, SFC; and 15D2, Mickleham,
lane to Juniper Hill, planted in woods, some young bushes apparently regenerated, 1967, DPY. Likely to become naturalised but
this is hardly so at present.

BUXACEAE

Buxus L.
B. sempervirens L.   Box.   Native.   1610.
Rare, 20 (including planted trees).   Map 113.   (Plate 16).
Native on steep slopes on chalk, of which the river-cliff of Box Hill,
15D1, is a fine example. Similarly it is likely to be native on steep

slopes above Headley Lane, 15D2 and 15E2, and in Norbury Park, 15D3. It is tempting to suppose it native in Titsey Plantation, 45A3, 1950, RARC, but there are records of planting in the nineteenth century. Off the chalk it is always planted.

RHAMNACEAE

Rhamnus L.
R. catharticus L.   Buckthorn.   Native.   1763.
Locally common, 108.   Map 114.
Hedges, woods, copses and grassland on the chalk, elsewhere mainly by streams and ponds. The great majority of localities are on chalk, another group extends along the Thames where calcium carbonate is unlikely to be deficient; some of the others are by streams draining from the chalk.

Frangula Mill.
F. alnus Mill. (*Rhamnus frangula* L.)   Alder Buckthorn.
Native.   1732.   Locally common, 185.   Map 115.
On damp and rather peaty soils in damp copses, wet heathland, woods. Avoids the chalk and is absent from large areas of basic soils towards the NE.

VITACEAE

Vitis L.
V. vinifera L.   Grape Vine.   Persistent alien.
The pips from grapes sold by fruiterers germinate freely and the plants are hardy and compete successfully with native species. First record: 17E4 and 27A4, scattered along river wall between Kew and Richmond, 1935, RNP & JPMB (*Rep. BEC*, **11**, 26). 16B4, E. Molesey, 6 plants climbing on hawthorns and covering distance of about 20 yards by footpath by R. Ember, 1964, JES; 17E3, Sheen Common, 1950 onwards, a huge plant in 1956, BW; 17E4, by Thames on both sides of Kew Bridge, 1950 onwards, BW.

Parthenocissus Planch.
P. quinquefolia (L.) Planch.   Virginia Creeper.
Naturalised alien.
Commons and roadsides (as well as refuse tips) sometimes thoroughly established well away from houses. For example, 94C4, Compton Heath, by old sandpit covering several square yards over

shrubs and trees, 1968, SFC; 95D3, Rickford Common, large plant climbing over tree far from houses, 1968, JES & WEW; 16B3, W of Royal Mills, Esher, between footpath and railway, 1959, BW; 17D3, Richmond, climbing over hawthorn between Old Deer Park and towpath, 1959, BW; 17E2, Petersham Common, also Ham Common tip, 1959, BW.

LEGUMINOSAE

Lupinus L.
L. arboreus Sims.   Tree Lupin.
84C4, Farnham by-pass, no doubt originally planted but thoroughly established from seed by 1959, VML, onwards; 84E4, in quantity at base of Seale Lodge sandpit, 1965, DPY.

Laburnum Medic.
L. anagyroides Medic.   Laburnum.
Seedlings have been observed near planted trees in 15C1, Denbies Drive, 1960, JEL, and 15D2, roadside bank S of Juniper Hall, Mickleham, 1969, RMB.

Genista L.
G. tinctoria L. spp. *tinctoria.*   Dyer's Greenweed.
Native.   1836.   Locally frequent, 42.   Map 116.
Rough fields and commons, hedgebanks and wood borders, mainly on clay soils. Undoubtedly native in the S of the county, but the status in most of the other localities is doubtful. It is a species of old pastures, such as the one in which it grows in 85D5 with *Sanguisorba officinalis*, and few of these have survived ploughing. In some it may be a relic of nurseries, or gardens.

G. anglica L.   Petty Whin.   Native.   1548.
Locally frequent, 89.   Map 117.
Rough and somewhat damp ground on commons and heaths. Still quite common on the London commons, including Mitcham Common, where it was recorded by Martyn in 1763, often with, or near, *Ulex minor.*

Ulex L.
U. europaeus L.   Common Gorse.   Native.   1753.
Common throughout the county.
Commons, heaths, hedgerows, less common on chalk.

Plate 14   *Rusty Back*. Asplenium ceterach, *on brickwork, Woldingham Station*

Plate 15   *Juniper,* Juniperus communis, *6m tall, from a fine colony on Hackhurst Downs (1968)*

Plate 16 *Box*, Buxus sempervirens, *from one of the finest colonies in Britain on the River Cliff, Box Hill (1974)*

U. gallii Planch.   Western Gorse.   Native.
1936, Britton, *J. Bot.* **74,** 355.   Very rare, 2.
Heathy commons. Britton said his specimen came from 'near
Haslemere' and the precise locality is not known. The following
year E. C. Wallace found it in 83D4, Golden Valley, Hindhead,
1937, where he showed it to me a year or two later; it disappeared
during the war probably due to military activities. Another small
clump of perhaps 12 plants was found on Thursley Common, 94A1,
by S. B. Chapman in 1967. They grow on slightly raised ground in a
very wet area; the identification has been confirmed by M. C. F.
Proctor, and they are still there.

U. minor Roth (*U. nanus* T. F. Forst.)   Dwarf Gorse.
Native.   1814.   Locally common, 180.   Map 118.
Rough grassy or heathy ground on commons. Perhaps especially
common on Bagshot Beds and Lower Greensand. This species
thrives on poor gravelly ground unproductive for agriculture and
its habitats tended to be left as 'waste of the manor' and hence
preserved as common land.

Cytisus L.
C. multiflorus (Aiton) Sweet.   White Broom.   Garden alien.
96D3, small seedling plant, Chobham Common, 1970, EJC; 15D1,
railway bank near Dorking (= S end Pixham Lane), D. McClintock,
Supplement (to) the Pocket Guide to Wild Flowers, 11, 1957.

Sarothamnus Wimm.
S. scoparius (L.) Wimm. ex Koch (*Cytisus scoparius* (L.) Koch)
Broom.   Native.   1836.   Frequent.
Sandy and gravelly places on commons, heaths, hedgebanks and
railway banks.

Ononis L.
O. repens L.   Common Restharrow.   Native.   1830.
Locally common, 100.   Map 119.
Chalk grassland, pastures, roadside banks and tracksides. Mainly
on the chalk and Bargate Beds. Some of the outlying localities
may be on Paludina limestone, but it seems that it does occur
on neutral soils.

O. spinosa L.   Spiny Restharrow.   Native.   1827.
Rare, 21.   Map 120.
Rough pastures and coarse grassland on clay soils. Sometimes confused with the spiny form of *O. repens*, but we have one report of *O. spinosa* from the chalk—25, Epsom Downs, 1956, EM det. DPY.

Medicago L.
M. falcata L.   Sickle Medick.   Naturalised alien.
1830.   Rare, 17.   Map 121.
Gravelly towpaths, roadsides by new roads, rough ground on commons, chalk scrub. 84B2, 94B2, 94E4, 17D1, 25A2, 25B1, 25C1, 26C1, 26D1, 26E4, 27A4, 27B1, 27B2, 27B3, 27B4, 36A4, 38C3. *M. falcata* is usually regarded as a native in E. Anglia; in Surrey it is an alien infrequently introduced but persistent when it is. The means of introduction is not known with certainty, but there is association with places where horses are, or have been, used, and its seeds may be imported with those of lucerne. The following are a few examples of persistence: 17D1, Teddington Lock, c1920, CEB, 1957 onwards, BW; 25B1, Reigate Heath, 1933, JEL, 1960, BMCM; 26C1, Banstead Downs, c1920, CEB, 1970, SFC; 27B4, Barnes Common, NE corner, 1942, Mrs M. Whitehouse, 1971, SFC.

M. x varia Martyn (*M. falcata x M. sativa*).
Naturalised alien.   1931.   Rare, 4.
Waste places, etc, introduced with parents. 16B5, towpath by Hurst Park Paddock, 'flowers greenish to dirty purple', 1944, BW, 1954–8, LJJ; 26E2, edge of disused Croydon airport, 1962, RSRF; 36A4, with *M. falcata* in field behind Beddington Lane dump, 1956, BW; 38C3, disused Surrey Commercial Docks, Rotherhithe, 1972, JEL & LMPS.

M. sativa L.   Lucerne.   Naturalised alien.
1836.   Frequent.
Persisting as a relic of cultivation by the sides of fields and tracks, also on railway banks and waste places.

M. lupulina L.   Black Medick.
Native (but has been sown for fodder).   1836.   Common.
Short grassland on chalk, sand and gravel, roadsides and waste places.

M. polymorpha L. (*M. hispida* Gaertn., *M. denticulata* Willd.)
Toothed Medick.   Casual.   1796.
The localities listed in Salmon, 1931, are probably all grain introductions. The only recent reports are: 04B4, Chilworth Railway
Station, with other aliens, 1958, JCG; 04C4, potato field between
Lockners and Postford Farms, 1963, JCG. At this time wool
shoddy was being unloaded at Chilworth Station for distribution
to farmers for use as manure, and no doubt this species and *M.
laciniata* L., which occurred with it in 1963, were introduced in
this way.

M. arabica (L.) Huds. (*M. maculata* Sibth.)
Spotted Medick.   Colonist.   1657.   Rare, 44.   Map 121.
Arable fields, nurseries, gravel pits, sandy tracks and roadsides.
Except that they confirm the preference for light soils, our records
bear little relation to those given by Salmon, 1931, and in most
cases they represent introductions of short-term persistence. It
appeared with other wool aliens at 04B4, Chilworth Railway
Station, 1958, JCG.

Melilotus Mill.
M. altissima Thuill. (*M. officinalis* auct.).   Tall Melilot.
Alien.   1836.   Rare, 44.   Map 122.
Mainly by the sides of tracks on the chalk and especially on the
escarpment E and W of Betchworth, 25, where it has been known
since 1836. Also by railways, field borders and waste places.

M. officinalis (L.) Pall. (*M. arvensis* Wallr.).
Ribbed Melilot.   Alien.   1886.   Common, 151.   Map 123.
Waste ground, near railways, sand and gravel pits, field borders and
roadsides. This species has shown a great increase since Salmon,
1931. Our workers have had great difficulty in recording this species
and the last, due solely to confusion over scientific names, especially
the use of '*M. officinalis*' for both. The plants are distinct enough;
it is only botanists who have created confusion and made it impossible to rely on records unless they are permanently backed by
herbarium material.

M. alba Medic. (*M. alba* Desr.)   White Melilot.   Alien.
1831.   Frequent, 76.   Map 124.
Waste places, railway yards, sandpits, roadsides and field borders.

Tends to appear on disturbed ground, increase rapidly and then decrease as the area becomes grassed. There is no clear relationship between the localities recorded by Salmon, 1931, and the recent ones.

M. indica (L.) All. (*M. parviflora* Desf.)   Small Melilot.
Alien.   1852.   Rare.
Waste ground, towpaths, refuse tips. In the past this has sometimes become established for a time in Surrey, as, for example, by the towpath between Putney and Kew, CEB, but in recent years it has been more of a casual. We have records from centrads 04, 16, 25, 27, 35, 36 & 38.

Trifolium L.
T. ornithopodioides L. (*Trigonella ornithopodioides* (L.) DC.)
Fenugreek.   Native.   1805.   Very rare, 8.   Map 125.
Gravelly or sandy places on commons, mainly on the London Clay gravels. 83E5, Thursley Cricket Pitch, 1965, DCK; 16B3, Esher Green, 1958, RAB; 25B1, Reigate Heath, 1950, 1962, BMCM; 26E4, Mitcham Common, 1957, RCW, JEL; 26E5, Mitcham Common, 1933, JEL, persisted until c1960, JEL; 27B1, Wimbledon Common, 1953, AWJ, 1961, DWF; 36C3, Addington Hills, 1948, LJJ; 36D3, Spring Park, 1954, RARC; Wickham, nr top of Spout Hill, 1958, LJJ. Comparison of our records with those in Salmon, 1931, shows remarkable persistence. The Thursley Cricket Pitch, Reigate Heath, Mitcham Common, Wimbledon Common and Addington records date from at least 1900, 1931, 1931, 1836 and 1923 respectively, while Esher Green cannot be far from Borrer's Esher record of 1805. Nevertheless this is a decreasing species at great risk. There are few plants left in Surrey and the habitats are all at risk.

T. pratense L.   Red Clover.   Native.   1724.   Common.
Chalk grassland, meadows, pastures, waste ground. The native plant (var. *sylvestre* Syme), with toothed leaflets and usually solid stems, is common on undisturbed grassland. Larger cultivated forms (var. *sativum* (Crome) Schreb.), with entire leaflets and fistular stems, occur usually in cultivated fields, where they are grown for hay, or on field borders, roadsides and waste ground.

[*T. ochroleucon* Huds. Grew on Duppas Hill, Croydon, 1778–1880, in 36A3, and Salmon, 1931, gives later records as a casual. Not seen recently.]

T. medium L.   Zigzag Clover.   Native.   1838.
Frequent, 147.   Map 126.
Wood borders, meadows, railway banks. Scarce on the chalk and
on Bagshot Beds in the NW, but otherwise generally distributed
and most common on heavy soils.

T. incarnatum L.   Crimson Clover.   Casual.   1838.
This handsome plant, which formerly frequently persisted for a
time as an escape from cultivation, is now seldom grown and hence
exceedingly rare. It occurred in 93C2 and 93D2 in a dry pasture
along the county boundary, 1963, JCG.

T. arvense L.   Haresfoot Clover.   Native.   1666.
Locally common, 138.   Map 127.
Sandy or gravelly places on commons, in pits and round the edges
of arable fields. Especially common, even abundant, on the light
soils of the Lower Greensand and Bargate Beds in the W of the
county.

T. striatum L.   Knotted Clover.   Native.   1763.
Frequent, 61.   Map 128.
Sandy and gravelly places on commons and by tracks and pits.
Quite common on the commons in the SW. Still in two places on
Mitcham Common, 26E4, whence it was recorded by Martyn in
1763.

T. scabrum L.   Rough Clover.   Native.   1794.
Rare, 6.   Map 129.
On gravel or sand on commons, paths and pits. 84D4, Moor Park,
1966, DCK; 84E4, sandpit, Seale, 1962, WEW; 94B4, c1960, OVP;
96C4, Puttenham Heath, 1968, DWB; 94D4, c1960, OVP; 17D1,
towpath, Ham, 1953, BW. All the localities, except the last, occur
in a line just S of the Hog's Back.

T. subterraneum L.   Subterranean Clover.   Native.
1696.   Rare, 31.   Map 130.
Short turf on sand or gravel, commons, greens, meadows, roadside
banks, football and cricket pitches. Also occasionally on refuse
tips. A persistent plant still, for example, on Kew Green and in
Kew Gardens, 17E4.

T. glomeratum L.   Clustered Clover.   Native.   1805.
Very rare and perhaps extinct, 1.
On and by a sandy track. All our records come from a sandy track in the parish of Artington, 94D5, where it formerly occurred in some plenty and could be seen until c1964, when the field was fenced and used for pigs.

[*T. suffocatum* L. Suffocated Clover. Error. Recorded from 26E4, Mitcham Common, bare ground by tip, one plant, 1951, RCW, conf. DPY, Hb. Croydon Nat. Hist. & Sci. Soc. (*Lond. Nat.*, **38**, 18). The specimen is a poor one, but it is not this species.]

T. hybridum L.   Alsike Clover.   Alien.   1762.
Formerly common as an escape from cultivation, this used to persist on field borders and roadsides but is now seldom seen. More frequent now is the smaller, pink-flowered plant with solid stem, *T. elegans* Savi, which is connected with *T. hybridum* by a series of intermediates, and may be the wild form from which it was bred. This is found fairly often on waste places, refuse tips and roadsides introduced with bird-seed and grain.

T. repens L.   White Clover.   Native.   1836.
Very common throughout the county. Lawns, pastures, meadows, commons and roadsides.

T. fragiferum L.   Strawberry Clover.   Native.   1827.
Rare, 35.   Map 131.
Meadows and grassy places by the Thames and Wey Navigation and elsewhere, roadsides and brickpit, usually on clay. Widely distributed, but most frequent on alluvium by the Thames, and on Weald Clay in the SE.

T. campestre Schreb. (*T. procumbens* auct.)   Hop Trefoil.
Native.   1838.   Common.
Dry grassland, commons, roadside banks, especially on chalk and sand.

T. aureum Poll. (*T. agrarium* auct.)   1838.
Never more than a casual introduced with clover and grass seed, this has been reported only from 25C1, garden, probably introduced with birdseed, 1966, BMCM.

T. dubium Sibth.   Lesser Trefoil.   Native.   1838.   Common.
Dry grassland, meadows, chalk downs, commons and roadsides.

T. micranthum Viv. (*T. filiforme* L.)   Slender Trefoil.
1724.   Rather rare.
Short turf of lawns, commons and pastures, usually on sand or gravel. Mainly on the gravelly commons of London Clay and Bagshot Beds, and Lower Greensand where it may be locally common.

Anthyllis L.
A. vulneraria L.   Kidney Vetch.   Native.   1836.
Locally common, 63.   Map 132.
On the chalk in grassland, roadsides, chalkpits and waste ground. Elsewhere rare. In 16A5, Chelsea & Lambeth Reservoirs on banks of imported chalk; in 26E4, at a cement works; also on roadsides off the chalk at 95A4, on sand; 96D3, on road shingle, and 83C4.

Dorycnium Mill.
D. rectum (L.) Ser.
15E1, Brockham Hill 'bomb crater', 1948, 1949, 1950, JEL, Hb. L, Kent & Lousley, *Handlist*, 64, 1952; huge plants in flower, still well established, 1967, BWu.

Lotus L.
L. corniculatus L.   Common Birdsfoot Trefoil.   Native.
1836.   Common throughout the county.
Chalk grassland, roadside banks, commons, waste places.

L. tenuis Waldst. & Kit.   Narrow-leaved Birdsfoot Trefoil.
Native, 1863.   Rare, 16 (probably under-recorded).
Map 133.
Usually on clay soils, especially on London and Weald Clays, in rough grassland and in such places probably native, but also on disturbed ground where clay has been dumped in 96D3, and 16E2 and in brickworks in 34C5. In 35C4 it grows in chalk grassland to W of Caterham by-pass, where it may have survived since F. C. S. Roper recorded it from Caterham about a century ago.

L. uliginosus Schkuhr   Marsh Birdsfoot Trefoil.   Native.
1836.   Common.
Marshes, wet fields, ditches and wet heaths.

L. angustissimus L.   Slender Birdsfoot Trefoil.
Established alien.   1931.   Very rare, 2.
Sandy arable fields when fallow. Native in SW England. 84D3,
Tilford, rough sandy field, 1951, NYS (*Watsonia* **2,** 339), unable to
refind, 1964, NYS & DPY. 04B3, SW of Blackheath, 1930, J. G.
Lawn (*Rep. BEC*, **9,** 341); near Wonersh, Pugsley, Watts & Still,
1935, (*Rep. BEC*, **11,** 26); 1936, 1944, JEL; 1953, 1957, BW; 1958,
JCG; field very overgrown, unable to find, 1969, B.Wu.

Galega L.
G. officinalis L.   Goat's-rue.   Naturalised alien.
First record: 1952, Kent & Lousley, *Handlist*, 65.
Frequent, 34 (probably under-recorded).   Map 134.
Waste ground, roadsides, chalk downs. This species first attracted
attention about 1935 and has spread rapidly since then. Originating
from gardens, it is sometimes abundant from self-sown seed.

Robinia L.
R. pseudacacia L.   False Acacia.   Naturalised alien.
First record: 1931, Bishop, Robbins & Spooner, 31.
Common, 137.   Map 135.
Commons, woods, plantations and lanesides. Common over most
of the county, except perhaps on chalk, from self-sown seeds, and
spreading.

Colutea L.
C. arborescens L.   Bladder Senna.   Naturalised alien.
First record: 1952, Kent & Lousley, *Handlist*, 66. Although less
spectacular in Surrey than in other vice-counties in the London
conurbation, this species is now thoroughly established in the NE.
The spread started along railway banks, but has now extended to
waste ground elsewhere. 17D2, Ham Pits, 1960, JLG; 17E4,
railway bank, Kew to Gunnersbury, PHC, Hb. LNHS, 1951, DHK;
26E3, River Gardens, 1957, RCW; 26E4, Beddington Lane &
Beddington Corner, 1957, RCW; 27E4, nr Battersea Park, 1965,
RARC; 35D4, Nore Hill Gravel Pit, 1964, FGC; 37B5, Railside,
Bermondsey, PHC, Hb. LNHS; 38C1, Surrey Commercial Docks,
1965, RARC.

Astragalus
A. *glycyphyllos* L.   Wild Liquorice.   Native.   1805.
Rare, 9.   Map 136.
Rough scrubby fields, roadside banks, and edges of woods. On warm, south-facing slopes along the chalk escarpment. The localities fall into three groups: in the W, below the Hog's Back from Seale (where it is in a sandpit) to above Shoelands; in the E, from near Merstham to above Tandridge; and, in the centre, below Brockham limeworks, 25A1, 1950, BMCM and Headley Warren Nature Reserve, 1970, HM-P. At the reserve it appeared suddenly and in quantity in Stanton's Field. It is possible that it is not native, as JHPS sowed seed elsewhere on the reserve a few years earlier and did not observe any plants as a result, but there is a pre-1863 record from Headley Lane which is near.

Ornithopus L.
O. *perpusillus* L.   Birdsfoot.   Native.   1763.
Locally common, 185.   Map 137.
Dry sandy and gravelly places on commons, heaths and roadsides, weed on light arable land. Especially common on Bagshot Beds and Lower Greensand in the W. Still on Mitcham Common, 26E3, whence it was first recorded for Surrey in 1763.

Coronilla L.
C. *varia* L.   Crownvetch.   Established alien.
1931, Salmon.
Rather frequent on railway banks, waste ground and woodland clearings in the London suburbs, and occasionally elsewhere.

Hippocrepis L.
H. *comosa* L.   Horseshoe Vetch.   Native.   1778.
Locally frequent, 53.   Map 138.
Chalk grassland, banks and slopes on the chalk.

Onobrychis Mill.
O. *viciifolia* Scop.   Sainfoin.   Native.   1666.
Frequent on the chalk. Chalk grassland, slopes in chalkpits, chalky roadside banks. Sainfoin was a fashionable crop used for the reclamation of chalk grassland on the Berkshire Downs, and no doubt elsewhere at the end of the eighteenth and beginning of the nineteenth centuries, and often survives as a relic of cultivation round the edges of fields and sides of tracks in the S of England.

Such relics sometimes take the form of the Giant or French Sainfoin, which is much larger than the native forms and thus easily recognised, though as its life-cycle is shorter it can only persist where ground is disturbed to allow reproduction from seed every few years. In Surrey there is no reason to question the established view that it is native on the chalk, at least on the slopes of the escarpment, and there are numerous early records to support this. Merrett in 1666 wrote 'In some mountainous parts of Surrey, plentifully', and the 'mountainous' parts are likely to have been the chalk. Martyn in 1793 mentioned places where it was cultivated and added 'with us it was remarked in a wild state, before it was adopted for cultivation, on many of our chalk downs' (*Fl. Rustica*, **2**, 47). Not all the plants which persist from cultivation are large; some are hardly distinguishable from natives except by the situations in which they grow.

Vicia L.
V. hirsuta L.   Hairy Tare.   Native.   1827.
Common throughout the county. Hedgebanks, cornfields and other cultivated ground, waste places. Still fairly evenly distributed.

V. tetrasperma (L.) Schreb.   Smooth Tare.   Native.
1832.   Frequent.
Hedgebanks, arable fields, commons and waste ground.

V. tenuissima (Bieb.) Schinz & Thell. (*V. gracilis* Lois.)   Native.
Found by E. C. Wallace in 1927 amongst grass by Denbies Drive, Dorking, 15C1 (*Rep. Watson BEC*, **3**, 430), 1935, JEL, Hb. L. This attractive plant with glabrous pods with 5 or more seeds, and 4 or more flowers on slender peduncles which much exceeded the leaves, persisted, in the company of *Galium pumilum*, until after 1950, but has not been seen recently. It was reported earlier by H. C. Watson from a wheatfield at Hook, 16E3 (?), before 1874, where it was probably a casual; Salmon, 1931, 251.

V. cracca L.   Tufted Vetch.   Native.   c1785.
Very common throughout the county. Bushy places, hedgebanks, wood borders and waste places.

V. tenuifolia Roth.   Established alien.   1932 (Beadell, 28).
16D2, Ham Pits, established for many years, 1942, 1945, JEL, Hb. L, 1960, DPY, Hb. Young; 35D4, rough bank near Warren Barn,

Warlingham, Beadell, 1932; track near Warren Barn, 1943, JEL, Hb. L. Persisted at both stations after the last date given.

V. villosa Roth subsp. *villosa* and subsp. *varia* (Host) Corb. (*V. dasycarpa* auct.) occur occasionally as casuals on refuse tips or in weedy arable fields. The species was first recorded by Irvine, 1857–8 (*Ill. Handb. Br. Plants*). The two subspecies, which grade into one another, have each occurred several times, the most recent record being subsp. *villosa*, 24D5, waste ground near St John's Church, Earlswood, 1971, BMCM, EMCI, ! JEL.

[*V. sylvatica* L. Wood Vetch. Recorded from near Fairoak Lane, Chessington, 16D1, 1934, M. L. Wedgwood (*Rep. BEC*, **10**, 824), but was probably either an error or a casual introduction.]

V. sepium. Bush Vetch. Native. 1836.
Very common.
Borders of woods and copses, hedgerows, rarely in open grassy places.

[*V. lutea* L. Occasionally on refuse tips. Probably never more than a casual, and not recorded by us.]

V. sativa L. Common Vetch.
Escape from cultivation. Now rare, and decreasing. Field borders, waste ground.

V. angustifolia L. (*V. bobartii* E. Forst.).
Narrow-leaved Vetch. Native. 1800. Common.
Commons and heaths, especially on sand and gravel, roadsides and grassy places.

V. lathyroides L. Spring Vetch. Native. 1805.
Rare, 13. Map 139.
Sandy ground on commons, roadsides and by tracks. This species is most at home on the sandy commons of the SW, such as those of Hankley, Thursley and Puttenham, in Windsor Park, about St Martha's Hill and Farley Heath, and it was reported from Wimbledon Common by Mrs B. Welch in 1957.

[*V. bithynica* (L.) L. Recorded from Woking and Weybridge by Lady Davy (*Rep BEC* **7**, 180, 1924; *ib.* **9**, 687, 1932) and no doubt a casual.]

Lathyrus L.

L. aphaca L.   Yellow Vetchling.   Alien.   1805.
Very rare, 4.
This Mediterranean species may be native in some southern counties, but in Surrey it is now, and probably always has been, a grain alien which does not persist for more than a year or two. 15E5, Ashtead Common, 1970, BRR; 16D2, Chessington on chalk railway embankment, plentiful, 1962, JES & BW; 17D2, Ham Pits, 1963, PCH, abundant, 1973, JEL et al.; 17E4 (?), Kew, derelict garden with other aliens, JPMB; 25D5, Banstead, in a garden, 1964, Mrs L. Reddick.

L. nissolia L.   Grass Vetchling.   Native.   1724.
Locally common, 99.   Map 140.
Rough grassy fields and commons. A difficult distribution to explain; it thrives on clay, but also occurs on sand and chalk. Varies greatly from year to year so that it may be abundant one year and scarce or absent the next on the same date, but even on this it is hard to provide reliable statistics as it is so much more conspicuous for a short period when the crimson flowers are at their best. There is a colony of plants with white flowers on Mitcham Common, 26E4, RCW.

L. hirsutus L.   Hairy Vetchling.
This alien, probably introduced with seed-corn, was found in 1858 by W. Robinson thoroughly established at the top of a sloping cornfield on Halliloo Farm, Warlingham, 35C4 (Brewer, 1863, 67). It was seen here at intervals until after 1932 (Beadell, 1932, 31), but has not been reported recently from here, or from two new stations found in the vicinity by A. Bennett in 1874. Salmon, 1931, gave other localities where it was a casual and we have it reported from 94C1, Enton, a few plants on a poultry farm, 1959, 1961, 1963, GSE; 17D2, Ham, KMM; 36D2, Selsdon Vale, a patch 6 yards long, 1960, RARC—also on refuse tips.

L. pratensis L.   Meadow Vetchling.   Native.   1814.
Common.
Hedges and grassy places.

L. tuberosus L.   Tuberous Pea.
An alien, often introduced with chicken food, which is persistent in other counties, has been found: 26C1, Banstead Downs in

spinney, rarely flowering but known for several years, 1959, EM; 35D4, Woldingham, Lunghurst Valley in derelict garden, 1966, RARC; 35D4, Nore Hill, among hawthorn, 1966, RARC; 17D2, Ham, 1966, KMM.

L. grandiflorus Sibth. & Sm.
This handsome garden alien is well established at 93A3, Grayswood, on railway embankment near cottages, 1965, SFC; 25A1, top of railway cutting E of Betchworth Station, covering a wide area, 1957, 1963, DPY; 27B4, Barn Elms reservoir, on Thames-side in quantity, 1962, LMPS.

[*L. palustris* L. There are specimens of this fen-loving species from several seventeenth-century botanists from 'Peckham Field', as reported first in 1666 by Merrett (*Pinax*, 70). It has long been extinct in the county.]

L. sylvestris L.   Narrow-leaved Everlasting-pea.   Native.
1724.   Locally frequent, 13.   Map 141.
Thickets, hedges, railway cuttings, rough fields. This species has its headquarters in the Buckland—Wingate Hill—Merstham—Hooley —Quarry Hangers area, where it is frequent. Outside this area it is usually a garden escape and it has sometimes been confused with narrow-leaved forms of the more common garden plant, *L. latifolius*.

L. latifolius L.   Broad-leaved Everlasting-pea.
Naturalised garden alien.   1855, Irvine (*Phyt*, **3** (NS), *Suppl.*, 167). Common, 93.   Map 142.
Railway banks and sometimes hedges, road cuttings and waste ground. Very common indeed on railway banks in suburban areas and, as the map shows, it follows the Brighton line, Oxted branch, Redhill to Guildford, main Woking and other railway lines. Elsewhere it is usually introduced with garden rubbish, but once established it is persistent and spreads. Occasionally it appears in places far from houses, such as the Headley Warren Reserve, in 15E3, where its introduction is difficult to explain.

L. montanus Bernh. (*L. linifolius* (Reichard) Bässler var. *montanus* (Bernh.) Bässler).   Bitter Vetchling.   Native.   1548.
Locally common, 142.   Map 143.
Woods, hedgerows and sometimes rough, rather acid, pastures. Most common on the Weald Clay, which was formerly heavily

wooded, rare or absent from lighter soils where oakwoods are uncommon.

ROSACEAE

Spiraea L.

[*S. salicifolia* L. is a garden escape well established in Wales and Scotland, but we have no recent records for Surrey.] It is often confused with a group of N. American species and their hybrids which sometimes become well established on Surrey commons and for which we have the following records:

S. menziesii Hook. 96C4, Sunningdale end of Chobham Common, established bush, 1966, DPY; 96D2, Lightwater Bog, 1957, JEL; 14E3, Holmwood Common, nr Four Wents Pond, 1961, JEL.

S. x billiardii Herincq 15E5, Ashtead Common, one small bush not far from gardens, 1964, EJC.

S. x fulvescens Dipp. 83E3, Hindhead, S of Gibbet Hill, site of old camp, 1961, BMCM.

S. douglasii Hook. 06E3, Walton, 1963, DMcC; 14D5, Dorking sandpit, 1961, BMCM; 15D5, 1963, RARC.

S. tomentosa L. 83E3, Hindhead, below Gibbet Hill, on site of army camp, 1961, 1970, BMCM.

Filipendula Mill.

F. vulgaris Moench (*Spiraea filipendula* L.; *Filipendula hexapetala* Gilib.) Dropwort. Native. 1838.
Locally frequent, 30.   Map 144.
Downs and dry pastures on chalk; pastures, sometimes damp, on alluvial soils by the Thames. Mainly along the chalk escarpment, with an outlying locality in the W on a small outcrop of chalk on Compton Common, 94C4. Also in the alluvial meadows near the Thames and 16B4, nr Esher Mills, 1959, BW & JES, and 16E3 in a damp meadow at Tolworth, 1963, JES. Perhaps extinct in Ham Meadows, 17D2, where Mrs B. Welch was unable to find it after 1944.

F. ulmaria (L.) Maxim. (*Spiraea ulmaria* L.)   Meadowsweet.
Native.   1838.   Common and generally distributed.
Marshes, stream sides, wet woods.

Rubus L.
R. parviflorus Nutt. (*R. nutkanus* Moç.)
Established garden alien.
94E4, Shalford, established on a railway embankment for about
40 yards, 1967, EMCI, det. B. Miles.

R. phoenicolasius Maxim.   Japanese Wineberry.
Established garden alien.
15D2, Norbury Park, slope above R. Mole, 1951, BMCM, 1953,
WHS, 1958, B. Sturdy. The area has now been replanted with trees
and the Wineberry is probably destroyed.

R. idaeus L.   Raspberry.   Native.   First record: 1838,
Irvine, (*London Fl.*, 193) or 1838, Luxford (*Fl. Reigate*, 44).
Frequent.
Woods, copses and commons.

R. spectabilis Pursh.   Established garden alien.
93A1 & 93A2, in quantity along lane SW of Haste Hill, 1959, DPY.

R. caesius L.   Dewberry.   Native.   First record: 1838,
Luxford (*Fl. Reigate*, 45).   Frequent, 133.   Map 145.
Hedges, wood margins and lanesides. Common on the chalk
and Weald Clay and also by the Thames; scattered records elsewhere.

R. fruticosus L. *sensu lato*.   Blackberry, Bramble.
Native.   Very common.
Commons, woods, lanes and waste places, abundant when allowed
to develop on the metropolitan commons, and self-sown in London
gardens from seed carried by birds.

The microspecies of *R. fruticosus*
by Alan Newton

The following account is based on the field and herbarium records
of A. Newton, some of B. A. Miles, a few of E. S. Edees, and on
specimens seen in the herbaria of the British Museum (Natural

History); the Botany School, Cambridge; Manchester University; South London Botanical Institute and Liverpool University. While it cannot be claimed that all the taxa represented in Surrey have been thoroughly mapped, or that more intensive fieldwork would not reveal many new facts about bramble distribution, the scope of the survey has been fairly comprehensive and has included a significant portion of the material collected by the leading batologists of the past who have interested themselves in the Surrey *Rubi*. Due to its proximity to London, Surrey has always been a popular hunting ground since the days of H. C. Watson; many of the specimens published by Rogers, Linton and Murray in the *Set of British Rubi* (1892–5) were collected in Surrey, and many exsiccata were widely distributed through the Exchange Clubs, so that the recognition of the taxa present in the county is, despite their number, not the pilgrimage into the unknown which faces the investigator in other less well-worked areas. Thanks to the considerable attention paid to the genus by previous botanists* the proportion of plants met with, which can be recognised and confidently named, may be as high as 95 per cent. There are, of course, as elsewhere, clumps and bushes either of isolated individuals or of rather wider distribution which cannot be satisfactorily named; W. Watson described some of these as new species and they are useful for the Surrey list even when of very local distribution. Less fortunately in other instances he assigned continental names both to widespread Surrey brambles and to isolated bushes on the basis of written descriptions or Sudre's plates; current research into authentic syntype material of these taxa reveals a disappointingly high rate of false identity. Further, Watson's own specimens are sometimes not homogenous and credibility is further weakened thereby.

The list given on pp 185–9 uses the name which can be justified unequivocally by reference to type or syntype material and synonyms are given where useful. If there is doubt as to the correct nomenclature the usage is appropriately qualified in cases where the plant has a wide distribution in the county; incorrect attributions of continental names to isolated plants have been omitted. References to original descriptions and localities are given for *Rubi* described from Surrey.

---

*The following collectors' gatherings have been seen, at least in part: C. Avery, J. G. Baker, W. H. Beeby, C. E. Britton, E. F. Linton, E. S. Marshall, B. A. Miles, W. H. Mills, G. Nicholson, H. J. Riddelsdell, W. M. Rogers, C. E. Salmon, H. C. Watson, W. C. R. Watson, A. H. Wolley-Dod.

There is much ground in Surrey favourable to the growth of brambles, but by far the greatest number and variety of plants occurs either on gravel terraces or on the Lower Greensand formation, which give rise to light, permeable and agriculturally unrewarding soils. The fringes of such areas which must for a long period have survived as heath, scrub or open woodland are most conducive to the optimum development of *Rubi*, though disturbance to old heath surfaces, chiefly by burning, is now widespread and has resulted in the luxuriant spread of bracken in many places, effectively confining the bramble communities to narrow marginal strips. Perhaps due to under-exploration of the Weald Clay areas in the south, there is a relative dearth of Sect. *Triviales* species which favour damp clay soils and of which each district appears to have its own often undescribed representatives.

In the following list the distributions are given in centrads of the National Grid (not tetrads as used elsewhere in this *Flora*). The place of publication is added for taxa described first from Surrey, and the type locality is given:

Sect. *Suberecti* P. J. Muell.
  R. nessensis W. Hall   84, 04, 14, 25, 34
  R. scissus W. C. R. Wats.   83, 84
  R. opacus Focke ex Bertram   93, 94, 34
  R. plicatus Weihe & Nees   84, 94–6, 04, 17, 25, 34
  R. affinis Weihe & Nees   93, 94, 27
  R. nobilissimus (W. C. R. Wats.) Pearsall   06, 16—(*Rep. BEC*, **10**, 485, 1934) from Abrook Common.
  R. divaricatus P. J. Muell.   83, 84, 93, 94, 06, 16
  R. integribasis P. J. Muell.   95, 96, 06, 16

Sect. *Triviales* P. J. Muell.
  R. sublustris Lees   84, 94–6, 04, 16, 25, 27
  R. balfourianus Blox. ex Bab.   84, 93, 94, 05, 06, 15, 16, 45
  R. britannicus Rogers   94—(*J. Bot.*, **32**, 49, 1894) from Munstead.

Sect. *Sylvatici* P. J. Muell.
  R. gratus Focke   93, 96, 04, 06, 25, 26, 35, 37, 45
  R. confertiflorus W. C. R. Wats.   83, 84, 93–5, 04, 16—(*J. Bot.*, **73**, 194, 1935) from Witley.
  R. averyanus W. C. R. Wats.   06, 14, 15, 25—(*Lond. Nat.*, **31**, *Suppl.*, 97, 1952) from Walton Heath.

R. calvatus Lees ex Bloxam 84, 93, 94, 25

R. nitidoides W. C. R. Wats. 25, 27, 36—(*Rep. BEC*, **8,** 786, 1929) from 'N.E. Surrey'.

R. crespignyanus W. C. R. Wats. (*R. simulatus* W. C. R. Wats. non P. J. Muell.) 36—(*Watsonia*, **3,** 286, 1956) from Addington Hills.

R. carpinifolius Weihe & Nees 83, 84, 93–5, 04, 06, 15, 16, 25, 27

R. glanduliger W. C. R. Wats. 84

R. selmeri Lindeb. 27, 45

R. laciniatus Willd. Alien from gardens. 94, 95, 04, 16, 17, 26, 27, 37. Has been known in Surrey for at least 60 years (Salmon, 1931, 281) and is frequently bird-sown on commons.

R. oxyanchus Sudre 05, 06, 27

R. lindleianus Lees 83, 84, 93–5, 04, 06, 15, 16, 25, 27

R. macrophyllus Weihe & Nees 84, 94, 16, 27, 45

R. subinermoides Druce 93, 94, 06, 15, 16, 25, 27, 36

R. bakeranus Barton & Riddelsd. 95, 96, 17, 27—(*J. Bot.*, **73,** 128, 1935) from Wimbledon Common.

R. silvaticus Weihe & Nees 95, 06

R. pyramidalis Kalt. 94, 04, 06

R. mollisissimus Rogers 84, 94, 96

R. crudelis W. C. R. Wats. 84, 93, 94, 05, 06—(*J. Bot.*, **71,** 228, 1933) from Witley Common.

R. londinensis (Rogers) W. C. R. Wats. (*R. imbricatus* var. *londinensis* Rogers) 15, 16, 27, 37—(*J. Bot.*, **41,** 89, 1903) from Barnes, Wandsworth and Wimbledon Commons.

R. milfordensis Edees 83, 93–6, 04, 16—(*Watsonia* **9,** 250–1, 1973)

R. incurvatus Bab. 14

R. polyanthemus Lindeb. 84, 93–6, 04, 06, 16, 17, 25, 27

R. rubritinctus W. Watson 93, 94

R. rhombifolius Weihe ex Boenn, (*R. incurvatus* var. *sub-carpinifolius* Rogers ex Riddelsd.) 94, 06, 16, 17, 25, 27

R. rhombifolius var. *megastachys* W.-Dod. 06, 16—(*J. Bot.*, **44,** 64, 1906) from Walton Common.

R. cissburiensis Barton & Riddelsd. 84, 94–6, 04–6, 15, 16, 25, 27, 36, 37

R. cardiophyllus Muell. & Lefèv. 83, 84, 93–6, 04, 06, 15–7, 27

R. imbricatus Hort (*R. neomalacus* Sudre sec W. Wats.) 83, 84, 93, 94, 04, 06, 15, 16

R. sprengelii Weihe 84, 93–5, 04, 05, 14, 25

R. permundus W. C. R. Wats. (*R. mundiflorus* W. C. R. Wats. non Sudre) 04—(*Rep. BEC*, **11**, 445, 1938) from Netley Heath.

Sect. *Disclores* P. J. Muell.

R. ulmifolius Schott 04, 15–7, 25–7, 36–7

R. pseudobifrons Sudre sec W. C. R. Wats. 84, 94–6, 04, 06, 15–7, 25, 27, 36, 37

R. procerus P. J. Muell. Himalayan Giant. 95–6, 04, 15, 27 This garden escape is becoming increasingly established on commons and waste places.

R. hylophilus Rip. ex Genev. (*R. brittonii* Barton & Riddelsd.) 95, 06, 16, 25

Sect. *Appendiculati* (Genev.) Sudre

R. surrejanus Barton & Riddelsd. (*R. hirtior* W. C. R. Wats.) 93–6, 04–6, 15, 16, 25—(*J. Bot.*, **70**, 189, 1932) from Walton Heath.

R. schmidelyanus var. *breviglandulosus* Sudre 05

R. vestitus Weihe & Nees 94, 04, 06, 15, 25, 27

R. leucostachys Schleich. ex Sm. 93, 16, 27, 37

R. conspersus W. C. R. Wats. 83, 93–5, 37—(*J. Bot.*, **73**, 255, 1935) from Witley.

R. macrothyrsus Lange 93, 94, 36

R. leyanus Rogers 27

R. chaerophyllus Sag. & Schultze 45

R. ahenifolius W. C. R. Wats. 93, 95, 96, 04–6, 27—(*Watsonia*, **3**, 288, 1956) from Farley Heath.

R. iodnephes W. C. R. Wats. 27—(*Lond. Nat.* **31**, *Suppl.*, 99, 1952) from Barnes Common.

R. radula Weihe ex Boenn. 16, 17, 27, 36

R. sectiramus W. C. R. Wats. 27, 36—(*Lond. Nat.*, **13**, 60, 1933) from Putney Heath

R. echinatus Lindl. 83, 84, 93–6, 04–6, 15, 16, 27

R. echinatoides (Rogers) Sudre 84, 93, 94, 15, 37

R. rudis Weihe & Nees 16, 25

R. flexuosus Muell. & Lefèv. 84, 93, 94, 04, 15, 16, 25, 27

R. bloxamii Lees 94

R. trichodes W. C. R. Wats. 93—(*Watsonia*, **3,** 289, 1956) from Witley.

R. largificus W. C. R. Wats. 25, 27, 35, 36—(*Rep. BEC*, **8,** 507, 1928) from Worm's Heath (Warlingham).

R. adamsii Sudre 04, 06

R. glareosus Rogers 83, 84, 93, 94, 04, 14, 17—(*J. Bot.*, **50,** 309, 1912) from Witley, Tilford, etc.

R. euryanthemus W. C. R. Wats. (*R. pallidus* var. *leptopetalus* Rogers) 05, 27, 36

R. insectifolius Muell. & Lefèv. (*R. nuticeps* Barton & Riddelsd.) 06, 16, 25

R. putneiensis W. C. R. Wats. 27—(*Watsonia*, **3,** 289, 1956) from Putney Heath.

R. scaber Weihe & Nees 93, 25, 37

R. praetextus Sudre 04

R. rufescens Muell. & Lefèv. (*R. rosaceus* var. *infecundus* Rogers) 94, 06, 15, 16, 25, 36

R. anglosaxonicus Gelert. 16, 27

R. phaeocarpus W. C. R. Wats. 84, 93, 94, 06

R. tardus W. C. R. Wats. 26, 27, 36

R. moylei Barton & Riddelsd. 84, 94, 96, 04, 06, 27

R. diversus W. C. R. Wats. 96

R. wedgwoodiae Barton & Riddelsd. 83, 84, 93, 94—(*Proc. Cottesw. Nat. Field Club*,. **24,** 212, 1933) from Witley.

R. rosaceus Weihe & Nees sec. W. C. R. Wats. 93, 94

R. formidabilis Muell. & Lefèv. 06, 16, 27

R. murrayi Sudre 93, 25, 27

R. pygmaeopsis Focke sec W. C. R. Wats. 36, 45

R. coronatus var. *cinerascens* W. C. R. Wats. 27—(*Lond. Nat.* **13,** 65, 1933) from Putney Heath.

R. bercheriensis (Druce ex Rogers) Rogers (*R. apricus* var. *sparsipilus* W. C. R. Wats., nom. illegit. superf.) 94, 06, 16

R. infestus Weihe ex Boenn. sec W. C. R. Wats. 84, 93–6, 04–6

R. milesii A. Newton (*R. adenolobus* W. C. R. Wats. nom. inedit.) 84, 94, 27

R. dasyphyllus (Rogers) E. S. Marshall 84, 96, 16, 27

R. marshallii Focke & Rogers 84, 93–5, 04, 05, 14, 27— (*J. Bot.*, **33,** 103, 1895) from Witley and Munstead.

R. lapeyrousianus Sudre sec W. C. R. Watson 96, 06

R. atrebatum A. Newton (*R. koehleri* var. *cognatus* Rogers p.p.) 83, 84, 94, 95, 04, 26—(*Watsonia*, **10,** 23, 1974) from Milford.

Sect. *Glandulosi* P. J. Muell.
 R. hylonomus Muell. & Lefèv.

Potentilla L.
 P. palustris (L.) Scop. (*Comarum palustre* L.)   Native.
 1666.   Rare, 14.   Map 146.
Spongy bogs and shallow water in acid ponds. This is now practi-
cally restricted to the commons of the SW: Hankley—Thursley—
Elstead—Puttenham, etc.

P. sterilis (L.) Garcke.   Barren Strawberry.   Native.
 1836.   Common.
Round the edges of woods, on banks and dry pastures.

P. anserina L.   Silverweed.   Native.   1718.   Common.
Roadsides, commons, sandpits, waste ground, field borders and
pondsides. Thrives where it has little competition on bare ground
so that it can extend its runners to form a carpet.

P. argentea L.   Hoary Cinquefoil.   Native.   1814.
Rather rare, 41.   Map 147.
Gravelly and sandy banks, commons and pits. Mainly on Lower
Greensand in the W and alluvial gravels by the Thames; scattered
on gravels and sands elsewhere.

P. recta L.   Sulphur Cinquefoil.   Established alien.
 1866.   Rare.
Commons, old wall, roadside, waste ground. 94B1, Witley Com-
mon, old camp site, 1955, OVP; 94D5, Guildford By-pass, 1974,
JFL, ACL; 04A4, Shalford Common, for several years past, 1969,
JFL; 05A2, N of Jacobswell, 1961, KMM; 15E2, 15E2, Headley
Lane, on old wall, 1905 Hb. SLBI, 1925, 1935, 1949, JEL; 26B1,
Cuddington, roadside verge, 1961, DPY; 36A2, rough grassland
by Croydon Airport, 1960, JEL & DPY.

P. norvegica L.   Norwegian Cinquefoil.   Established alien.
 1897.   Rare, 3.
Gravelly commons, gravel-pit, garden. 94B1, Witley Common,
1955, OVP, 1964, EJC; 35B3, Stanstead gravel-pits, 1963, DCK;
36A5, Thornton Heath, garden in Galpin's Road, 1956, HWP.

P. intermedia L.   Established alien.   1894.   Rare, 3.
Gravel-pits, roadsides, and waste ground. 86E1, Camberley, a
good-sized colony on roadside, 1965, JCG, CPP & JES, Hb. JEL;
17D2, Ham Tips, 1958, BW (from here back to 1921, Hb. K); 17E3,
Richmond, waste ground by railway station, 1945, JEL, Hb. L.

P. erecta (L.) Rausch (*P. tormentilla* Neck.)   Tormentil.
Native.   1836.   Very common.
Commons, heaths, open woods, dry pastures.

P. anglica Laichard (*P. procumbens* Sibth.)
Trailing Tormentil.   Native.   1763.   Uncommon.
Roadside banks, borders of woods, heaths.

P. reptans L.   Creeping Cinquefoil.   Native.   1836.
Very common.
Commons, roadsides, hedgebanks, waste places.

Hybrids between the last three species occur in Surrey and are
probably not rare. It is also possible that the plants referred to as
P. anglica are themselves of hybrid origin. Salmon, 1931, 283–4,
listed:
> *P. erecta* x *reptans* = *P.* x *italica* Lehm.
> *P. anglica* x *erecta* = *P.* x *suberecta* Zimmet.
> *P. anglica* x *reptans* = *P.* x *mixta* Nolte ex Reichb.

These were supported by records from E. S. Marshall, C. E. Britton,
and others. None of our workers has paid special attention to
this small critical group.

Fragaria L.
F. vesca L.   Wild Strawberry.   Native.   1836.   Common.
Woods, hedgebanks and scrub, avoiding acid soils and especially
common on the chalk.

F. moschata Duchesne (*F. elatior* Ehrh.).
Hautbois Strawberry.   Naturalised alien.   1835.
Very rare, 2.
Our two records suggest that this is a very persistent species: 03A5,
Hascombe, roadside S of White Horse, 1961, JCG. Probably the
same locality as Beeby's reported in Salmon, 1931. 05E4, W.
Horsley, beside track S of church, never fruits, 1965, DPY. This is
the place where it was found by E. Step, recorded in Salmon, 1931,
and it was found there again in 1944, NYS.

F. ananassa Duchesne (*F. chiloensis* auct.)   Garden Strawberry.
Persists occasionally on railway banks and roadsides. C. E. Britton
knew it for at least ten years by a roadside near Ottershaw (*J. Bot.*,
70, 317, 1932).

Geum L.
   G. urbanum L.   Herb Bennet.   Native.   1836.   Common.
Woods, hedgebanks and shady places.

[*G. rivale* L. Water Avens. The only Surrey record is from near Chertsey
by Miss M. Tulk in 1919. Despite repeated search we have failed to
refind it and the species must be regarded as extinct.]

Agrimonia L.
   A. eupatoria L.   Agrimony.   Native   1664.   Common.
Hedgebanks, copses, roadsides, and edges of fields.

   A. procera Wallr. (*A. odorata* (Gouan) Mill.).
Fragrant Agrimony.   Native.   1867.   Frequent, 106.
Map 148.
Wood borders and hedges, avoiding calcareous soils. Especially
frequent on the Weald Clay but showing no preference for clay
soils generally.

Alchemilla L.
   A. filicaulis Buser subsp. *vestita* (Buser) M. E. Bradshaw
Native.   Rare, 8.
Tracks and pastures. 93A1, nr Lythe Hill Lake, 1962, JEL & JES,
Hb. L; 93B1, nr Lythe Hill Lake, edge of conifer nursery, 1960,
BW; 05C1, Clandon Downs, 1965, CPP, 1966, JEL, Hb. L; 13E4,
Capel, nr Lower Gages Farm, 1944, BW, 1967, SFC; 14A5,
hedgebank nr Abinger Common, 1929, ECW, Hb. L; 14B4, Wotton,
meadow by Tillingbourne, 1964, CPP; 14D3, Holmwood Common,
known for many years, 1958, BMCM; 17D4, towpath nr Kew
Gardens Gate, 1 plant, adventive, 1946, BW.

[*A. monticola* Opiz. Probably alien, and no recent records. Found in
quantity in a dry meadow used as a camp, Betchworth Hills, 15E1, in
1931, JEL, Hb. L, det. SMW (*Watsonia* **1**, 12, 1949). Possibly introduced
on the wheels of caravans or by horses, but neither would seem very
likely. Also collected near Woking in 1906 and 1908 by Miss M. Saunders,
Hb. BM, Walters, *op cit.*. This was by White Rose Lane, 05A4, in a
sandy field. Not refound in either locality.]

A. xanthochlora Rothm.   Native.   Very rare, 2.
25C4, Banstead Heath, growing near *Listera ovata* and *Ophioglossum vulgatum*, 1959, Mr & Mrs F. G. Fuller, JEL, Hb. L. 27B1, Wimbledon Common, 1949, BW, Hb. BM, 1950, AWJ, perhaps adventive, but J. F. Young (1796–1860) collected it on the Common, Hb. BM.

A. mollis (Buser) Rothm.   Garden escape.
15D1, nr 'Stepping Stones' below Box Hill, 2 plants, 1948, WHS (*Handlist*, 103); 2 plants, 1956, DPY; 1 plant, much overgrown, with few leaves, 1960, DPY.

Aphanes, L.
A. arvensis L. (*Alchemilla arvensis* (L.) Scop.)   Parsley Piert.
Native.   1640 (for aggregate species).
Very common, 220.   Map 149.
Borders of arable fields, open habitats on commons and roadsides, common on the chalk and on neutral soils, but rare or absent from very acid soils.

A. microcarpa (Boiss. & Reut.) Rothm.   Native.
1949, Walters, *Watsonia*, **1**, 167.   Common, 181.   Map 150.
Borders of arable fields, open habitats in sandy or gravelly places on heaths and commons. Almost absent from the chalk, common on the acid sands in the W of the county.

Sanguisorba L.
S. officinalis L. (*Poterium officinale* (L.) A. Gray).   Native.
1548.   Very rare, 2.
Marshy meadows. 85D5, old meadowland nr R. Blackwater, 1957, RWD, 1966, JES & WEW (first record from here 1884); 26A2, Ewell, marshy ground to N of ox-bow of Hogsmill river, amongst *Filipendula ulmaria*, 1966, CBA.

Poterium L.
P. sanguisorba L. (*Sanguisorba minor* Scop. subsp. *minor*).
Salad Burnet.   Native.   1746.   Locally common, 134.
Map 151.
Mainly on short chalk grassland, especially on the escarpment, but also on the Bargate Beds; on the better-drained parts of alluvial meadows by the Thames which are not deficient in calcium carbonate; and in neutral areas of the Lower Greensand and Bagshot

Sands in well-drained habitats where railway embankments and roadsides have been made up with chalk or with neutral soils. G. M. Ash, who was a farmer, informed us that this species, rather than *P. polygamum*, was grown as a fodder crop about Grayswood, and this may explain some of the records off calcareous soils.

P. polygamum Waldst. & Kit. (*Sanguisorba minor* Scop. subsp. *muricata* (Spach) Nordb.)   Fodder Burnet.
Agricultural relic.   Uncommon, 26.
A fodder crop grown mainly on the chalk, where it persists on road and trackside banks and edges of arable fields. Usually much larger than the last species, but only reliably separated by the fruit.

Acaena Mutis ex L.
A. novae-zelandiae Kirk (*A. anserinifolia* auct. Brit. non Druce)
Pirri-pirri Burr.   Naturalised alien.   Rare.
Thoroughly established on a common. 93C5, Enton, 1961, GSE, no further information. 94B1, Witley Common, known here to OVP and others since 1939 or earlier, 1959, JEL, Hb. L, at least four colonies scattered about the common SW and S of Borough Farm; 94E1, Tuesley, 1960, GSE. It is not known whether the species was introduced here from a garden or as a wool alien, but once introduced it is likely to be spread widely by the fruits attaching themselves to the coats of rabbits, dogs and other animals.

Rosa L.
No special study has been made of Surrey roses in recent years and in any case the numerous varieties which account for most of the 26 pages in the two accounts in Salmon, 1931, are now regarded as hybrids or of little significance. A great deal of work is needed on this difficult group.

R. arvensis Huds.   Field Rose.   Native.   1836, Cooper.
Common.
Hedgebanks, woods and scrub.

R. pimpinellifolia L. (*R. spinosissima* auct.)   Burnet Rose.
Native.   1666, Merrett.   Very rare, 3.
On a common where it has been known for over three centuries, probably alien elsewhere. 84C1, Churt, on the side of a track NE of Crosswater, probably introduced, 1964, JEL; 15C1, Polesden

Lacy, by drive under trees, probably planted, 1966, SFC; 27B3, Barnes Common, 1955, BW, five large patches, 1971, SFC. It was on Ham Common, 17E1, until at least 1944, BW. There were formerly many stations in N. Surrey but the others have all been lost.

R. rugosa Thunb.   Established alien.   1931, Salmon.
Rare, but increasing.
Well established, for example, at: 04B4, Chilworth swamp, 1969, RARC; 16E2, Chessington, in rough grassy field, 1965, EJC; 17D2, Ham pits, 1 large shrub, 1964, EJC; 35D1, Godstone, at intervals along hedge along Oxted Road, planted but naturalised.

R. stylosa Desv.   Native.   1836, Cooper.   Hedges.   Rare.
Apparently always scarce in Surrey, this species has probably been under-recorded recently. Our only current reliable record is from 15E1, Brockham on chalky roadside bank (f. *lanceolata* (Lindl.) W-D.), 1968, EJC, det. RM.

R. canina L.   Dog Rose.   Native.   Very common.
Hedges, woods, scrub, waste ground. Despite the collection of hips for rose-hip syrup, destruction of hedges, and building, Dog Rose is still plentiful and grows as far into London as Wimbledon Common. Various segregates occur: *R. dumalis* Bechst., *R. obtusifolia* Desv., etc.

R. tomentosa Sm.   Native.   1856, Brewer.   Rare.
In hedges, probably scattered in small numbers of bushes over much of the county. For example: 14D2, footpath to Trout's Farm, nr Beare Green, 1964, CPP; 36E2, Addington, Castlehill Ruffs, 1962, DPY, det. RM.

R. sherardii Davies   Native.
First record: 1724, Sherard in Ray, *Synopsis*, Ed. 3, 478, and in Hb. Rand (Thameside between Richmond and Kingston). Rare. 26C1, Banstead Downs, 1955, RCW, det. RM, a large patch, 1967, EJC.

R. rubiginosa L.   Sweet Briar.   Native.
Cooper, 1836, 33.   Locally frequent.
Grassland and scrub on the chalk, where it is frequent; rare on other soils.

R. micrantha Borrer.   Native.   1856, Brewer, 69.
Rare, 16.   Map 152.
Chalk grassland and scrub. Almost restricted to the chalk but in a
hedge on clay soil at 95C1, Wanborough, near Flexford, 1963,
AB & WEW.

R. agrestis Savi   Native.
Very rare and mainly round Thursley, Witley and Puttenham.
Not reported recently but I have seen it: 94A5, Lane on Hog's
Back W of Puttenham, 1945, and 04A1 (?), near Hascombe,
1944, with E. B. Bishop.

Hybrids reported include:
   R. canina x pimpinellifolia = *R. x hibernica* Templeton = *R. x
   glabra* W.-Dod. 27B3, Barnes Common, 1948, JEL, Hb. L,
   1951, LNHS. Known here for many years under various
   names; I failed to refind it when I last searched.
   R. canina x rubiginosa. 36B2, Sanderstead, edge of small wood
   behind Arkwright Wood, 1949, DPY, det. RM.

Prunus L.
P. spinosa L.   Blackthorn, Sloe.   Native.   1838.
Common throughout the county.
Hedges, woods and thickets of all soils, sometimes forming dense
scrub which is difficult to penetrate.

P. domestica L. subsp. *domestica.*   Wild Plum.
Naturalised alien.   1837.   Frequent, 133.   Map 153.
Hedges or wood borders. Usually as isolated trees bird-sown from
orchards or grown from dropped plum stones. For example:
95A2, Ash Ranges, small plant near target, probably from a plum
dropped by troops, 1966, DPY; 26B4, N. Cheam, relic of very old
hedge in a green lane, 1957, DPY; 35B2, Bletchingley Place Farm,
det. EFW.
subsp. *insititia* (L.) C. K. Schneid. (*P. insititia* L.).   Bullace.
Naturalised alien.   1856.   Rare.
Hedges and wood borders. Occasional plants are found as relics in
a hedge, as at 05E2, W. Horsley, near Place Farm, at intervals,
1958, EAB, BW. Specimens from 35B5, Kenley, very old trees in a
spinney, were used for *Fl. Br. Isles Illustr.*, fig 619, 1950, APC, det.
EFW, but Dr Young collected further material from the same
locality in 1960, and found that it did not correspond fully with

EFW's description. We have had considerable difficulty in naming some of our material of these two subspecies.

P. cerasifera Ehrh.   Cherry Plum.   Planted alien.
Frequent, 23.
Widespread in hedges. This species has been much confused with *P. spinosa*, and especially with the var. *macrocarpa* Wallr. which also has rather large flowers and flowers which appear with the leaves, and there has also been some confusion with *P. domestica*. It starts to flower earlier than *P. spinosa* and probably accounts for most of the 'early-flowering blackthorn' records in the county. It is certainly much commoner than our records suggest, but we have no evidence that it spreads from seed.

P. avium (L.) L.   Wild Cherry.   Native.   1718.   Frequent.
Woods and hedges. Common in woods on the chalk and often associated with beech; frequent elsewhere in oak and other woods on a wide range of soils.

P. cerasus L.   Sour Cherry.   Escape from cultivation.
1838.   Rare.
Abandoned gardens and orchards, hedges and woods. Sour Cherries are less often cultivated than formerly, but many of them are self-fertile and may be spread by birds. They are free-suckering and tend to form a thicket of bushes not more than 7m tall. Unfortunately, *P. avium* also suckers and sometimes flowers when quite small and it is on such plants that most records of *P. cerasus* are based. In 50 years I have only seen *P. cerasus* in Surrey once, but we have the following recent records: 25A2, Walton, Little Heath, status doubtful, 1946, ECW; 35C2, Godstone, Flower Lane, 1961, DPY; 35D4, Woldingham, in old neglected estate, 1963, RARC; 35D5, Chelsham, narrow copse, 1963, RARC; 35E2, Oxted, 1963, RARC; 35E4, Lumberdine Wood, 1962, RARC; 36E2, Addington, Rowdown Wood, many bushes over a wide area, 1962, DPY.

P. padus L.   Bird Cherry.   Planted tree.   1838.   Very rare.
Hedges and copses. *P. padus*, abundant as a native elsewhere in Britain, has a very poor status in Surrey. The first record was from the yard of a Reigate inn, others were also of planted trees, and some arose through confusion with *P. serotina*. We have received

the following records: 94C2, Eashing Moor, copse by R. Wey, 1 tree, 1965, GSE; 94C3, Slowley Copse, 1950, GSE.

P. serotina Ehrh.   Rum Cherry.   Naturalised alien.
1853, Loudon (*Encycl. Trees & Shrubs*, 291–2).   Rare.
Woods and heaths. 84C1, Frensham Common, 1942, WEF, Hb. BM; NW of Little Pond, several planted trees reproducing aggressively into native coppice, 1963, DPY; 94B2(?), Peper Harrow, 1925, W. A. Shaw; 94B3, Shackleford Heath, sandy soil, seemingly well established, 1965, GSE, 1966, AJS & WEW; 06D2, Brooklands, 1962, SFC. See Lousley, 1963, *Proc. BSBI*, **5**, 122–123. 15D4, Leatherhead, in hedge, 1974, BRR; 15E2, 25A2, 25A3, Headley Heath, scattered trees, 1974, BRR; 25B3, Withybed Corner, self-sown, 1974, BRR. No doubt under-recorded.

P. laurocerasus L.   Cherry Laurel.   Naturalised alien.
Self-sown in parks, by drives and woods. 93B5, Witley Park, frequent seedlings on roadside, 1959, SFC; 14A3, near Felday, seedlings, 1960, BW; 15D2, edge of yewwood, above drive above, inn, seedlings, 1960, BW; 26A3, Worcester Park Wood, naturalised, 1957, BW.

P. lusitanica L.
Portugal Laurel has been reported as regenerating by path to Mickleham Downs opposite Juniper Hall, 15D2, 1963 B.Wu.

Cotoneaster Medic.
C. simonsii Bak.   Himalayan Cotoneaster.
Naturalised alien.   1917, Leith (*Rep. BEC*, **4**, 485).
Rare, 24.   Map 154.
Bushy places, edges of woodland, hedges. Bird-sown on a wide variety of soils.

C. horizontalis Decne.   Wall Cotoneaster.   Established alien.
1954, McClintock ex Kent (*Proc. BSBI*, **1**, 156).
Rare, 12.   Map 155.
Railway banks, chalk pits, woodland edges.

C. microphyllus Wall. ex Lindl.   Small-leaved Cotoneaster.
Established alien.   1917, Whitchurch, Caterham, Surrey, Mrs Hanbury Tracy (*Rep. BEC*, **4**, 485).   Rare, 9.
Bird-sown on chalky slopes. In Surrey this is never abundant, as

it is on limestone in Wales and SW England, but occurs usually as single bushes.

C. frigidus Wall. ex Lindl.
15E1, Box Hill, 1 tree in dense native scrub, 1960, ECW & CCT, 1966, EMCI, 1967, BWu—there may be more than 1 tree. A hybrid of *C. frigidus* occurs in a hedge at 35C2, Hill Top, Caterham.

Pyracantha M. J. Roem.
P. coccinea M. J. Roem.   Established alien.
Bird-sown in a clump of bushes in the centre of Banstead Downs, 26C1, 1958 onwards, JEL, and doubtless elsewhere.

Crataegus L.
C. laevigata (Poir.) DC. (*C. oxyacanthoides* Thuill.; *C. oxyacantha* auct.).   Midland Hawthorn.   Native.   1870.
Locally common, 156.   Map 156.
Woods on clay, often well inside the wood where it can withstand considerable shade; less often in copses and hedges. This is the most common hawthorn on the Weald and Atherfield Clays in the S, well spread but less abundant on the Clay-with-Flints, London Clay; frequent on the Gault.
x *C. monogyna* = *C. media* Bechst.   Probably not rare on clay.
It is recorded from 84B3, Knowles Lane, Farnham, 1966, DPY; 97E2, Runnymede, just N of path to Kennedy Memorial, 1965, DPY; 26B4, Motspur Park Wood, 1957, RCW—but has been little studied.

C. monogyna Jacq.   Hawthorn.   Native.   1836.
Abundant throughout the county, 499.   Map 157.
Woods, and especially the edges, hedges, commons, chalk downs and in scrub everywhere. Hawthorn was spread widely by planting as 'Quickset' in vast numbers in the eighteenth and early nineteenth centuries by people who received allotments under the Enclosure Acts on condition that they enclosed them. From these hedges birds spread the seeds. It may well be our most widespread woody species since it occurs in built-up areas as well as in the country and on all soils.

Mespilus L.
M. germanica L. (*Pyrus germanica* (L.) Hook. f.).   Medlar.
Relic of ancient cultivation.   1831.   Very rare, 1.
This still grows in a hedge near the top of Redstone Hill, Redhill,

25E3, where it was first recorded by John Stuart Mill in 1831 (Hooker, W. J., *Br. Flora*, ed. 2, 220). Building and road alterations have made great changes in the habitat since I knew it first about 1925, and the 'hedge' is now more of an 'island'. Isolated bushes of Medlar have been reported elsewhere in the county but are non-spinous recent relics of cultivation.

Amelanchier Medic.
A. lamarckii F.-G. Schroeder (*A. canadensis* auct. non Medic; *A. intermedia* auct. *A. confusa* auct. non Hyland; *A. laevis* Clapham, Tutin & Warburg non Wiegand). Juneberry.
Naturalised alien. 1893, Dunn (*Fl. S-W Surrey*, 26).
Locally common, 64. Map 158.
Heathland and open heathy woodland, usually on sandy soils, often associated with *Betula* spp. Thoroughly naturalised around (a) Chobham, Bagshot, Camberley and Pirbright; (b) Frensham, Churt and Haslemere; (c) Hurtwood Common, extending SW to Hascombe and Hambledon; and, less plentifully, (d) round Ockham Common. Dunn's first record of 1893 gave it as 'plentifully naturalised' in the Hurtwood, and many other localities in the Tillingbourne Valley, and copses about Witley and Thursley. It is still so, and yet Salmon in 1931 evidently failed to appreciate the extent to which this tree is established far away from houses. This plant is a lovely sight at the end of April when it is in full flower for a brief period, and again in September when the leaves turn a glorious crimson before dropping. During the summer it is inconspicuous and often overlooked. It is thought that only one species of *Amelanchier* is naturalised in Britain though this has been given many names (Schroeder, 1972). The fruits appear to ripen in mid-summer and to be distributed by birds.

Sorbus L.
S. aucuparia L. Rowan. Native. 1763.
Locally common.
Woods and heathland, mainly on sandy, rather acid soils. This species is perhaps most characteristic of the Greensand hills from Leith Hill, through the Hurtwood and Winterfold to Hambledon, where it thrives on the slopes. Probably native elsewhere on some sandy commons but often bird-sown from planted trees.

S. intermedia (Ehrh.) Pers. (*S. scandica* (L.) Fr.)
Occasionally found as a planted tree. It reproduces from seed occasionally, though less frequently in Surrey than in N. Kent.

S. aria (L.) Crantz   Common White-beam.   Native.
1640.   Locally common, 185.   Map 159.
Common and characteristic of the chalk escarpment but scattered trees are not uncommon on well-drained slopes on the Lower Greensand hills, and occur occasionally on other soils.
x *S. aucuparia* = *S.* x *thuringiaca* (Ilse) Fritsch. The following, all named by Dr Young, no doubt originated from garden trees: 93A3, Weydown Common, 1965, EMH; 15D1, near Stepping Stones, below Box Hill, 1964, EJC; 25B5, Epsom Downs, a small seedling, parent not seen, 1965, DPY.

S. latifolia (Lam.) Pers. *sensu lato*.
Occasionally planted. 35E3, between Botley Hill and top of Oxted chalkpit, 1950, RARC.

S. torminalis (L.) Crantz   Wild Service Tree.   Native.
1664.   Rather rare, 82.   Map 160.
Woods and hedgerows on clay. This is very much a plant of the Weald Clay, where in some woods and districts it is quite frequent. It may perhaps also be native on London Clay, and especially in 93A1 and 95D1. Elsewhere it is almost certainly planted, or from planted trees, and indeed it is sometimes planted on Weald Clay as, for example, 1 tree at Vann Lake Reserve, 13C5.

Pyrus L.
P. communis L.   Wild Pear.   Alien of garden origin.
1835.   Uncommon.   Map 161.
Roadside hedges, commons, sandpits, wood borders, usually as isolated trees. Our map shows a random scatter of localities over the county with a greater frequency in the commuter areas on the west side. There are minor differences in characters, such as variations in the onset of flowering and the hairiness of the leaves, and in their shape and that of the fruit. It is probable that all our plants have originated from pear cores thrown away by farm workers or walkers, though birds may sometimes play a part in distribution.

opposite Early Marsh Orchid (*Dactylorhiza incarnata* subsp. *pulchella*), Thursley Bog (*Photo by D. M. Turner Ettlinger*). Published in memory of Dr D. P. Young with funds provided by the BSBI

J. E. Lousley 1907-76

Malus Mill.
  M. sylvestris Mill.   Apple.   Native or orchard-descended.
  1838.   Common.
  Widespread in woods, copses and hedgerows. Divided into:
    subsp. *sylvestris* (*M. sylvestris* Mill. *sensu stricto; M. pumila*
      auct.)   Native.   Fairly common.
    subsp. *mitis* (Wallr.) Mansf. (*M. domestica* Borkh.). Descended
      from cultivated trees.   Common.
  The introduced plant is rarely thorny, has the leaves persistently
  woolly beneath, and the pedicels, receptacle and outside of the
  calyx tomentose, while the fruit is sweeter. The apple has been in
  cultivation from very early times; over the centuries pips from
  garden fruit have become distributed over the countryside and
  there has probably been crossing with wild trees. This would explain
  why we have been unable to make a clear-cut separation of the
  two subspecies in Surrey.

CRASSULACEAE

Sedum L.
  S. telephium L.   Orpine.   Native.   1763.
  Uncommon, 111.   Map 162.
  Hedgebanks and wood borders, most frequent on sandy banks on
  the Lower Greensand and Weald Clay. It is sometimes difficult to
  distinguish escaped garden plants from natives.

  S. spurium Bieb.
  A garden plant, it is occasionally established—as, for example,
  06A4, Lyne Sewage Farm, a fine colony for many years along brick
  edge of sewage tank, 1965, EJC; 14A4, Raikes Lane, on bank of
  lane across road from Dean Cottages, 1962, BW & JCG.

  S. album L.   White Stonecrop.
  A garden plant occasionally established on wall-tops in villages or
  on the brickwork of reservoirs, but never far from gardens in Surrey.

  S. acre L.   Biting Stonecrop.   Native.
  Rare, 47 (+28 of garden origin).   Map 163.
  Dry grassland on heaths and downs, also on walls. Regarded as
  native on basic soils, such as 15D2, Mickleham Downs, and 15E2,

Juniper Top, and on neutral light soils on the commons of SW
Surrey, and on light soils round Chertsey, Weybridge and Walton.
We also have records from 28 tetrads from walls, cemeteries,
churchyards, etc, where it is likely to be a garden escape.

S. reflexum L. is reported occasionally as a garden escape as, for
example, 35A1, Nutfield Marsh, 1957, BAK & WH; 45B2, West-
wood Farm, 1957, RARC.

[*Sempervivum* L.
S. tectorum L.   Houseleek. 1836. Formerly planted extensively on roofs
as a supposed protection against lightning and for medicinal purposes.
Brewer, 1863, gave it as frequent in the county (and Luxford, 1838, and
Brewer, 1856, said the same for the area round Reigate), but it gradually
became rare so that by Salmon, 1931, it was only 'Scattered here and
there throughout the county'. It is now extremely rare—I am able to cite
only two records: 45B2, Limpsfield, escaped but hardly naturalised on a
steep bank beside old cottages, Moorhouse Bank, 1961, DPY; 95D3,
Fox Corner, the 'Fox', on the roof of one of the usual outbuildings,
1974, JFL, AL. A planted species with no proper place in our *Flora*.]

Crassula L.
C. tillaea L.-Garland (*Tillaea muscosa* L.)   Mossy Stonecrop.
Native.   1954 (*Proc. BSBI*, **1**, 169).   Very rare, 2.
On sandy tracks and nursery beds. 96A2, a troublesome weed for
10 or more years in Waterer's Nursery, Bagshot, 1932, B. Schafer,
Hb. Mus. Brit; *loc. cit.*, apparently not abundant, 1962, BW; 97D1,
on Bagshot Sand near Virginia Water, 1952, 1953, SH, Hb. Mus.
Brit; not seen since.

Umbilicus DC.
U. rupestris (Salisb.) Dandy (*Cotyledon umbilicus-veneris* auct.)
Wall Pennywort.   Native.   1850.   Rare and local, 6.
Map 164.
Sandy banks in shade, and on stone walls. Occurs in 83C5, 83D4,
83D5, 84B1 and 93A1 in a fairly compact area round Frensham,
Churt and Haslemere in the extreme SW of the county, and usually
in small numbers. We have not refound it in 94A1 near the Half
Moon Inn, or near Thursley, where there are old records. An
interesting new locality a long way from the others is 35A1,
Pendell House, on old stone walls of the garden, 1964, JES.

SAXIFRAGACEAE

Saxifraga L.
  S. tridactylites L.   Rue-leaved Saxifrage.   Native.
  1827.   Rare, 20.   Map 165.
Sandy fields and banks, walls. Luxford and Brewer regarded this
as common, and even Salmon, 1931, regarded it as so common
that there was no need to give localities. In my own boyhood it was
a frequent plant, but now it is rather seldom found. The reason
is partly the destruction of old walls (cf. *Erophila verna*), but this is
not the full explanation. This may be associated with smoke
pollution.

  S. granulata L.   Meadow Saxifrage.   Native.   1597.
  Local, 37.   Map 166.
Sandy or gravelly fields, cemeteries, usually in well-drained places
but also in basic alluvial meadows near the Thames. Still persists
in the turf of Kew Gardens, 17E4, and in meadows at Ham, 17D2.

Chrysosplenium L.
  C. oppositifolium L.   Opposite-leaved Golden-saxifrage.
  Native.   1836.   Locally common, 133.   Map 167.
Stream-sides, wet woods, by ponds. With the exception of one
colony near Virginia Water, 96E5, restricted to S of the chalk.
Common on the Lower Greensand, Weald Clay and Hastings Beds.

  C. alternifolium L.   Alternate-leaved Golden-saxifrage.
  Native.   1837.   Rare, 14.   Map 168.
Wet woods and swamps, and especially alder holts, and by streams.

GROSSULARIACEAE

Ribes L.
  R. rubrum L. (*R. sylvestre* (Lam.) Mert. & Koch)
  Red Currant.   Native.   1666.   Frequent.
Woods, hedges and banks of streams. Commonly bird-sown from
gardens and, as this has no doubt gone on for centuries, it is no
longer possible to separate native localities.

  R. nigrum L.   Black Currant.   Native.   1837.
  Uncommon, 120.   Map 169.
By streams, canals and ponds, damp woods. Sometimes bird-sown

from gardens, but the native status of this species is more convincing than that of the last.

R. sanguineum Pursh   Flowering Currant.   Garden alien.
14D1, Bennett's Wood, N of Coles' Lane, 1 bush in cut-down wood believed to be bird-sown, 1958, BMCM & BW.

[*R. alpinum* L.   Mountain Currant. Occasionally planted as at 35E2, Barrow Green, roadside hedge, 1958, CNHSS.]

R. uva-crispa L.   Gooseberry.   Native.   1793.   Common.
Hedges, woods, streamsides, in pollarded trees. Commonly bird-sown, and often brought from gardens.

DROSERACEAE

Drosera L.
D. rotundifolia L.   Common Sundew.   Native.
1718.   Locally common, 60.   Map 170.
On wet peat by paths and ditches on heaths and in spongy bogs. Surviving on Wimbledon Common, 27B1, and Reigate Heath, 25B1, but locally common on the great areas of heathland towards the W of the county.

D. intermedia Hayne (*D. longifolia* auct.)
Long-leaved Sundew.   Native.   1696.
Locally frequent, 34.   Map 171.
Spongy bogs and wet peat, often growing with *D. rotundifolia*, but more common in the wettest places.

SARRACENIACEAE

Sarracenia L.
S. purpurea L.   Pitcher Plant.
96D4, Chobham Common, on Sphagnum. Presumably deliberately planted but the habitat was chosen with considerable skill. First found in 1968 by A. C. Withers and communicated by Dr S. Waters, of Bedford College, and Dr E. Lodge, of Royal Holloway College. Spread over about 1 sq m with offspring, but somewhat decreased by 1974, JEL.

LYTHRACEAE

Lythrum L.

L. salicaria L.   Purple Loosestrife.   Native.   1597.
Common, 200.   Map 172.
By rivers, ditches and ponds, in reed-swamps, and marshy meadows.
Almost absent from the chalk and from acid heathy areas.

[*L. hyssopifolia* L. was formerly well established at Mitcham but there
are no current records. *L. junceum* Banks & Sol., which is commonly
mistaken for it, is often introduced with food for cage birds.]

L. portula (L.) D. A. Webb (*Peplis portula* L.).
Water Purslane.   Native.   1633.   Frequent, 92.   Map 173.
Pond margins, woodland rides, damp places in sandpits, etc.

THYMELAEACEAE

Daphne L.

D. mezereum L.   Mezereon.   Native?
1835.   Very rare, 1.
Woods on the chalk. 05E1, W. Horsley, 1967, MLC. This locality
is in scrub which has grown up on chalk grassland and I saw 3
plants in 1969. There are old records for five places on chalk in
Surrey which Salmon accepted as native. It still grows in similar
places on the Chilterns and in Sussex; seed might be brought from
there by birds, and also from Surrey gardens.

D. laureola L.   Spurge Laurel.   Native.   1762.
Locally frequent, 56?   Map 174.
Woods, mainly on calcareous soils. This is usually regarded as
characteristic of chalk and limestone, but in Surrey nearly a third
of the localities are off the chalk. It is sometimes planted for the
evergreen foliage.

ONAGRACEAE

Epilobium L.

E. hirsutum L.   Great Willowherb.   Native.   1832.
Common throughout the county.
Streambanks, marshes, by ponds, roadsides, also frequent in drier
places on commons and waste ground.
x *E. montanum*=*E.* x *erroneum* Hausskn. 15D2, Mickleham Downs,
1960, TDP.

E. parviflorum Schreb.   Hoary Willowherb.   Native.   1836.
Frequent, and scattered throughout the county, 125.   Map 175.
Streambanks, marshes, pits and damp workings.

E. montanum L.   Broad-leaved Willowherb.   Native.
1633.   Very common.
Woods, hedgerows, walls and garden weed. Common also in the
built-up areas.
x *E. obscurum* = *E. x aggregatum* Cĕlak. 93B4, Brook brickworks,
1955, OVP.
x *E. roseum* = *E. x mutabile* Boiss. & Reut. 37B2, Dulwich Woods,
1958, JEL.

E. lanceolatum Seb. & Mauri.   Spear-leaved Willowherb.
Native.   1852.   Rare, 13.   Map 176.
Sandy and gravelly banks, walls, roadside slopes, and brickworks.
83D5, Churt, DPY; 84B3, River Lane, Wrecclesham, DCK; 93B4,
Brook birchwoods, OVP; 94C3, Charterhouse Copse, OVP;
94D5, Guildford By-pass on gravel dumps, TDP; 94E2, Coombe
Lane, Bramley, AB, EJC, CPP; 03A5, lane Hascombe to Place
Farm, DPY; 17E4, Kew, by Herbarium, PHR (*Watsonia* 6, 36);
24D4, BMCM; 25B1, Lawrence Lane, Buckland, BMCM; 25E1,
Redhill, the Moors, RARC; 26C4, St Helier, railway, RCW;
35B5, Kenley, 2 plants, DPY.
x *E. montanum* = *E. x neogradense* Borbás. 93B4, Brook brickworks,
1955, OVP.

E. roseum Schreb.   Pale Willowherb.   Native.   1798.
Rather rare, 42.   Map 177.
A weed in gardens, shady passages, the base of walls, and occa-
sionally by streams.

E. adenocaulon Hausskn.   American Willowherb.
Naturalised alien.   1934, Ash (*Rep. Watson BEC* 4, 218–9).
Abundant throughout the county.
Damp woods, roadsides, gardens, waste places, railway banks,
almost ubiquitous. This North American species was first brought
to the notice of British botanists from Rice's Lane, Witley, 93C5?
by G. M. Ash in 1934; his account the following year (*J. Bot.* 73,
177–184, 1935) was based mainly on Surrey specimens and showed
that it has been in the county undetected since at least 1921 when
J. Fraser found it at Woking (Hb. Kew). The species had been

much confused with *E. tetragonum* and *E. obscurum* and is now
known to have been in Britain since 1891. It first appeared in
Scandinavia about 1900, and Poland about 1917, and has spread
rapidly to Germany, Luxemburg, Belgium, Holland and western
France. Its spread in Surrey, as in southern England generally, has
been phenomenally fast and it is now by far our commonest
willowherb.

x *E. hirsutum*. 96E5, Callowhill, 1968, DPY.

x *E. montanum*. 93B4, Brook birchwoods, 1949, GMA, 1955, OVP;
25B1, Buckland, garden weed, 1965, BMCM & EJC.

x *E. obscurum*. 93B4, Brook brickworks, 1949, GMA, 1955, OVP;
06A4, Lyne, 1965, EJC; 06A5, nr Whitehall Farm, Stroude, 1965,
EJC; 24D5, Earlswood Common, 1965, EJC.

x *E. parviflorum*. 93B4, Brook birchwoods, 1949, GMA, 1955, OVP.

x *E. roseum*. 26A3, Worcester Park, 1956, RCW.

E. tetragonum L. subsp. *tetragonum* (*E. adnatum* Griseb.).
Square-stalked Willowherb.   Native.   1821.
Frequent, 116.   Map 178.
Damp ground, woodland clearings, roadsides, ditchsides and
cultivated ground.
Subsp. *lamyi* (F. W. Schultz) Nyman (*E. lamyi* F. W. Schultz).
We have been unable to separate satisfactorily from subsp. *tetragonum* although Salmon, 1931, and the late G. M. Ash recognised
it as a separate species.
x *E. montanum* = *E.* x *beckhausii* Hausskn. 93D5, Hambledon,
1949, GMA.

E. obscurum Schreb.   Short-fruited Willowherb.   Native.
1863.   Frequent, 111.   Map 179.
Marshes, ditchsides, moist ground in woodland, roadsides.
x *E. parviflorum* = *E. dacicum* Borbás. 34B1, Burstow, 1961, DPY.

E. palustre L.   Marsh Willowherb.   Native.   1836.
Rather rare, 51.   Map 180.
Heathy ponds and acid marshes.

Reference should be made to Salmon, 1931, 325–32, for additional
hybrids between the above species. These were reported before the
recognition of *E. adenocaulon* and some of them were in fact that
species.

[*E. nerterioides* Cunn.   New Zealand Willowherb. This is not known to be established in the wild in the county but is an aggressive weed in a few gardens: 83E2, Haslemere, on a garden wall, King's Road, 1963, DWF; 83D4, Hindhead, spreading from rockery in garden of 'Hillside', Tilford Road, 1965, DCK; 05B5, in a neglected garden at Pyrford Court, 1968, WEW.]

E. angustifolium L. (*Chamaenerion angustifolium* (L.) Scop.)
Rosebay.   Native.   1763.   Common throughout the county.
Waste ground, railway banks, roadsides, woods, heaths and commons. By Brewer, 1863, this species was widespread in the county, and Salmon, 1931, gave it as 'Frequent and often locally abundant' in four of his districts and rare in the other six. The further increase in the last 40 years has been spectacular and it is now one of our most abundant plants. White flowers are rare, they are reported from: 05E1, Effingham Forest, Green Hill, 1957, BW; 37D5 Surrey Commercial Docks, 1972, JEL & LMPS.

Oenothera L.
O. erythrosepala Borbás   Evening Primrose.
Naturalised alien.   1931, Salmon, but no doubt earlier.
Frequent and widespread, especially on sandy soils. This is our commonest Evening Primrose; we have records from 27 tetrads, which are clearly very incomplete.

O. parviflora L.
Recorded from 06E2, Walton Common, 1961, DP det. PHR; 26E4, waste ground by Mitcham Junction, 1963, EJC.

Circaea L.
C. lutetiana L.   Enchanter's Nightshade.   Native.
1792.   Common and widespread.
Woods, damp shady places and shady places in gardens, on base-rich soils.

HALORAGACEAE

Myriophyllum L.
M. verticillatum L.   Whorled Water-milfoil.   Native.
1666.   Very rare, 4.
Ponds and a slow-running river. 84B3, R. Wey between Passmore Bridge and River Lane, Farnham, 1966, DCK; 06B5, Eastley End, the Fleet and stream below it, 1965, EJC, and Thorpe, pond in old

gravel-pit, 1958, BMCM; 07A1 & 07A2, Langham Pond, 1969, JES & WEW (known here for over 40 years). The records in Salmon, 1931, cover a very long period and the species is probably no rarer than it was earlier.

M. spicatum L.   Spiked Water-milfoil.   Native.
1763.   Uncommon, 34.   Map 181.
Canals, ponds, ditches and slow rivers.

M. alterniflorum DC.   Alternate Water-milfoil.   Native.
1850.   Rare, 10.   Map 182.
Ponds, lakes, canals. This still persists in a short stretch of the long-disused Wey & Arun Canal N of Alford in 03B4, 03C4, and 03C5 where W. H. Beeby found it over 60 years ago.

HIPPURIDACEAE

Hippuris L.
H. vulgaris L.   Marestail.   Native.   1814.   Very rare, 4.
Ponds and canal. 85E2, canal, Ash, 1963, WEW; 85E3, Mychett, canal, 1963, WEW and Mychett Lake, 1972, SFC; 05B1, Clandon Park, Upper Lake, 1957, BW (recorded from here by John Stuart Mill in 1841, and by C. E. Britton, *J. Bot.* **70**, 317, 1932); 15C5, Fetcham Millpond, 1948–58, LJJ, a locality dating back to 1838, very well known to many contemporary workers, and destroyed when the pond was drained for water-board work about 1962; 25E2, New Pond, Merstham, 1947, JDL—Dr Young was unable to refind this.

CALLITRICHACEAE

Callitriche L.
C. stagnalis Scop.   Common Water-starwort.   Native.
1633.   Common and widely distributed.
Characteristically a small-leaved creeping plant on mud in puddles in woodland rides (var. *serpyllifolia* Lönnr.), but also common in streams, ditches, marshes, ponds and an alder-bog. As our helpers collected only material in fruit which they could send for identification, this and other species of *Callitriche* are usually heavily under-recorded.

C. platycarpa Kütz. (*C. polymorpha* Lönnr.)   Native.
1891.   Rare, 11.   Map 183.
Field-ponds, streams, ditches. All the dots on the map are based

on determinations by Dr J. P. Savidge, and reflect the enthusiasm of several workers in the eastern part of the county.

C. obtusangula Le Gall   Native.   1876.   Rare, 9.
Ditches and ponds. Recorded from 94E4, 24B5, *26D5, *26E3, *26E4, *26E5 and *34E1 (* = det. JPS). Still fairly plentiful on Mitcham Common whence it was first recorded in 1876.

C. intermedia Hoffm.   Native.   1792.
Common, 74.   Map 184.
Rivers, canals, ditches, ponds. Apart from *C. stagnalis* this is our most common *Callitriche* and, because it is easily recognised by the bicycle-spanner-like leaf apices, it is the most frequently recorded. Even so our records do not fully reflect the distribution.

[*C. palustris* L. This European species, which grows mainly in mountainous areas, was reported from Britain by Dr. H. D. Schotsman in 1954 citing a specimen from Petersham, Surrey. This was an error for *C. stagnalis* Scop., Meikle & Sandwith, *Proc. BSBI*, **2**, 135–6, 1956.]

LORANTHACEAE

Viscum L.
V. album L.   Mistletoe.   Native.   1718.
Frequent, 34.   Map 185.
Probably most common on apple, mainly in gardens, but our most numerous reports were on lime, *Tilia x europaea* ,and poplar, *Populus x canadensis*, which are more conspicuous. Other frequently reported hosts include hawthorn, *Crataegus monogyna*, acacia, *Robinia pseudoacacia*, whitebeam, *Sorbus aria*, and maple, *Acer campestris*. We have failed to refind mistletoe on oak, including the famous mistletoe-oak at Dunsfold where Salmon saw it as late as 1917.

SANTALACEAE

Thesium L.
T. humifusum DC.   Bastard Toadflax.   Native.   1778.
Rare, 17.   Map 186.
Chalk grassland, a plant of short, and usually grazed, turf. Restricted to the chalk from Pewley Down in the W to Purley Downs and Riddlesdown in the E.

CORNACEAE

Swida Opiz
S. sanguinea (L.) Opiz (*Cornus sanguinea* L.; *Thelycrania sanguinea* (L.) Fourr.).   Dogwood.   Native.   1832.
Common generally and locally too abundant.
Woods, scrub and hedgerows, especially on calcareous soils. The rapid increase of Dogwood following the withdrawal of grazing on chalk grassland is a major threat to the amenity and natural history interest of these areas, and has necessitated, and will continue to demand, much of the effort of conservation corps.

S. sericea (L.) Holub. (*Thelycrania sericea* (L.) Dandy; *Cornus stolonifera* Michx.)
A garden shrub which persists once it is established, and is just out of gardens in a few places, such as Enton, 94C1.

ARALIACEAE

Hedera L.
H. helix L.   Ivy.   Native.   1836.
Very common throughout the county.
Woods, hedges and walls, almost as common in the built-up areas as in the country.

UMBELLIFERAE

Hydrocotyle L.
H. vulgaris L.   Marsh Pennywort.   Native.   1793.
Locally common.
Bogs and marshes, usually a sign of acid soil conditions.

Sanicula L.
S. europaea L.   Sanicle.   Native.   1763.
Common, 181,   Map 187.
Common in beechwoods along the chalk and in oakwoods on loam on Weald Clay. Rare on Lower Greensand and very rare on London Clay; absent from Bagshot Beds.

Chaerophyllum L.
C. temulentum L.   Rough Chervil.   Native.   1836.
Common throughout the county.
Hedgebanks, roadsides, wood borders.

Anthriscus Pers.

A. caucalis Bieb. (*Anthriscus vulgaris* Pers.; *Chaerophyllum anthriscus* (L.) Crantz)   Bur Chervil.   1836.   Rare, 19.   Map 188. Commons, roadside banks, margins of arable fields, usually on loose sandy soils.

A. sylvestris (L.) Hoffm. (*Chaerophyllum sylvestre* L.) Cow Parsley.   Native.   1836.   Common. Margins of woods, hedgebanks, copses and waste places. Still abundant along many Surrey lanesides, and especially beautiful where the soil is moist and it gets a little shade, but decreasing owing to the widening of roads and the use of pesticides.

[*A. cerefolium* (L.) Hoffm.   Garden Chervil. A garden escape. 1838. Although Salmon, 1931, was able to report seven more or less contemporary records, this has not been reported recently. I knew it for several years on a bank by the road between Reigate Heath and the town, 25C1, (1935, H. S. Redgrove, Hb. Lousley), but last saw it there about 1940.]

Scandix L.

S. pecten-veneris L.   Shepherd's-needle.   Native.   1836. Rare, 19.   Map 189. Weed of arable crops and especially corn. Like Salmon (1931, 343) I would have expected to find this 'chiefly on the chalk', but we have more records from the Weald and London Clays. A decreasing species owing to modern agricultural practice.

Myrrhis Mill.

M. odorata (L.) Scop.   Sweet Cicely.   1843. Garden escape occasionally established for a time, especially in the Puttenham and Milford area. For example: 94C2, Lower Eashing, side of Godalming By-pass, 1953, AWW, and 94C4, Wanborough Common, wood near garden, 1966, OVP.

Torilis Adans.

T. japonica (Houtt.) DC. (*Caucalis anthriscus* (L.) Huds.; *Torilis anthriscus* (L.) C. C. Gmel., non Gaertn.) Upright Hedge-parsley.   Native.   1832. Common throughout the county. Mainly a hedgerow plant, but also wood borders and waste ground.

T. arvensis (Huds.) Link. (*Caucalis arvensis* Huds.)
Spreading Bur-parsley.   Native.   1814.   Very rare, 4.
Arable fields. Reported since 1950 from: 95E3, Woking, Pile Hill, 2 plants in sandy cornfield 1964, WEW; 25D1, Redhill Common, by access road crossing N part, 1963, FDSR; 35E5, on arable gravel (Clay-with-Flints) by footpath from Chelsham Church to Fickleshole Road, in stubble, 1962, RARC; 36B1, Sanderstead, field by Mitchley Hill, 1958, LJJ. Formerly frequent on a wide range of soils, and still not particularly rare in the 1940–50 decade, this species has shown one of the most spectacular decreases of all the arable weeds. It may belong to the group with *Agrostemma githago* which depended on periodic reintroduction of their seed with foreign seed-corn.

T. nodosa (L.) Gaertn. (*Caucalis nodosa* (L.) Scop.).
Knotted Bur-parsley.   1832.   Native.   Very rare, 2.
Dry places, towpath and hedge. 17D2, towpath below Ham Dock, 1946, BW, 1948–52, LJJ, and Petersham Meadows, 1950, BW; 36A5, Norbury, hedge of recreation ground, Pollards Hill, 1962, DPY. Salmon, 1931, described this as 'uncommon'; it is now practically extinct. Formerly it grew in a number of places on gravel by the Thames, of which the localities in 17D2 are survivals. It also grew in fields and roadsides on the chalk; these include Epsom Downs in 25B5, where I knew it below the Grandstand and near Drift Bridge until July, 1944. The rapid decrease of this species is difficult to explain, but the British distribution suggests that it requires warm summer temperatures, and the reason may be climatic.

Caucalis L
[*C. platycarpos* L. (*C. daucoides* L.). Colonist, 1746. Formerly a rare cornfield weed, of which we have no recent records.]

Coriandrum L.
C. sativum L.   Alien.   1814.
The seeds are imported for several purposes and this is still a frequent casual on refuse tips.

Smyrnium L.
S. olusatrum L.   Alexanders.   Established alien.
1724.   Rare, 8.
Roadside, hedgebanks, Thames towpath. 95D2, Worplesdon, S of

churchyard, 1956, LJJ; 05D2, alley W of Cranmore, 1967, DPY; 05E2, W. Horsley, N side of A246, an old and well-known locality, where it persisted after road improvement; 16A4, 16A5, by the Thames nr Sunbury Lock, recorded from here by J. S. Mill in Brewer, 1863, and still abundant in 1957, JEL; 17D2, on river-wall, 1953, BW; by Thames at Ham Dock, 1958, LJJ—possibly carried down from the large Middlesex colony at Hampton Court; 36B1, nr top of Riddlesdown Road, 1949, LJJ; 36B2, hedgebank, Selsdon Road, Croydon, 1947, DPY, gone by 1957. This mainly Mediterranean species is probably a relic of former use as a pot herb in Britain.

S. perfoliatum.   Naturalised alien.
1953 (Kent & Lousley, *Handlist*, 127).   Rare, but spreading.
Under shrubs in parks and elsewhere. 17E3, Richmond, bushes by Kew Road, escape from Kew Gardens, 1948–53, BW; 27E4, flowerbed weed, Battersea Park, 1939, Mrs G. M. Gibson—still there and spreading, JEL.

Conium L.
C. maculatum L.   Hemlock.   Native.   1827.
Frequent, 143.   Map 190.
Damp places by rivers, streams and ditches, wood borders and waste places. Hemlock is most at home in damp hollows by rivers and about half our Surrey localities are associated with the rivers Wey and Mole.

Bupleurum L.
[*B. rotundifolium* L. Thorow-wax. Colonist. 1746. This was probably never established in Surrey in cornfields for more than a few years, and may have been extinct before Salmon, 1931. *B. lancifolium* Hornem. (*B. subovatum* auct.), commonly introduced with cage-bird seed, and in other ways, is found in gardens and on refuse tips and often mistaken for *B. rotundifolium*. It has rather narrower, more pointed, leaves, fewer (2–3) rays to the umbel, and larger rough-surfaced fruits.]

[*B. tenuissimum* L., which Salmon, 1931, says was first recorded in 1650, was found in several places on London Clay commons. It has not been reported since 1916, when C. E. Salmon collected it on a roadside near Epsom, Hb. Mus. Brit, Hb. Kew.]

[*B. falcatum* L., found in 1856 at the W end of Reigate Heath, Brewer, 1863, lasted only a few years, and has not been recorded recently.]

Apium L.

[*A. graveolens* L. Wild Celery. 1793. Formerly native by the Thames but not reported during our survey.]

A. nodiflorum (L.) Lag.   Procumbent Marshwort.   Native. 1799.   Frequent throughout the county.
By ponds, ditches and streamsides.

A. inundatum (L.) Reichb.f.   Lesser Marshwort.   Native. 1666.   Rare, 31.   Map 191.
Shallow pools and ditches, and canal. In the Basingstoke Canal from Brookwood to Ash Vale, otherwise scattered over the county in small shallow pools.

Petroselinum Hill.

[*P. crispum* (Mill.) Airy Shaw. Wild Parsley. Found occasionally as an obvious garden escape but not recorded as established in the county.]

P. segetum (L.) Koch   Corn Parsley.   Native.   1666.
Rare, 10.   Map 192.
Edges of arable fields, roadsides, hedgebanks. This has always been rare in Surrey and has not decreased significantly.

Sison L.

S. amomum L.   Stone Parsley.   Native.   1548.
Locally common, 175.   Map 193.
Edges of woods, hedgebanks, and roadsides. Mainly on clay soils, especially common on the Weald and Atherfield Clays, also on London Clay and Gault.

Carum L.

C. verticillatum (L.) Koch   Whorled Caraway.   Native. 1908.   Very rare, or extinct, 2.
Marshy, turfy places. 05A5, Horsell Common, by small pond near canal, 1906–46, 6 plants in flower, 1946, WEW; 06A1, Horsell Common, by track, 1929–50, 1 plant in flower, 1950, WEW. Dunford Bridge, 1907–67, this well-known locality has undergone great changes in recent years arising from bridge widening, tinkers' encampments, and quite recently scrub development. It is now very much less suitable and the last record is of 1 plant, 1967, JES. The Surrey localities for *C. verticillatum* were over 60 miles E of the otherwise most easterly localities for this species, in Dorset.

C. carvi (L.)   Caraway.   Alien.   Rare.
Waste ground and roadsides. 17, Ham Common, roadside, TBR, 1974.

Conopodium Koch
 C. majus (Gouan) Koch (*Conopodium denudatum* Koch).
Pignut.   Native.   1710. · Common.
Woodlands with deep litter, heathland and pastures on neutral or somewhat acid soils.

Pimpinella L.
 P. saxifraga L.   Lesser Burnet-saxifrage.   Native.
1793.   Common.
Chalk grassland, roadsides and well-drained pastures, mainly on basic or neutral soils. Very variable in leaf shape.

 P. major (L.) Huds.   Greater Burnet-saxifrage.   Native.
1746.   Locally frequent, 42.   Map 194.
Grassy places along margins of woods and copses, roadsides and hedges. Mainly a plant of the chalk but with a few localities on Weald Clay in the eastern part of the county.

Aegopodium L.
 A. podagraria L.   Goutweed.   Naturalised alien.
1793.   Common throughout the county.
Near houses in gardens, waste ground, roadsides and hedges. An ancient introduction which is now a serious weed in gardens both in town and country. The far-creeping rhizomes are very difficult to eradicate from clay soils, and the plant is known to gardeners as 'Ground Elder'.

Sium L.
 S. latifolium L.   Greater Water-parsnip.   Native.
1724.   Very rare, 3.
Now only by ponds. 97D1, Windsor Great Park, Obelisk Pond, 1965, EJC; 07A1, Egham, pond by by-pass, 1956, AWW; 07A1, 07A2, Runnymede, by a pond, a locality at least 50 years old, where it persisted in 1972. Elsewhere in Britain this fine plant is associated with fen-like conditions and it still grew in such places in the NE of the county in the eighteenth century. More recently it was known from a number of places by the Thames along almost its whole length, but the embankments, anglers and boating have destroyed it. In 1945 I found a fine colony in 06B2, by a pond at Eastly End, Thorpe. Here it occurred in two strongly contrasting forms—with broad leaflets and with narrow—but extended gravel-winning

activities seem to have destroyed it. It may yet be refound in the vicinity of Thorpe.

Berula Koch
B. erecta (Huds.) Coville   Lesser Water-parsnip.   Native.
1633.   Rather rare, 26.   Map 195.
Ditches, streams, ponds, canal, and marshy ground. Most common along the partially dried-up Basingstoke Canal; rather frequent on wet commons round Elstead, Puttenham and Godalming; scattered localities elsewhere.

Oenanthe L.
O. fistulosa L.   Tubular Water-dropwort.   Native.   1633.
Rather rare, 31.   Map 196.
Marshy places and shallow water of ponds, ditches and canals; thrives especially on spongy marsh. Now mainly in the W of the county, where the Basingstoke Canal accounts for 8 tetrads. In the E still at the old localities at Mitcham Common, 26E4, and New Pond, Merstham, 25E2, but gone, for example, from many of the old localities round Esher and Molesey.

O. pimpinelloides L.   Corky-fruited Water-dropwort.
Native.   1931.   Very rare, 5.
Commons and other rough places on clay. 95C1, Worplesdon, Backside Common nr Wood Street, covering a ¼ acre, 1962, WEW & AJS; 16E1, Epsom Common, near Stamford Green in rough ground probably ploughed during 1939–45 war and also on plateau above the Stew Pond; 16E2, Chessington, clayey meadows S of church, 1959, K. Page comm. A. Hitch; 24E5, Earlswood, field on E side of railway, 'flourishing and has increased considerably', 1961, FDSR; 35C5, Warlingham, S of Court Farm, 1 plant, 1951–4, LJJ.

O. silaifolia Bieb.   Narrow-leaved Water-dropwort.
Native.   1845.   Rare, 10.
Damp rich meadows, usually cut for hay. 06C4, 06D4, Chertsey Mead, an old locality, now restricted to two fields, 1962, SFC; 07A1, Egham, hedge and ditch adjoining Runnymede, 1956, AWW; 24B5, by Mole, Little Flanchford, 1950, FR; 24C4, nr Mole, E of Sidlow Bridge, 1964, DCK; 24D2, Hookwood by brickworks, 1964, RARC; 24D4, N of Kinnersley Manor, near Mole, 1958,

AWJ; 34E5, Oxted Brook meadows, 1962, RARC, N of Gibbs/ Oxted Brook junction, 1967, RARC, Gibbs Brook, S of Foyle Farm, 1963, 1967, RARC; 44A3, bank of R. Eden W of Cernes, 1964, ECW & JES; water meadows on Cernes/Haxted footpath, 1970, RARC; 44B3, S of Puttenden Manor, in water meadows, 1970, RARC.

In Surrey the localities fall into three groups: (a) water meadows by the Thames at Chertsey and Runnymede, (b) water meadows by the R. Mole S of Reigate and (c) water meadows near the R. Eden headwaters, from which it extends across the Kent border along the river to Tonbridge. This is a decreasing species due to drainage, ploughing and re-seeding, and the use of fertilisers. The species has decreased in (a) and (b), and disappeared from other areas given in Salmon, 1931, but the fine colonies in (c) had not been discovered at that time.

O. lachenalii C. C. Gmel. Parsley Water-dropwort.
Native. 1873. Very rare, 1.
Marshy ground by a pond and in hollows on clay on a common. 28E4, Mitcham Common. Found here by A. Bennett and W. H. Beeby in 1873 and known to Salmon before 1931, C. Avery, 1928, Hb. L. More recent records are 1957, JEL, Hb. L., 1958, two colonies, RCW and 1974, JEL. There are still about 20 plants in this, the only certain Surrey locality.

O. crocata L. Hemlock Water-dropwort. Native.
1548. Too common.
Wet places of all sorts—ponds, rivers, wet copses, marshes and ditches. An unpleasant and poisonous plant which persists even on the Thames embankments in central London.

O. aquatica (L.) Poir. (*O. phellandrium* Lam.) Native.
Rare, 23. Map 197.
Ponds, ditches, rivers, canal. Really rarer than the records suggest, as the plant is often in very small quantity in field ponds where cattle trample it out of existence for a time.

O. fluviatilis (Bab.) Colem. River Water-dropwort.
Native. 1865.
A species of fast-running water recorded from the Thames, and some nearby streams, at intervals from Staines to Kingston. It has

not been reported by any of our contributors and may be extinct. Aquatic vegetation in the Thames is greatly reduced as a result of being slashed about by the propellers of motor cruisers, but on the other hand *O. fluviatilis* is a shy flowerer and was rarely seen in flower in Salmon's time, and it may quite well be still there. Present or not, there cannot be many counties which can report all seven British species of *Oenanthe*.

Aethusa L.
A. cynapium L.   Fool's-parsley.   Native.
1836.   Very common.
A weed of gardens and arable, roadsides and waste places. The dwarf form of cornfields, about 4 or 5in tall, (var. *agrestis* Wallr.) is seldom seen nowadays.

Foeniculum Mill
F. vulgare Mill.   Fennel.   1793.
A garden escape persisting for a time on waste ground, refuse tips and railway banks, but not permanent.

Silaum Mill.
S. silaus (L.) Schinz & Thell. (*Silaus flavescens* Bernh.).
Native.   1793.   Frequent, 107.   Map 198.
Commons and roadsides on heavy clay soils. Mainly on Weald Clay, London Clay and Gault.

Angelica L.
A. sylvestris L.   Wild Angelica.   Native.   1640.
Frequent and widespread.
Marshy meadows, wet roadsides, open damp woods and streamsides.

A. archangelica L. (*Archangelica officinalis* Hoffm.)
Garden Angelica.   Established alien.   1771.
Locally frequent, 16.   Map 199.
Banks of the Thames and waste places. Anciently cultivated as celery or for candying the root, this has been known from by the Thames in Surrey for two centuries, and still occurs scattered along from Rotherhithe to Hampton Court. Elsewhere by a roadside and carrot field.

Pastinaca L.
P. sativa L. (*Peucedanum sativum* (L.) Benth. ex Hook. f.)
Wild Parsnip.　　Native.　　1793.
Very common on the chalk, frequent elsewhere. Chalk grassland and scrub, roadsides and rough commons.

Heracleum L.
H. sphondylium L.　　Hogweed.　　Native.　　1763.
Common throughout the county.
Roadsides, hedgerows and rough grassy places.

H. mantegazzianum Somm. & Levier.　　Giant Hogweed.
Naturalised alien.
This is the plant identified as *H. villosum* Fisch. (which some of it may be) in Salmon, 1931, who gives four localities. We have it reported from streamsides, riverbanks, roadsides and waste ground from 83C5, 83E4, 93C2, 94E5, 03A2, 03A5, 14A2, 14A3, 26B5, 35D1, but it is more frequent than this. While the number of localities has increased, Giant Hogweed shows no tendency for rapid spread along rivers as it does in some other counties.

Daucus L.
D. carota L. subsp. *carota*　　Wild Carrot.　　Native.　　1836.
Abundant on the chalk and frequent in the rest of the county.
Chalk grassland, roadsides and other grassy places.

CUCURBITACEAE

Bryonia L.
B. dioica Jacq.　　White Bryony.　　Native.　　1832.
Frequent and especially so on the chalk.
Hedgerows, copses and woodborders, round rabbit warrens on the chalk.

ARISTOLOCHIACEAE

Aristolochia L.
A. rotunda L.　　Dwarf Birthwort.　　Established alien.
1929, Burkill, H. J., *Lond. Nat.* **8**, 16–17; & 1930, *Rep. BEC*, **9**, 136.
35D2, Woldingham, South Hawke, in a spinney on chalk hillside. The earliest specimen of this I have seen was collected by F. A. Swain in 1925, Hb. L. In 1928 it was found by H. J. Burkill and R. W. Robbins as: 'a patch 6ft x 10½ft intermixed with *Viola*,

Nepeta, etc. and fruiting freely. Found by a lady and reported to HJB who found it after long searching. Was known to Beadell of Warlingham 10 or 12 years ago and reported to C. E. Salmon. Has every appearance of having been there for many years.' It had therefore been known since 1916–18. It was about this time that C. E. Britton reported it as found on the downs near Shoreham, Kent, for two or three seasons about 1901 (*Rep. BEC*, **4**, 428, 1917). When I first knew it there were two patches on open chalkdown and my photographs taken in 1939 show leaves of *Viola hirta* as described by Robbins earlier. By 1959 scrub had invaded the down and the habitat was practically woodland; in November of that year it was partly cleared by the Conservation Corps of the Council for Nature and the plant increased and flowered. It is still far too shaded and overgrown. The great interest of this plant is that it was found at about the same time in three places (two in Kent) on south-facing calcareous slopes similar to places where I have seen it in southern Europe. The European distribution is against it being claimed as native, but it is difficult to imagine ways in which it could have been introduced to these downs 55 to 70 years ago.

EUPHORBIACEAE

Mercurialis L.
M. perennis L.   Dog's Mercury.   Native.   1836.
Locally abundant.
Woods on good soils, especially in beechwoods on the chalk, less common in areas with acid soils and in the NW and NE of the county.

M. annua L.   Annual Mercury.   Colonist.   1827.
Locally common, 120.   Map 200.
Cultivated and waste places, roadsides and refuse tips. A steadily increasing alien. Salmon, 1931, had records from 26 localities; in 20 years we collected records from 120 tetrads. Mainly a garden weed and the seeds are probably spread with nurserymen's potted plants.

Euphorbia L.
E. lathyris L.   Caper Spurge.   Native?
1913.   Very rare, 1.
In a wood appearing only after the wood has been coppiced or felled. 35D5, Henley Wood, near the dene-hole, W. Turner (*J. Bot.*,

51, 225–6, 1913). Since then it has appeared at intervals, sometimes in abundance, as when I first saw it in 1923, sometimes in smaller quantity. Seeds can remain viable in the soil for many years (as in my garden) and, although it has not been reported during the present survey, it is likely to reappear when conditions are right. Caper Spurge also occurs as a casual in gardens, allotments, churchyards, sewage farms and roadsides—we have 14 records.

E. dulcis L.
A park-plant, well established at 25D2, Gatton Park, in long grass, 1964, CWW.

E. platyphyllos L.   Broad-leaved Spurge.   Native.
1763.   Rare, 23.   Map 201.
Cornfield weed on heavy clay soils. Twenty-one of our records are from Weald Clay; one in 84B3 near Farnham is on Gault Clay; and one in 14B5, in thin mixed woodland on Ranmore Common, on Clay-with-Flints. It is on very chalky soil in 25B2, a cornfield below Buckland Hills, B.Wu, 1973, JEL. A decreasing species.

E. helioscopia L.   Sun Spurge.   Native.   1836.
Common throughout the county.
Gardens, arable fields and waste ground.

E. peplus L.   Petty Spurge.   Native.   1836.
Common throughout the county.
Mainly gardens and waste ground, and especially common in rather shady, dank London suburban gardens, but also in arable.

E. exigua L.   Dwarf Spurge.   Native.   1762.
Frequent, 87.   Map 202.
Most common on the chalk, and on neutral soil on the Weald Clay, and scattered elsewhere. In gardens and arable fields.

E. pseudovirgata (Schur) Sóo (*E. uralensis* Fisch. ex Link, *E. virgata* Waldst. & Kit. non Desf.)   Twiggy Spurge.
Alien.   1887.   Rare, 26.   Map 203.
Railway banks, commons, waste places. This is a very variable taxon thought to be of hybrid origin, and it includes most, and probably all, the plants recorded from Surrey as *E. esula* L. Between 1945 and 1955 I found seven clumps on Epsom racecourse, and they all differed in leaf-width and shape, the plants in each

clump being uniform owing to vegetative reproduction. Twiggy Spurge originated from Central Europe but may have come to us from Canada or the USA, where it is abundantly naturalised. The means by which it is spread is not known, but the distribution on Epsom racecourse suggested hay brought for the racehorses, and a great many of our localities are on railway banks or near railways.

[*E. esula* L. This, and *E. esula x cyparissias*, may occur in Surrey, but we have been unable to distinguish them from *E. pseudovirgata* in our fieldwork.]

E. cyparissias L.   Cypress Spurge.   Established alien.
1909.   Rare, 1.
Chalk scrub. 25B4, Walton Downs, known from here since 1909 and still there in 1974. In this locality it has been claimed as a native, but by comparison with other similar habitats elsewhere there can be little doubt that it was introduced from the Continent by racehorses. Cypress Spurge also occurs occasionally as a garden escape as, for example, 04D4, Shere, on footpath, 1965, EJC, and 26E2, Wallington Station, 1958, GIC.

E. amygdaloides L.   Wood Spurge.   Native.
1763.   Locally abundant.
Wood, copses and hedgerows. At its best in moist woods but common on the chalk.

POLYGONACEAE

Polygonum L.
P. aviculare L. *sensu lato*.   Common Knotweed.   Native.
1724.   Abundant throughout the county.
On tracks, waste places, arable ground and roadsides. The following segregates occur in Surrey:

P. aviculare L. *sensu stricto* (*P. heterophyllum* Lindm.) Very common in cultivated land, waste places, roadsides, etc.

P. rurivagum Jord. ex Bor. Rare, a decreasing cornfield weed, easily overlooked. Doubtless still to be found though not recorded during the present survey—see Kent & Lousley, 240.

P. arenastrum Bor. (*P. aequale* Lindm.) Frequent and widespread, especially characteristic of tracks on sandy and gravelly soils. A form characterised mainly by small narrow leaves (*P. microspermum* Jord.; *P. calcatum* Lindm.) has attracted much interest in Surrey. Recorded from 26E5, Mitcham Common, ABe (*Rep. BEC*, **1**, 18, 1880 & 55, 1882); 26D4, Mitcham, nr the 'Goat', 1957, Hb. DPY; 16B2, Esher, margin of pond on West End Common, CEB (*Rep. BEC*, **6**, 575, 1921 & 207, 1923; *op cit*, **10**, 447–9, 1932; *Rep. Watson BEC*, **4**, 229, 1934); and 25A2, Headley Heath, 1946, ECW (*Rep. BEC*, **13**, 367, 1948). This may deserve further study.

[*P. cognatum* Meisn. Alien. Established near Westerley Weir, Kew Green, 17E4 from before 1872 (*J. Bot* **10**, 339, 1872) until it was destroyed by the construction of the tennis courts in 1923 (*Rep. BEC*, **8**, 924–5, 1929). For further details see Lousley, *Watsonia*, **1**, 319–20, 1950.]

P. bistorta L.   Bistort.   Native.   1666.
Rare, 35.   Map 204.
Wet alluvial meadows and roadsides, rarely in woods. In addition to the native localities, in some of which it has a long history, Bistort is sometimes found as a garden escape. Mainly on Upper Greensand.

P. amplexicaule D. Don
A garden plant which is sometimes established: 83C5, Churt, bank of Barford Stream, 1959, IES; 35C4, Warlingham, chalk slope above Stuart Rd., 1953, PG.

P. amphibium L.   Amphibious Bistort.   Native.
1666.   Frequent, 134.   Map 205.
Floating in ponds, rivers and ditches; the terrestrial form on pond margins, marshy meadows, towpaths, damp cultivated fields and railway banks. Most frequent on Bagshot Sands and London Clay in the N, but widely scattered elsewhere.

P. persicaria L.   Red Shank.   Native.   1836.   Very common.
Pondsides, ditches, cultivated ground, waste places.

P. lapathifolium L. (including *P. nodosum* Pers, *P. maculatum* (Gray) Dyer ex Bab., *P. petecticale* (Stokes) Druce)
Pale Persicaria.   Native.   1666.
Very common throughout the county.
Cultivated fields, waste places, pond margins and roadsides.

P. hydropiper L.   Common Water-pepper.   Native.
1836.   Frequent.
Ditches, by streams, marshy meadows, and damp places in woodland rides.
[*x P. persicaria* has been found at 05D5, Wisley, by Wey, 1932,
ECW, *Rep. Watson BEC,* **4,** 186, 1933.]

P. mite Schrank (*P. laxiflorum* Weihe).
Tasteless Water-pepper.   Native.   1845.
Rare, 13.   Map 206.
Shallow ditches and wet hollows, in bed of canal, damp meadows.
A decreasing and often wrongly identified species. Formerly at
intervals along the Thames from Battersea to Chertsey Meads.

P. minus Huds.   Small Water-pepper.   Native.
1763.   Very rare, 4.
Wet hollows and on horse-ride. 94E4, Shalford Common for many
years, last seen c1950, JEL; 16B2, Esher, West End Common,
1963, JES; 17E1& 17E2, Ham Common, last record 1955, BW.
The rapid decrease in this species is partly explained by the drying
up of ponds and ditches, but the main cause is unknown.
x P. persicaria. 17E1, Ham Common, 1955, BW, Hb. Mus. Brit.;
17E2, Ham Common, N corner, 1944, JEL, Hb. L.

P. convolvulus L. (*Fallopia convolvulus* (L.) A. Löve).
Black Bindweed.   Native.   1836.   Common.
Cultivated ground, hedgebanks, waste ground.

P. dumetorum L. (*Fallopia dumetorum* (L.) Holub)
Copse Bindweed.   Native.   1836.   Rare, 14.   Map 207.
Scrambling over hedges in lanes, and sides of woods and copses.
Very irregular in appearance; it sometimes comes up in great
abundance after trees have been felled or a hedge heavily cut back,
persists in decreasing quantity for a year or two, and then is not
seen again for many years until appropriate conditions recur.
The precise habitat also varies and some of our records are perhaps
50 or 100m down the lane from an earlier report. The persistence
and irregularity of appearance of this species is well shown by the
history of the locality at Copse Wood, Wimbledon, 27B1, where it
was first found in 1834 and recorded for the first time in Britain in
1836, Babington (*Trans. Linn. Soc.* **17,** 460). It now appears fairly
regularly because the garden hedges are cut back regularly, but is

found in several different spots within a small area. The localities are all on light, well-drained soils, possibly because heavy wet soils would not permit the seeds to remain viable for long periods. With the exception of Wimbledon all our current records are from the W of the county. It has not been reported recently from (a) round Esher and Cobham and (b) Redhill-Reigate-Dorking, where there are numerous old records. In some of these localities it will probably reappear.

P. aubertii L. Henry (*Reynoutria aubertii* (L. Henry) Moldenke; *P. baldschuanicum* Regel).   Russian Vine.
First record: 1955, Kent & Lousley. A garden alien which can hardly be ignored. Common and conspicuous in garden hedges where it is very persistent, and sometimes spreading on to adjacent commons or wooded waste ground.

P. cuspidatum Sieb. & Zucc. (*Reynoutria japonica* Houtt.)
Japanese Knotweed.   Established alien.
1931 (Salmon, *Fl.*, 564).   Common, 286.
Commons, roadsides, railway banks, waste places. The increase in this aggressive garden plant has been rapid; from a single record in the *Flora* of 1931 it has become common almost throughout the county. Whereas in 1930 it was news to report it from Mitcham as 'a well-established alien' in that district (Lousley, *Rep. BEC*, **9**, 839, 1932), this is now true almost everywhere. The increase arises from gardeners, misguided enough to grow it, throwing out roots which grow very easily in competition with all native plants except trees and the taller shrubs. We have no evidence that it spreads from seed.

P. sachalinense F. Schmidt (*Reynoutria sachalinensis* (F. Schmidt) Nakai)   Giant Knotweed.   Established alien.
First record: this *Flora*.   Frequent, 34.
Commons, roadsides, railway banks, sites of refuse tips. As persistent as the last species, and spread in the same way, but perhaps extending its ground less rapidly.

P. polystachyum Wall. ex Meisn. (*Aconogonum polystachyum* (Wall.) Kral).   Established alien.   First record: this *Flora*.
Rare, 3.
Roadsides. 83D4, Hindhead, above Woodcock Bottom, 1962, WEW; 04D3, between Hound House & Farley Green, 1954, AWJ;

06E2, Walton Common, on roadside verge, 1965, EJC. Established from roots thrown out from gardens.

P. campanulatum Hook. f.   Established alien.
1956 (*Proc. BSBI*, **2**, 137).   Rare, 2.
In swamps near houses. 27A1, Wimbledon Common, near Warren Farm, 1952, CA (first record), 1972, JEL; 34D1, Woodcock Hill, 1963, RARC.

Rumex L.
R. acetosella L.   Sheep's Sorrel.   Native.   1836.   Common.
Dry commons and heaths, cultivated land, usually on poor acid soils. The aggregate species is now divided into three segregates: the common widely distributed plant is *R. acetosella* L. *sensu stricto; R. angiocarpus* Murb. is not known to occur, but we have:
    R. tenuiflorus (Wallr.) Löve.   Native.
First record: this *Flora*.   Locally frequent, 34.   Map 210.
Dry sandy places on commons, and in arable fields. Restricted, as at present recorded, to the Lower Greensand and Bagshot Beds.

R. acetosa L.   Common Sorrel.   Native.   1836.   Common.
Meadows, pastures, roadsides, woodland rides; at its best in rich meadowland.

R. hydrolapathum Huds.   Water Dock.   Native.   1770.
Locally frequent, 55.   Map 211.
By rivers, canals and streams, and in adjacent marshes, also by ponds and lakes where it is often planted for ornamental purposes. The distribution of this fine plant follows the Thames, and then extends down the R. Wey and along its tributaries, including the Tillingbourne, the Basingstoke Canal, and the course of the long-disused Wey and Arun Canal. It is rare by the R. Mole, and occurs also by lakes. In recent years it has become much scarcer by the Thames, owing to embankments. motor cruisers, and the use of the plant as a platform by anglers. On the other hand it has increased on the derelict canals which are no longer cleared out at frequent intervals.
[x *R. obtusifolius*=*R. x weberi* Fisch.-Benz. Salmon, 1931, 570–1, gave two records of this hybrid. The first, from 94A3, Cut Mill Pond, 1874, Warren, was probably incorrectly identified; the second, from 'Bourne

Brook basin lying between Egham and Chertsey', Beeby, 1886, may be correct, but I have not seen a specimen. We have no current records.

[*R. confertus* Willd. Established alien. 35A4, Old Coulsdon, 1955, HBr & JEL (*Lond. Nat.*, **34**, 4, 1955). This grew in a rough field where Mr Britten had known it since 1942, and it had survived ploughing. It has not been seen since about 1960, but the steep slope has become extremely difficult to search owing to the growth of dense scrub and it is possible that the plants survive.]

[x *R. crispus* = *R. skofitzii* Blocki and *R. confertus* x *obtusifolius* = *R. borbasii* Blocki. 35A4, Old Coulsdon, one clump of each, with parents, JEL & HBr, Hb.L.]

[*R. cristatus* DC. (*R. graecus* Boiss. & Heldr.) Has been reported once: 17E4, bank of Thames by Kew Bridge, 1938, JEL, Hb. L.; destroyed by reconstruction of the river-wall a few years later.]

R. patientia L.　　　Established alien.
27C3, Wandsworth, railway embankment, 1967, DPY. subsp. *orientalis* (Bernh.) Danser: 17D2, Ham Gravel Pits, plentiful, 1942, JEL (*Rep. BEC*, **12**, 577, 1944), towpath by Ham Pits, 1964, EJC; 27A4, bank of Thames nr Barnes Bridge, 1933, JEL, Hb. L. *Rep. BEC*, **12**, 147, 1939).

R. crispus L.　　　Curled Dock.　　　Native.　　　1666.　　　Abundant.
Roadsides, waste places, cultivated ground, and shores of ponds and lakes. Usually in rather open habitats.
x *R. obtusifolius* = *R. x acutus* L. (*R. pratensis* Mert. & Koch) The commonest of all *Rumex* hybrids. Frequent in all areas where the parents grow together.
x *R. sanguineus* = *R. x sagorskii* Hausskn. Probably not uncommon when searched for. 06E4, Walton-on-Thames, Thamesbank, 1942, JEL, Hb. L; 35D5, Chelsham, border of Holt Wood, 1942, JEL, Hb. L.

R. obtusifolius L.　　　Broad-leaved Dock.　　　Native.
1836.　　　Too common throughout the county.
Waste ground, roadsides and field borders. An agricultural pest which requires open ground for the establishment of seedlings but persists in closed communities. The common plant is subsp. *obtusifolius* (subsp. *agrestis* (Fries) Danser). In addition the following have occurred:

　　　subsp. *transiens* (Simonk.) Reching.f. An alien, formerly abundant on the banks of the Thames from Putney to Kew. To this belong the records of 'var. *sylvestris* Wallr.' in Salmon,

1931, 567; see Lousley, 1939 (*Rep BEC*, **12,** 126–7). It is now frequent by the R. Wandle from Beddington Park to Earlsfield in 26D4, D5, E3, E4 & 27C2; 26D5, Wandle at Merton & Morden Hall Park, 1957, RCW & JEL, Hb. L.

subsp. *sylvestris* Wallr. Rechinger. Alien from eastern Europe. Was found in 26D4, Mitcham Junction, in a gravel pit, 1938, JEL (*op cit.* 127). The locality has since been built over.

subsp. *obtusifolius* x *palustris* = *R.* x *steinii* Becker. Only recent record: 26E4, Beddington Sewage Farm, in gravel pit with parents, 1956, JEL, Hb. L.

subsp. *obtusifolius* x *pulcher* subsp. *pulcher* = *R.* x *ogulinensis* Borbás has occurred at 27B3, Barnes Common, 1942, JEL, Hb. L.

subsp. *obtusifolius* x *sanguineus* = *R.* x *dufftii* Hausskn. Probably quite frequent on wood borders where the parents grow together. 15E1, Boxhill, by woodland ride, 1970, JEL, Hb. L; 35D4, Warlingham, Slines Oak, 1942, JEL, Hb. L; 35D5, Chelsham, Holt Wood, 1942, JEL, Hb. L; 37B2, Dulwich Woods, 1958, JEL, Hb. L.

R. pulcher L.   Fiddle Dock.   Native.   Rare, 9.   Map 212.

Dry commons and pastures, usually on sandy soils. Formerly frequent on alluvial soils near the Thames about Kew, Richmond and Putney, now greatly decreased, but probably still on Barnes Common, 27B4, and perhaps Kew, 17E4, and Richmond, 17E2. Still on Shalford Common, 04A4, an old station, Esher Green, 16B3, and Ditton Common, 16C4, where it was first found by H. C. Watson (1804–81), and in a few places near Godalming. The native Fiddle Dock is subsp. *eu-pulcher* to which the above records refer.

subsp. *divaricatus* (L.) Murb. An alien introduced from the Mediterranean. 24D4, refuse tip nr Dover's Green, 1961, BMCM, Hb. L.

R. sanguineus L. (*R. viridis* (Sibth.) Druce)   Wood Dock.

Native.   1780.   Common.

Woodland rides, copses, lanes, usually in rather shaded places but sometimes in full sun on roadsides and waste places. The common plant has green veins, but the typical species (var. *sanguineus*) with brilliant purplish-red veins on the leaves, which has been distributed throughout Europe in cultivation, formerly occurred. The most recent record is from 93B4, Brook, nr brickworks, where it was found by W. C. R. Watson in 1934 and persisted until at least 1949, GMA—see Lousley, *Rep. BEC*, **12**, 128–31, 1939.

R. conglomeratus Murr.   Clustered Dock.   Native.
1836.   Frequent, 197.   Map 213.
Wet commons on heavy soils, ditch banks, river and stream sides, marshes.
x *R. crispus* = *R.* x *schulzei* Hausskn. 35A1, Nutfield Marsh, 1951, JEL, Hb. L.—see also Salmon, 1931, 565.
x *R. maritimus* = *R.* x *knafii* Čelak. Very rare. 35C1, Godstone, Ivy Millpond, 1949, JEL, Hb. L.
x *R. obtusifolius* = *R.* x *abortivus* Ruhmer. No recent records, but see Salmon, 1931, 565.
x *R. pulcher* = *R.* x *muretii* Hausskn. No recent records, but see Salmon, 1931, 565.

R. palustris Sm.   Marsh Dock.   Native.   Rare, 7.
Gravel pits, gravel roadways, ditches. 05A2, Jacob's Well, ditch on edge of refuse tip, 1968, JFL, 1969, JFL & JEL, Hb. L.; 16A5, Walton, nr Sunbury Weir, and sandy roadways, 1956, BW; 17D4, river wall between Kew & Richmond, 1951, RAB; 26D4, Mitcham Junction, gravel pits, 1933 to c1944, JEL, Hb. L; 26E4, Beddington Corner, gravel pits, 1956, JEL; 27D1, Collier's Wood, Wandle Valley Sewage Works on sludge-beds, 1958, RCW; 35D4, Nore Hill, on tipped soil in quantity, 1961, RARC & DPY. A decreasing species.

R. maritimus L.   Golden Dock.   Native.   Very rare, 1.
Pond-sides. 84C1, Frensham Great Pond, on drained bed, 1941, ECW, Hb. L; 35C1, Godstone, Ivy Millpond, 1949, CDP, 1964, RARC; Godstone, Town = Bay Pond, 1942, JEL, Hb. L, dominant on muddy bed, 1959, BAK & DPY, 1964, RARC; Leigh Millpond, 1934, JEL, Hb. L; N of Tilburstow Hill, on disturbed ground, 1964, RARC. Records during our survey restricted to the vicinity of Godstone.

R. triangulivalvis (Danser) Reching. f. (*R. salicifolius* auct.) Alien from N. America sometimes established for a few years. 26E5, Mitcham, Eastfields, 1934, JEL, Hb. L, (*Rep. BEC.* **10**, 985, 1935); 27A1, Wimbledon Common, near Beverley Brook, 1957, JEL, Hb. L; 37C5?, Surrey Commercial Docks, Barnard's Wharf, 1967, RCP, Hb. L.

URTICACEAE

Parietaria L.

P. judaica L. (*P. ramiflora* Moench; *P. diffusa* Mert. & Koch)
Pellitory-of-the-wall.   Native.   1801, Garry (*J. Bot.* **41,**
Suppl., 168, 1903).   Rather rare, 55.   Map 214.
Old walls, churchyards, hedgebanks. The distribution follows the
Thames from central London to Kingston, the R. Wey and, less
convincingly, the rivers Mole and Wandle. In 24C5 on imported
chalk.

Soleirolia

S. soleirolii (Req.) Dandy   Mind-your-own-business.
Alien.   1930, Haslemere, Miss M. Drummon (*Rep. BEC.* **9,** 136).
Rare.
Churchyards, walls and occasionally in turf. 93, Haslemere, as
above; 94E5, Guildford, wall of river towpath, 1969, JFL; 04A4,
Clinthurst House, 1958, JCG; 04A3, Snowdenham Hall, 1959,
JCG; 06B5, Thorpe churchyard; 15B5, Stoke d'Abernon church-
yard, 1965, JES; 26D3, Carshalton churchyard, 1958, RCW; 36A3,
Croydon churchyard, 1958, RCW; 36C3, Shirley Hills, in turf,
1965, RARC.

Urtica L.

U. urens L.   Small Nettle.   Native.   1836.
Common, 286.   Map 215.
Gardens and other cultivated ground and waste places. Especially
abundant on light soils.

U. dioica L.   Stinging Nettle.   Native.   1836.   Very common.
Hedgebanks, about farms, wood borders and waste ground.

U. pilulifera L.   Roman Nettle.   Alien.   1837.
Garden weed. 27C3. Wandsworth, in a garden, 1960, I. C. Price,
1961, JEL & BHSR. This persisted until at least 1966, DPY.

CANNABACEAE

Humulus L.

H. lupulus L.   Hop.   Native.   1666.   Locally frequent.
Hedges, thickets and especially in damp places. Accepted as likely
to be native generally, but also occasionally a relic of cultivation
round Farnham where hops have long been cultivated.

ULMACEAE

Ulmus L.

U. glabra Huds. (*U. scabra* Mill.; *U. montana* Stokes).
Wych Elm.   Native.   1633.
Frequent, 173 (+14 recognised as 'planted').   Map 216.
Woods and hedges on basic soils and by streams. This species is most convincing as a native in ravines where Ragstone or other basic rocks are exposed; as, for example: 83E5, ravine N of Devils Punchbowl, 1969, FR; 94E2, Godalming, Cizden's Copse on Bargate Beds, 1957, FR; 04B5, Albury, valley on Bargate Beds at foot of Colyers Hanger, 1957, FR; 13D4, Capel, ghyll S of Holbrook's Farm, 1949, CNHS; 14C2, Nuttold Copse and Great Copse, by stream, 1961, SFC; 45B1, Limpsfield, Tenchley's Park, wood on Ragstone escarpment, 1957, FR. Elsewhere it often grows in places where it appears to be 'wild', but it has been commonly planted both on and off basic soils.
x plotii = *U. x elegantissima* Horwood. Planted tree. 36E1, Fairchildes, in hedge nr several *U. plotii* of same age, 1966, DPY det. RM.

U. procera Salisb. (*U. campestris* auct.; *U. sativa* auct.)
English Elm.   Native.   1832.   Common.
Roadsides and hedges. Our commonest elm is usually a planted tree which is replaced from suckers when the old trunk is felled or decays. Its rapid growth is dramatically demonstrated in 37C3, Nunhead Cemetery, where tombs only 30 years old are buried in tall elm woods, tilted, and split open.

U. angustifolia (Weston) Weston (*U stricta* (Ait.) Lindley).
Cornish Elm.   Occasionally planted.

U. coritana Melville. Occasionally planted as at 26E3, Beddington Park, 1957, RCW det. RM.

U. x hollandica Mill. (*U. major* Sm.)   Dutch Elm.
A hybrid of uncertain parentage, commonly planted near London, and less frequently elsewhere.

opposite Green-flowered Helleborine (*Epipactis phyllanthes*), Riddlesdown. (*Photo by P. Wakely*). Published in memory of Dr D. P. Young with funds provided by the BSBI

U. carpinifolia Gled. (*U. nitens* Moench) Planted here and there, mainly in parks. 15, Leatherhead By-pass, 1949, AEE, det. RM.

x U. glabra x plotii. Planted; rare. 34E3, Lingfield Common, edge of field, 1961, DPY det RM; 36C3, Shirley Park Golf-course, 1961, DPY, det RM.

U. plotii Druce.   Plot's Elm.   Very rare, planted.
36E1, Chelsham, about 12 large trees in hedgerows between church and Fairchildes, 1946, HKAS (*Watsonia*, 1, 53, 1954); S of Fickleshole, 1967, DPY conf. RM, 1971, SFC.

MORACEAE

Ficus L.
F. carica L.   Fig.   Established alien.   1936 (see below).   Rare.
Embankments, railway embankments, etc. Derived from fruiterers' figs and distributed by birds, water and perhaps other agencies. Our first evidence is from 16D5, Kingston, at junction of Hogsmill River and R. Thames, 1918, LJT, but the first printed record noted is 17E4, several bushes on the river-wall by the Thames between Kew and Mortlake, RNP & JPMB (*Rep. BEC*, 11, 41, 1936). Still there in 1947 and later. Recent records: 94A2, Elstead, brick wall opposite 'Golden Fleece' car park, 1967, BWu; 26D3, Carshalton churchyard, 1957, RCW; 26E4, Mitcham, railway cutting at Willow Lane bridge, a sizeable bush, 1940, DPY, a big tree, 1952, DPY, 12ft tall in scrub, 1957, RCW; 37A4, South Lambeth, 1965, AB. See also Lousley, 1948, '*Ficus carica* in Britain,' *Rep BEC*, 13, 330–3.

JUGLANDACEAE

Juglans L.
J. regia L.   Walnut.   Established alien.
1955, Kent & Lousley (*Handbook*, 256).
Locally frequent as self-sown tree.
Hedges, woods and lanes. Increasing. 06B5, Thorpe, self-sown by pond in old gravel pit, 1965, EJC; 15B1, Ranmore, seedlings in lane N of Hogden Farm, 1960, JEL; 15D1, railway embankment, 1961, BMCM; 15D2, E of Mickleham Church, 1958, BW; 15D2, nr Headley Lane, 1961, BW; 15D2, Mickleham, by path to Downs,

opposite Narrow-lipped Helleborine (*Epipactis leptochila*), Horsley. (*Photo by P. Wakely*). Published in memory of Dr D. P. Young with funds provided by the BSBI

1925–54, JEL; 15D3, Norbury Park, AEE, 1963, BWu, Icehouse Coombe, several trees, 1960, BW; 25B5, Epsom Downs, in scrubby copse, 1961, EJC; 35B5, Riddlesdown, in hedge, 1961, JEL.

MYRICACEAE

Myrica L.

M. gale L.   Bog Myrtle.   Native.   1718.

Locally plentiful, 18.   Map 217.

Bogs and wet heaths. Often abundant over limited areas in the NW round Camberley, Bagshot Heath, Pirbright, Brookwood, Woking, Lightwater and Virginia Water on acid heathland, less so on Puttenham Common and near Elstead. It is still abundant at Lightwater (in Folly Bog, 96B1, and Lightwater Bog, 96B2) where John Aubrey recorded its 'very grateful smell' in 1718. There is an isolated locality in 14C3, S of Broadmoor, 1954, RAB & FR (*SE Nat.* **1959,** 24); 1 large bush and 2–3 smaller ones on grassy path, 1965, RAB & JES; beside public footpath along boundary of a garden, here before the garden but too dry for a natural habitat, 1962, CPP.

PLATANACEAE

Platanus L.

P. x hybrida Brot. (*P. acerifolia* (Ait.) Willd.)

London Plane.   Naturalised alien.

1955, Kent & Lousley (*Handlist*, 253).   Rare.

Self-sown at intervals on embankment or towpath of R. Thames. 16A4, between Walton on Thames and Sunbury Lock, 2 self-sown plants on river-wall ¾ mile apart, 1957, JEL; 17E4, river-wall, small plant nr Brentford Gate of Kew Gardens, tree 10ft tall and smaller one just W of Kew Railway Bridge, another and seedlings just E of railway bridge; all 1958, BW; 17D3, on river-wall above Richmond Bridge, 1962, BW; 27A4, Mortlake, Thameside, 1939, AEE; 27B4, several up to 30ft on river-wall, 1960, BW. The evidence for a hybrid origin of London Plane is not convincing.

BETULACEAE

Betula L.

B. pendula Roth (*B. verrucosa* Ehrh.; *B. alba* auct.).

Silver Birch.   Native.   1836.   Very common.

Woods, heathland, commons and especially abundant on sandy soils. Increasing on many commons owing to withdrawal of grazing, and ability to grow again from the stump after heath fires.

x *B. pubescens* = *B.* x *aurata* Borkh. Plants with mixed characters are not infrequent in secondary birch scrub and are presumed to be this hybrid. The two species are not always easy to separate in Surrey.

B. pubescens Ehrh.   Downy Birch.   Native.   1893.
Frequent, 268.   Map 218.
In wetter situations than the last and at its best on heathland and commons on damp peaty soil. Pure stands are local and rather uncommon, and the frequency figure is somewhat misleading as records for many tetrads are based on a few trees growing with the much commoner *B. pendula*.

Alnus Mill.
A. glutinosa (L.) Gaertn. (*A. rotundifolia* Mill.)   Alder.
Native.   1791.   Frequent.
By rivers and streams and wet places in woods. Sometimes forming pure woods as 'alder holts' of which there are good examples at Colyer's Hanger, 04B5, and several near Reigate Heath, 25B1.
x *A. incana* = *A.* x *pubescens* Tausch. 24E3, Langshott wood with parents, 1949, JPMB, 1949, JEL, Hb. L, 1950, BMCM. Believed now destroyed.

A. incana (L.) Moench   Grey Alder.
Naturalised alien.   First record: this *Flora*.   Rare.
Woodland. Usually planted but appears to have regenerated in 34A2. 24E1, Horley, Horleyland, in mixed wood, presumably planted, 1964, RARC; Buckland Alders, presumably planted but very well established, 1968, BMCM; 34A2, Langshott (Brook) Wood, on bank of brook, 1949, BMCM & JEL, 1960, DPY.

CORYLACEAE
Carpinus L.
C. betulus L.   Hornbeam.   Native.   1725.
Frequent, 221.   Map 219.
Woods, copses, hedgerows, especially on loamy clays of the Weald Clay series in the SE. In addition to native occurrences, it is commonly planted and we have endeavoured to exclude these from our records. The distribution suggests that Hornbeam is likely to be native on the Weald Clay but elsewhere usually introduced.

Corylus L.
C. avellana L.   Hazel.   Native.   1718.   Very common.

Hedges, woods, copses. Common in hedges on the chalk on dry basic soils and equally at home on wet neutral or slightly acid soils.

FAGACEAE

Fagus L.

F. sylvatica L.   Beech.   Native.   1813.

Common, but sometimes planted.

Forming pure and magnificent woods on the chalk escarpment and Clay-with-Flints above (Plate 5). There are also fine examples on the Lower Greensand. Beechwoods have a characteristic ground flora which includes a number of the county's most interesting species. Beeches are also scattered over the better-drained soils throughout the county and it is sometimes difficult to distinguish those that have been planted.

Castanea Mill.

C. sativa Mill.   Spanish Chestnut.   Naturalised alien.

1830.   Locally frequent.

Planted as woods for coppicing, especially on the Lower Greensand, and occasionally reproducing from seed.

Quercus L.

Q. cerris L.   Turkey Oak.   Naturalised alien.

1931 (Salmon, *Flora*).   Locally abundant, 219.   Map 220.

Thoroughly established on sandy slightly acid soils, and especially on the commons in the W of the county. Seedlings are commonly seen and we have endeavoured to include only regenerating trees on our map. Spreading rapidly and variable in foliage.

Q. ilex L.   Holm Oak.

Occasionally planted but also self-sown on the chalk as at 15D2, Norbury Park, 1916, EBB; seedlings on downland above Mickleham Tunnel, 1958, BW; regenerating by path to Mickleham Downs opposite Juniper Hall, 1963, BWu.

Q. robur L.   Common Oak.   Native.   1718.   Common.

Woods, copses, hedgerows. The distribution produced by grid-recording shows the Oak as evenly distributed throughout the county, but this is misleading. Scattered trees are everywhere, some of them planted, but on neutral or basic clays and loams it is the natural dominant tree. On the Weald Clay especially it formerly formed almost continuous cover, and the view looking S from Leith Hill and other viewpoints shows fields cut out from oak woodland.

Q. petraea (Mattusschka) Liebl. (*Q. sessiliflora* Salisb.)
Durmast Oak.   Native.   1792.   Rather rare, 91.   Map 221.
Woods, usually on rather acid soils, and often on slopes. The Great
North Wood, which extended formerly from Selhurst to Nunhead
on the slopes of the Crystal Palace ridge, was mainly *Q. petraea*
(Lousley 1959 & 1960) and parts of it can still be seen. Salmon's
first record for *Q. robur* was Aubrey's reference in 1718 to the
Vicar's Oak which was part of the Great North Wood and may
belong to *Q. petraea*. His 1792 first record for the latter is certainly
correct and was also from the Great North Wood.
x *Q. robur* = *Q. x rosacea* Bechst. Salmon, 1931 cites six records
for this hybrid and our observers have often noticed trees with
intermediate characters. It may be that hybrids are not infrequent,
but both species are variable and much of this variation may be due
to normal intraspecific variation. See Jones, E. W., 1968, *Proc.
BSBI*, **7**, 183–4.

Q. rubra L. sec Du Roi   Red Oak.   Alien.
86E1, nr Camberley, 'well naturalised', 1966, JES & WEW, det.
DPY; 94C3, Shackleford Heath, 1 large tree, 1954, OVP. Several
other cultivated oaks persist where they have been planted on
commons, etc.

SALICACEAE

Populus L.
P. alba L.   White Poplar.   Alien.   1863.
Often planted but we have no records of this species regenerating.

P. canescens (*Ait.*) Sm.   Grey Poplar.
Possibly native in wet woods.   1863.
Widely distributed, the following seem to be localities less obviously
introduced: 06A2, Ottershaw Park, edge of woodland nr Little
Blackmole Pond, 1964, EJC; 35D2, Robins Grove wood, in wet
woodland, 1957, RARC.

P. tremula L.   Aspen.   Native.   1863.
Common, 226.   Map 222.
Woods and commons, often forming thickets by its spread from
suckers, mostly on rather poor moist soils.

P. nigra L.   Black Poplar.   Doubtfully native.   1852.   Rare.
This is still by the Thames at intervals from Putney, where J. Boswell Syme found it in 1852, to Barnes and Mortlake, 27A4—27C3.
There are old records from by the R. Mole at intervals from Sidlow Bridge, 24C4; Burford Bridge to Mickleham, 15D1, and between Cobham and Hatchford, 15A5, which suggest possible native status, but these require confirmation.

P. x canadensis Moench (*P. serotina* Hartig, *P. lloydii* Henry, etc.)
Italian Poplar (including Lombardy Poplar).
Commonly planted but reproduced only from cuttings.

P. gileadensis Rouleau.   Balsam Poplar.
Sometimes planted but not reproducing naturally in Surrey.

P. trichocarpa Hook.   Black Cottonwood.   Planted tree.
26E4, Mitcham Common, 1958, DPY; 35A3, Alderstead Common, 1961, DCK, det. DPY.

Salix L.
S. pentandra L.   Bay Willow.   Planted tree.   1724.   Rare, 4.
94E4, Shalford, 1 large tree by Wey Navigation, 1967, BMCM & EMCI; 16, Leatherhead Road opposite Star Inn, 1 female tree, BW; 27B2, Wimbledon Common, 1913, Pugsley, nr Windmill Road, 1943, JEL, 1972, JEL; 35C4, Warlingham, 1960, RARC.

S. alba L.   White Willow.   Native.   1836.
Rather rare, 91.   Map 223.
By streams, ponds, lakes. The distribution follows the Thames and Wey with many other scattered records. The var. *vitellina* (L.) Stokes, with young twigs bright yellow or orange, is widely scattered. The Cricket-bat Willow, var. *coerulea* (Sm.) Sm. (*S. alba x fragilis?*) is apparently the prevailing variety but we have only received the following confirmed records: 16E3, Chessington North, by a stream 'very unlikely to have been planted here', 1964, EJC det. RDM; 17E4, Mortlake, Thames towpath, 1955, BSBI Exc. det. RDM.
x S. babylonica = *S. x sepulcralis* Simonk. An ornamental willow frequently planted as at 15D1, Burford Bridge, by the Mole, 1954, BSBI Exc., det. RDM. The *S. alba* involved is usually, perhaps always, var. *vitellina*.

x *S. fragilis* = *S. rubens* Schrank (*S. viridis* Fries.)   Rare.
35A2, Warwick's Wold, 1960, DPY det. RDM; 35B1, Brewer St,
Bletchingley, 1960, DPY, det. RDM; 44B5, Langhurst, 1945,
CNHSS, det RDM. See also *S. alba* var. *coerulea*.
x *S. pentandra* = *S. x ehrhartiana* Sm. 25E1, Chilmead Farm, nr
Redhill, planted, 1950, RDM & NYS.

S. babylonica L. x *fragilis* = *S. blanda* Anderss. Planted, occasional.
17D1, Ham, riverside towpath, 1955, BSBI Exc., det. RDM.

S. fragilis L.   Crack Willow.   Native.   1836.   Common.
Streamsides, wet copses and by roadside ditches. *S. fragilis* was
formerly grown commonly for withes used for fencing and was
spread widely by easily rooted cuttings, and by the use of stakes
which rooted almost equally easily. It has been divided into the
following segregates:
      S. fragilis L. sec Smith (*S. viridis* Fries)   Rare.
      05B4, Pyrford, 1960, DPY det. RDM; 26A2, nr Epsom,
      opposite St Ebba's Hospital, 1964, RDM. var. *latifolia*
      Moss 05B5, Pyrford, 1 large tree by stream, 1958, BSBI Exc.,
      det. RDM.
      S. russelliana Sm.   Bedford Willow.   Common.
      For example: 05B4, Pyrford, 1960, DPY; 17E4, Kew, Thames
      towpath, 1955, BSBI Exc.; 25B4, Burgh Heath, 1962, DPY;
      34B2, Horne, S of Homehouse farm, 1961, RARC; 35A2,
      Warwick Wold, 1960, DPY; 35E1, Broadham Green, 1961,
      DPY; 36B3, Croydon, Lloyd Park, 1961, DPY; 36D4, Shirley,
      Monk's Orchard, 1960, DPY.—all det. RDM.
      S. basfordiana Scaling ex Salter. Planted as an ornamental
      tree. Probably rather rare. 05B5, Pyrford, 1960, DPY; 36D4,
      Monk's Orchard, 1960, DPY—both det. RDM.
      S. decipiens Hoffm. (*S. fragilis* auct. mult). 25E1, Chilmead
      Farm, nr Redhill, planted, 1950, RDM & NYS.
S. fragilis x *triandra* = *S. x speciosa* Host (*S. alopecuroides* Reichb.)
A. Kerner. Fraser in Salmon, 1931, said that the male tree was
widely distributed in the county, but Mr Meikle suggests that this
was an error for *S. fragilis* L. var. *latifolia* Anderss.

S. triandra L.   Almond Willow.   Doubtfully native.
1670.   Rare, 24.   Map 224.
Banks of streams and canals, by ditches in hedgerows.
x *S. viminalis* = *S. x mollissima* Ehrh (*S. undulata* Ehrh., *S. hippo-*

phaefolia Thuill.). 17D2, Thames bank by Ham Pits, 1966, EJC; 17E4, Thames bank above Kew Bridge, 1930, JEL & J. Fraser, 1933, with August catkins, JEL, female bushes, 1955, DPY, det. RDM, 1966, EJC det. RDM. This also grew formerly by the Thames at Mortlake, and near Chertsey Bridge. The northomorph *nm undulatum* (Ehrh.) Wimmer is generally more common than *nm hippophaefolia* (Thuill.) Wimmer.

S. purpurea L.   Purple Willow.   Native.   1762.
Rare, 22.   Map 225.
Ditches in watermeadows, by rivers, streams and ponds. More convincing as a native than any of the above species, though even this is sometimes planted.
x *S. viminalis* = *S. x rubra* Huds. Salmon, 1931, gives seven records and some of these may persist.

S. daphnoides Vill.   Planted tree.
27A4, riverside near Barnes Bridge, 1942, NYS, 1947, DHK, 1966, BWu, 1967, JRP.

S. acutifolia Willd.
27A4, by the Thames, Barnes Bridge, 1969, BWu, 1967, JRP.

S. viminalis L.   Osier.   Native.   1836.   Frequent, 35.
By streams and ponds, osier holts, flooded gravel pits. Spread over almost the whole county, but often originally planted.

S. calodendron Wimm. (*S. dasyclados* auct.)   Planted.
96E4, Chobham Common, frequent about Longcross, 1960, RPHT.

S. caprea L. subsp. *caprea*.   Goat Willow.   Native.
1813.   Common.
Woods, scrub and hedges on a wide range of soils, including the chalk escarpment.
x *S. viminalis* = *S. x laurina* Sm.   Probably not rare.
05C3, Groveheath near Sendmarsh, beside pond, 1966, EJC, det. RDM; 26C3, nr entrance to N. Cheam cemetery, 1958, RCW; 26D4, Mitcham Watermeads, 1958, RCW, det. RDM.
[x *S. cinerea* = *S. x reichardtii* A. Kerner. Salmon, 1931, gave three localities on the authority of J. Fraser.]

S. cinerea L.   Grey Willow, Sallow.   Native.

1835.   Abundant throughout the county.

Woods, hedges, gravel pits, waste ground, chalk pits on wet and dry soils. The wind-dispersed seeds are almost ubiquitous and seedlings soon appear on any bit of waste ground, even in central London. The two subspecies, subsp. *cinerea* and subsp. *oleifolia* Macreight (*S. atrocinerea* Brot.), both occur, the latter very much more common than the former. The two intergrade as elsewhere in Britain.

x *S. viminalis*=*S.* x *smithiana* Willd. 05B3, Send, disused sandpit, 1966, EJC, det. RDM; 05C5, Pyrford, by Wey Navigation, 1968, BSBI Exc., det. RDM; 06C2, New Haw, canal towpath, 1964, EJC, det. RDM.

[x *S. repens*=*S.* x *subsericea* Doell. Salmon, 1931, gave this from Putney Heath on the authority of J. Fraser. Not seen since and the hybrid is likely to be rare on account of the difference in flowering times.

S. aurita L.   Eared Willow.   Native.   1724.

Rare, 26.   Map 226.

Woods, scrub, hedges. We cannot believe that this species was ever as widespread and common as Salmon, 1931, seems to suggest. Our experienced helpers have had difficulty in finding it at all, except for the Weald Clay in the S, where it is locally frequent in a few areas. Apart from this, it is in small quantity in a few places in the W and on Wimbledon Common, 27B1, 27B2, where it has long been known from by the little streams which trickle down to the Beverley Brook.

x *S. caprea*=*S.* x *capreola* J. Kerner ex Anderss.

x *S. cinerea*=*S.* x *multinervis* Doell.

x *S. repens*=*S.* x *ambigua* Ehrh.

x *S. viminalis*=*S.* x *fruticosa* Doell.

are all reported in Salmon, 1931, 592–3, but we have received no recent records.

[*S. nigricans* Sm. (*S. andersoniana* Sm.) Dark-leaved Willow, and its hybrid with *S. phylicifolia* L., were known for many years on the Thames bank outside Kew Gardens, 27B4, but destroyed when the embankment was rebuilt about 1948. Both were garden escapes.]

S. repens L.   Creeping Willow.   Native.   1724.

Locally frequent, 70.   Map 227.

Damp or wet peaty places on heaths and commons. A decreasing species, as Salmon recognised, but the fact that 40 years later this

moisture-loving species is still to be found in 70 tetrads indicates
how much good undrained country still remains.

ERICACEAE

Ledum L.

L. groenlandicum Oeder.   Labrador Tea.

Established alien.   1931 (see below).

95B2, wet margin of Henley Park Lake, Normandy, 1930, C. E.
Marks, G. C. Druce, 1931, *Rep BEC*, **9**, 361 & 1932, *op cit.*, **9**, 690;
1967, EP, JRP, JEL & JES, 1974, JEL. In the first report Druce
claimed that 'from the water's edge to about 6ft back is one solid
mass of it all around the island', but this is very difficult to believe.
In recent years it has been limited to a small area, extending along
perhaps 12ft of the margin of the lake and going back about 9ft.
Druce also claimed that Marks had found it in a second locality,
but this has not been refound.

Rhododendron L.

R. ponticum L.   Rhododendron.   Naturalised alien.

1954, Kent & Lousley (*Handlist*, 183).

Locally abundant, 201.   Map 228.

Spreading as a shrub layer in woods and on heathland on sandy
and peaty soils. Salmon, 1931, makes no mention of this species;
it has spread so rapidly that it is now a major pest. Rhododendron
grows freely from seed and seedlings are abundant. Its dense
evergreen shade eliminates all species of the herb layer and it is
immune to all ordinary herbicides. To control Rhododendron is
likely to prove extremely expensive in many areas and, if it is not
controlled, their conservation future is bleak.

R. luteum Sweet.   Established alien.

96D3, Chobham Common, 1 plant in a boggy valley thought to be
sown rather than planted, 1970, EJC; 95B2, Pirbright, Peatmoor,
1 big plant and 2 smaller on edge of woodland, 1969, JES & WEW.
This species is spreading in Berks and Bucks.

Kalmia L.

K. polifolia Wangenh. (*K. glauca* L'Hérit, ex Ait.)

Established alien.   Very rare, 1.

96D2, Chobham Common, in a wet bog, 1910, C. E. Britton (*J. Bot.*
48, 205, 1910 & *Rep. BEC*, **2**, 576, 1911); Chobham Common, in a

'dangerous bog', away from tracks and habitations and screened
by sallows and birches, 1913, CEB (*Rep. BEC.* **3**, 482, 1914); a
patch about 10 x 3m with outliers, 1956, JEL; 1958, BW & WEW;
still plentiful, 1972, JES & JEL. This fine plant flowers in early
May and the bog must have been exceptionally wet in 1913 for
Britton to have described it as 'dangerous'. The species also grows
on Flanders Moss, Perthshire, in a very similar situation (Scott,
Glasgow Nat., **18**, 196–7, 1962), where it is thought to have been
disseminated by birds. The habitats are similar to those in North
America, where it is native, and Chobham Common is not far
from nurseries where American shrubs are grown.

K. angustifolia L.   Established alien.
96C3, S end of Sunningdale Golf Course in 'neo-bog', 1968,
PFLeB; 96D3, Chobham Common, a well-established plant in
Calluna-Erica tetralix 'neo-bog', 1966, PFLeB, det. DPY. The
discovery of a second species of *Kalmia* on Chobham Common in
1966 at first caused some confusion until it was confirmed that the
locality was some distance from the one known since 1910. *K.
angustifolia* is established in several counties and was found by
Lt Col G. Watts in 1943 near Fleet, Hampshire, not far from the
Surrey border, and I saw it there in 1960. At Rixton Moss, S Lancashire, it had been known for many years in 1913 and was still
there in 1960. In SW Yorkshire it persisted for at least 28 years.
It is not known how it gets to these remote spots but, as with the
last species, it is thought to be distributed by birds.

Gaultheria L.
G. shallon Pursh   Partridge Berry.   Naturalised alien.
1914.   Locally abundant, 25.   Map 229.
Heathy places and open woods. The Leith Hill locality, 14C2 etc.,
is impressive and has attracted considerable attention. First
recorded as 'quite wild' by H. J. Riddelsdell (*J. Bot.* **52**, 250, 1914),
it was discussed at length by C. E. Moss a year later (*J. Bot.* **53**,
279, 1915). Recorded again from Leith Hill by Miss Flora Russell
(*Rep. BEC*, **10**, 533, 1934), the occurrence here was related to
records in other parts of Britain by Lousley (*Rep. BEC*, **10**, 973,
1935). E. F. Hale has described its spread at Lob's Wood, between
Farnham and Tilford (*Gard. Chron.*, **144**, 55, 1958). The spread of
this species in the last 40 years is remarkable; it is now common
over a large area of the Lower Greensand and there are a few
localities elsewhere.

Pernettya Gaudich.
P. mucronata (L. f.) Gaudich. ex Spreng.   Established alien.
83D4, Hindhead, a fine patch naturalised above Woodcock
Bottom, 1962, WEW; 84D3, Moor Park, alder holt by R. Wey,
1969, SFC; 96D3, little wood opposite Chobham Place, 1961,
EAB; 14B2, Leith Hill, N slopes, remote from houses, 1970, CPP;
45B2, Broomlands Sandpit, 1968, RARC. Increasing, and believed
to be bird-sown.

Arbutus L.
A. unedo L.   Strawberry Tree.
Alien, planted and perhaps spreading from self-sown seed.
15C1, Denbies, well established, 1966, BWu.

Calluna Salisb.
C. vulgaris (L.) Hull   Ling.   Native.   1696.
Locally abundant throughout the county.
Heaths, commons, woods on acid soils and rarely in leached areas
of the chalk. Lovely on Wimbledon Common and other commons
near London, and even more so on the great commons in the W of
the county.

Erica L.
E. tetralix L.   Cross-leaved Heath.   Native.   1763.
Locally common, 101.   Map 230.
Bogs and the wetter parts of peaty heathland. The distribution of
this species reflects the distribution of wet peatlands in the county
and especially the large commons in the NW and SW. Less abundant in the Leith Hill area; scarce in 25 centrad.

E. cinerea L.   Bell Heather.   Native.   1666.
Common, and found throughout the county.
Dry heaths and commons. Very rare on the chalk, as, for example,
15C3, on top of Fetcham Downs, 1952, BW.

Vaccinium L.
V. myrtillus L.   Bilberry.   Native.   1718.
Locally abundant, 115.   Map 231.
Heathy woods and commons. The great headquarters of this species
in Surrey is on the Lower Greensand hills round Leith Hill, Holmbury St Mary and Winterfold Heath; John Aubrey's statement of
1718 (*Nat. Hist. & Antiq. of Surrey*, **4,** 115) is still true: 'Southward

from this place (Abinger) the County abounds with Heath and Fern, and is full of Bilberries, which the Country People call Whortell-berries'. An alternative name 'Hurt' has no doubt given its name to the Hurtwood. Bilberries are also plentiful on the commons round Haslemere and Farnham, and again round Frimley and Camberley.

V. oxycoccus L. (*Oxycoccus quadripetalus* Gilib.)   Cranberry.
Native.   1666.   Very rare.
Creeping over bog-moss on tussocks in spongy bogs, and some-times on slightly less wet ground on adjoining slopes. 83E4, Devil's Punch Bowl, 'shown me many years ago', CEB (*J. Bot*, **70**, 333, 1932); 94A1, Ockley Common, in quantity, 1933, ECW (*Rep. BEC*, **10**, 532, 1934), still there in at least two places, 1972, JEL; 94A1, Thursley Common, approx ½ mile W of Warren Lodge, 1949, JEL—a short extension of the same station. Since Cranberry was rediscovered by Mr Wallace in 1933, it has been seen by many botanists on Ockley Common which is in the same district as old records for Borough Farm and Peper Harow. It is more plentiful near Oakhanger in N Hampshire less than 3 miles over the Surrey border.

V. macrocarpon Ait.   American Cranberry.
94A3, Puttenham Common, growing with *Calluna* in boggy scrub, 1966, WEW, 1967, WEW & JEL. An alien which is well naturalised elsewhere in Britain and probably bird-sown at Puttenham. It covers only about 1 sq m and the habitat is perhaps too shaded for it to spread.

PYROLACEAE

Pyrola L.
P. minor L.   Common Wintergreen.   Native.
1847.   Rare, 10.
Heathy woods, often under pines. 83D4, S of Green Farm, under Sweet Chestnuts, 1949, BW & BMCM; 84B2, Frensham Heights School, 1960, VL & JEL; 84C2, Frensham, Kennel Lane, 1968, CD; 84E3, Crooksbury Hill, 1961, VL; Crooksbury Common, 1970, JCG, several patches in this neighbourhood, JEL; 94B1, Rodborough, Witley Common, 1967, DWB; 95A3, Tunnel Hill, under pine/birch canopy, 1965, DCK; 96E3, nr Longcross in mixed wood, 1964, WEW; 04E2, Pitch Hill, under pines, 1962,

LNHS; 06B1, Sheerwater Bog, 1950, BW, this well-known locality now built over; 06B1, New Zealand Golf Course, 1962, JES & BW; 14B2, Leith Hill, 1963, ECW & JES.

MONOTROPACEAE

Monotropa L.
M. hypopitys L.   Yellow Birdsnest.   Native.   1821.
Rare, 31.   Map 232.
In deep leaf litter, usually under beech on the chalk, rarely under oak or pine on Lower Greensand. Salmon, 1931, tried to divide this species into glabrous and hairy segregates with unconvincing results. Dr Young, with knowledge of the 40 years' additional work, and that the two are said to have different chromosome numbers, made a great effort to plot the distribution using modern characters and got the following results:
    subsp. *hypopitys*—15 tetrads
    subsp. *hypophegea* (Wallr.) Soó—19 tetrads
The range of both 'subspecies' extended the whole length of the chalk outcrop; the first had also one station on Lower Greensand at 14A3, Tenninghook Wood, 1962, HG, and the second, one station on Lower Greensand at 94A1, Thursley, Warren Mere under oak, 1968, 1969, DWB. In some cases characters were mixed, and in at least one the identification changed in different years, while in four instances both were found in the same tetrad.

PLUMBAGINACEAE

Armeria Willd.
A. maritima (Mill.) Willd.   Thrift.
Garden escape. 04B5, St Martha's, naturalised, 1960, RARC; 27D1 & 27D2, widespread in cemeteries, 1965, RARC.

PRIMULACEAE

Primula L.
P. veris L.   Cowslip.   Native.   1827.
Locally frequent, 150.   Map 233.
Chalk grassland, pastures on well-drained basic or neutral soils. Mainly on the chalk and on neutral soils on the Weald Clay. A decreasing species owing to the depredations of town dwellers, and thus rare in the N of the county, but also to the ploughing of old pastures. It is difficult to realise that in 1827 our first record was

from near the Swan, Stockwell, and fields on Lavender Hill, yet less than 150 years later it has gone from the NE corner of the county except where protected.

x *P. vulgaris*—*P. variabilis* Goupil non Bast.   False Oxlip.
Not rare but usually soon removed to gardens. The following are given as examples: 05D2, Gazon Wood, 1963, AB, AJS, WEW; 13C5, Vann Woods, several very fine plants, 1967, SFC, JEL; 25B5, Great Burgh, beech woods on chalk, 1949, DPY.

P. vulgaris Huds.   Primrose.   Native.
1827.   Locally still common.
Woods, copses and hedgebanks, especially on the Weald Clay, rare on sand. At its best after woods have been coppiced or felled. The primrose was never common on the Bagshot Sands and is now probably extinct in centrad 96 except as a garden outcast. Within easy reach of London, and some other Surrey towns, almost all accessible primrose roots have been removed to gardens—usually to die. During the last 50 years I have seen this zone pushed steadily farther and farther out, until now the plant survives only on railway banks and in woods from which the public are excluded. It is at its best in the wet clay woods in the S where ecological conditions are more favourable and public pressure less great.

P. spectabilis Tratt., a lime-loving species of the SE Alps, has been found at 35C4, Woldingham, on a field bank at the bottom of Birch Wood, some distance from houses, 1959, JPSR. Some of the employees at a local nursery are known to have amused themselves by sowing or planting garden plants in woods in the district.

Hottonia L.
H. palustris L.   Water Violet.   Native.   1597.
Rare, 23.   Map 234.
Ditches in water meadows, canals, ponds in woods (where it seldom flowers) and slow streams. This handsome plant was formerly abundant in London near the Thames, about Esher and Walton and in other places, but is decreasing fast. Most of the current localities are by tributaries of the Wey, the Basingstoke Canal and the old Wey and Arun Canal.

Lysimachia L.
L. nemorum L.   Yellow Pimpernel.   Native.   1836.
Frequent but local.

Damp places in woods, often by rutted tracks where water stands during the winter and in shade.

L. nummularia L.   Creeping Jenny.   Native.   1832.
Frequent.
Moist hedgebanks, ditchsides, moist grassy places. Occasionally also found as a garden escape.

L. vulgaris L.   Yellow Loosestrife.   Native.   1548.
Locally common, 121.   Map 235.
Canal, river and pond banks, often in rather fen-like conditions. Common on the Bagshot Beds in the NW and especially by the R. Wey and Basingstoke Canal, frequent about Tilford and Milford and widely scattered over the rest of the county. Var. *klinggraeffii* Abromeit (var. *maculata* Druce) 06B1, Byfleet canal-side, 1923, G. C. Druce & Lady Davy (*Rep. BEC*, **7,** 396, 1924). By the Basingstoke Canal between Woking and Byfleet, most plentiful towards the first-named locality, CEB (*J. Bot.*, **70,** 333, 1932). This interesting plant has the corolla lobes with a brownish crimson spot at the base and grows along a considerable stretch of the canal.

L. punctata L.   Established alien.
13A5, Ewhurst, Buildings Wood, naturalised with *Campanula persicifolia* in dense woodland near house, 1965, RARC, 1966, DPY; 16D1, Chessington, Telegraph Hill, in roadside ditch remote from houses, 1966, EJC; 25E3, Greystone chalkpit, on spoil bank, 1961, DCK; 27B1, Wimbledon Common, pathside ditch far from houses, 1964, EJC.

Anagallis L.
A. tenella (L.) L.   Bog Pimpernel.   Native.   1780.
Rare, 22.   Map 236.
Bogs, peaty ditches, woodland tracks and wet peaty heathland. A decreasing species now lost in all localities in the E of the county and some in the W. Drainage is the usual reason, but in others the plant has been shaded out. For example, in 1957 and 1958 there was a lovely colony at 04D2, Wickets Well, Winterfold Forest, which became overgrown with sallows and birches, and the Pimpernel could not be found in 1960 and 1961.

A. arvensis L. subsp. *arvensis*.   Scarlet Pimpernel.
Native.   1793.   Very common throughout the county.
Weed of arable ground and gardens, waste places, rarely on sandy
commons.
subsp. *foemina* (Mill.) Schinz & Thell. (*A. coerulea* Schreb.)
Native.   Rare, 11.   Map 237.
Cornfields and cultivated ground on the chalk. Blue-flowered forms
of subsp. *arvensis* also occur. A hybrid between the two subspecies
has been reported from 25D4 (?), Chipstead Valley, CDP.

A. minima (L.) E. H. L. Krause (*Centunculus minimus* L.)
Chaffweed.   Native.   1814.   Rare, 16.   Map 238.
Damp peaty or sandy tracks, usually in woods and associated with
Radiola linoides, and *Lythrum portula*, damp open ground round
ponds or on heaths. An unusual habitat was wartime gun-pits in
15B4, Bookham Common, (*Lond. Nat.*, **33**, 39, 1954). Most of the
localities are on Weald Clay. An inconspicuous little plant which
may be in a few more localities than we have recorded.

Glaux L.
[*G. maritima* L. Was found in 1910 with other maritime species by the
estuarine Thames at 27C3, Putney by W. A. Todd & CEB (*J. Bot.* **48**,
186, 1910), but has not been reported since.]

Samolus L.
[*S. valerandi* L. Brookweed. Salmon, 1931, gives six records. It has not
been reported recently.]

BUDDLEJACEAE

Buddleja L.
B. davidii Franch.   Butterfly Bush.   Naturalised alien.
1954, Kent & Lousley (*Handbook*, 187–8).
Abundant on waste ground, bombed sites, walls and railway banks
in the London suburbs, locally in chalk pits, in railway cuttings
and on walls elsewhere, 137. Map 239. The earliest evidence we
have traced for Surrey is from 17E4, wall of Kew Gardens, Kew
Road, 1933, DHK. This native of China was brought into cultivation about 1890, but it was nearly 40 years before it was noticed
as established anywhere in England. The period of rapid extension
in London and its Surrey suburbs was from 1940 to 1950, by which
time it was common in most parts of centrads 16, 17, 26, 27, 36
and 37, and a little later in 25 and 35. Outside these centrads it is

much more local but still spreading. The seeds are distributed from gardens by the wind. Var. *nanhoensis* (Chitt.) Rehder, which was introduced into cultivation from China in 1914, is established at 25A1, old chalkpit, Betchworth Hills, 1961, BMCM & JEL, Lousley, *Proc. BSBI*, **4**, 416–7, 1962.

OLEACEAE

Fraxinus L.
F. excelsior L.   Ash.   Native.   1836.
Common throughout the county except on acid soils.
In mixed woodland, scrub and hedges on basic and neutral soils. Much less conspicuous as forming woods than it is in counties to the W and N.

Syringa L.
S. vulgaris L. Lilac. A common garden shrub occasionally established, as at 35C1 (?), Godstone, established in hedgerow.

Ligustrum L.
L. vulgare L.   Wild Privet.   Native.   1836.   Locally common.
Wood borders and scrub on chalk soils, more widely distributed in hedgerows on neutral soils.

L. ovalifolium Hassk. Garden Privet. Commonly planted in hedges though hardly naturalised. Often confused with the native species.

APOCYNACEAE

Vinca L.
V. minor L.   Lesser Periwinkle.   Doubtfully native.   1780.
Frequent, 118 (plus 12 as obvious escapes from gardens).   Map 240.
Copses, hedgebanks and lanesides. Widely distributed over the county. This species is a favourite plant of old cottage gardens and, in addition to the obvious escapes, most of the occurrences are under some suspicion of being introductions, as indeed are most of the reports in Brewer, 1863. The distribution map supports this view.

V. major L.   Greater Periwinkle.   Established garden escape.
1846.   Rare, 25.
Hedgerows, wood margins, railway banks, roadsides.

GENTIANACEAE

Centaurium Hill
C. pulchellum (Sw.) Druce (*Erythraea pulchella* (SW.) Fr.)
Lesser Centaury. Native. 1837. Rare, 33. Map 241.
Damp places where there is little competition, in pastures, track-sides, commons, rides in woods, and on chalk downs. The localities fall into two groups: those across the middle of the county on the chalk (except for one in 15E5 on London Clay), and those in the S on Weald Clay where the usual habitat is woodland rides.

C. erythraea Rafn. (*Erythraea centaurium* auct., *Centaurium minus* auct., *C. umbellatum* auct.) Common Centaury. Native. 1664. Common.
Dry grassland on downs, pastures, heaths and also in woodland rides.

Blackstonia Huds.
B. perfoliata (L.) Huds. (*Chlora perfoliata* (L.) L.).
Yellow-wort. Native. 1640. Frequent, 62. Map 242.
Chalk grassland, rarely in pastures and banks on other soils. As the map shows clearly, Yellow-wort is common along the chalk outcrop with a few records to the N and S. In 96D3 it grew on chalk brought in to build up the ground, and in 26E4 on chalky banks by the railway where the chalk was doubtless brought from elsewhere. In 34E2 it was on tipped rubble. The remaining records from the London Clay in the S are from natural habitats including woodland clearings.

Gentiana L.
G. pneumonanthe L. Marsh Gentian. Native. 1850. Very rare, 2.
Wet heathland. 95A3, Worplesdon, Whitmoor Common, 1967, AJS & WEW; 96D3, Chobham Common, 5 colonies, WEW, PFLeB, etc. The relative abundance of this handsome plant appears to be influenced by heath fires, destruction by picking (these both depend on the weather) and damage to the peaty habitat from trampling and vehicles.

Gentianella Moench
G. amarella (L.) Börner (*Gentiana amarella* L.)
Autumn Gentian. Native. 1837.

Locally common, 58.   Map 243.

Chalk grassland, especially on dry slopes and banks.

G. anglica (Pugsl.) E. F. Warb. (*Gentiana lingulata* C. A. Agardh. var. *praecox* Towns.)   Native.   1876.   Rare, 13.   Map 244. Short grassland on chalk, usually where the chalk is exposed. (Plate 17). This species varies widely in numbers from year to year, according to the availability of habitats which meet its exacting requirements. I have known it on Banstead Downs, 26C1, since 1921, and have witnessed wide fluctuations (some depending on whether the sun was in or out), but have no reason to suspect a decrease. At Riddlesdown (36B1) it appears in new places when the old become overgrown with tall vegetation or scrub. Salmon, 1931, gave records from eight localities covering a period of 50 years; the present survey found it in 13 tetrads in 20 years. In 05D1 a colony on a small piece of chalk grassland on Clandon Downs nr Fuller's Farm, known since 1903 and last seen by me in 1945, was destroyed by ploughing. In 1974 1 small plant was found at 14A5, White Downs, DWB.

MENYANTHACEAE

Menyanthes L.

M. trifoliata L.   Bogbean.   Native.   1746. Rare, 23.   Map 245.

Ponds, swamps and the wetter parts of bogs. This wonderfully beautiful flower is no longer to be found in many of its old stations. Salmon, 1931, blamed over-picking but, as it usually grows under such wet conditions and has such an extensively creeping rhizome, damage from this cause is most unlikely. It is still plentiful in Farm Bog, Wimbledon Common, 27B1, where it has been known since 1836. The usual reason for its loss is changes in the habitat, of which a recent example is the filling in with hard core and refuse of the Lower Stew Pond, Epsom Common, 16E1, soon after our last record of 1957.

Nymphoides Hill

N. peltata (S. G. Gmel.) Kuntze (*Limnanthemum peltatum* S. G. Gmel., *Villarsia nympaeoides* Vent.).   Fringed Water-lily. Native in some localities.   1724.   Rare, 8.   Map 246.

Slow rivers, backwaters, canals, ponds. This lovely plant has long been accepted as a native in the Thames, and formerly had a good many localities in backwaters and adjacent ditches; of these only one has been confirmed in recent years—06E4, plentiful by Thames, Walton-on-Thames, 1950, DHK. It is still plentiful in the Basingstoke Canal, where it occurs in six tetrads, and is likely to have been brought to the canal by barges from the Thames. Similarly, it is still (since 1852) in 37B4, Surrey Canal, Rotherhithe, 1956, LMPS, JEL, TGC, DHK, which communicates with the Thames. It has been planted in 03C5, Cranleigh, pond on the Green, 1956, BW; 14A4, pond by Tillingbourne S of Abinger Hall, 1972, KL, and 14C2, Broome Hall Lake, 1961, SFC.

BORAGINACEAE

Cynoglossum L.
C. officinale L.   Hound's-tongue.   Native.   1836.
Locally frequent, 37.   Map 247.
Chalk grassland and scrub, often round rabbit warrens, wood borders, and sandy commons. Mainly in two areas: along the chalk, and on the sandy commons round Puttenham, Milford and Thursley. An interesting outlier is 27B1, Wimbledon Common, whence it was first recorded in 1836 and still occurs regularly near Springwell Cottage.

C. germanicum Jacq. (*C. montanum* auct.).
Green Hound's-tongue.   Native.   1666.   Very rare, 3.
Woods and wood borders, mainly under or near beech trees. (Plate 18). 94C4, wood near Compton, 12 plants, 1948, WEW; a few, 1949, 1950; 6 plants, 1951, WEW; 1953, AHGA, Hb. Mus. Brit; wood felled and plants not found 1966, WEW & JEL; 15D2, Norbury Park, several places, seen every year in varying numbers, JEL et al; Headley Lane, a few plants every year, JEL; Mickleham Downs, several places above Mickleham village, for example, 1947, JEL & DHK, 1960, BW, 1953, RAB; 15D3, Norbury Park, Icehouse Combe, 1960, BW. The headquarters of this species in Surrey is Norbury Park where it was first recorded by Merrett in 1666. Like most woodland species it varies widely in numbers from year to year according to available light and, although the populations are at a fairly low level at present, this is related to the density of tree canopy. From time to time dogs and humans carry the spined fruits away to start new colonies for a while, and this is believed to account for many of the scattered records. It has not been seen recently in 15E4, Ashtead, near Thirty Acre Barn, where ECW showed it to JEL in 1927.

Symphytum L.

This very difficult genus is currently under revision by several botanists who are approaching it from different aspects and whose conclusions therefore do not always agree. We have relied mainly on the characters given by Perring, 1969 (*Proc. BSBI*, **7**, 553–6).

S. officinale L.		Common Comfrey.		Native.		1836.
Common, 125.		Map 248.
By rivers, stream and ponds. On the banks of the Thames, Wey and Mole and scattered over the rest of the county wherever there is water. Mainly with yellowish-white corollas.

S. asperum Lepech.		Alien.
26B2, Nonsuch Park, E of house, 1962, DMcC, det. AEW; 35A5, Coulsdon, above Old Lodge Lane, 1959, DPY, 'near *S. asperum*' det. AEW.
x *S. officinale* = *S. x uplandicum* Nyman. (*S. peregrinum* auct.)
Russian Comfrey.		Established alien.
Wallace, 1928, *Rep. Watson BEC.*, **3**, 436. Undated records in Salmon, 1931, 460, are earlier.		Common, 166.
Roadsides, edges of cultivated fields, waste places. Russian Comfrey was introduced as a forage plant and first attracted attention as an escape from cultivation in England about 1861. In Surrey it made a late start compared with other counties and Salmon, 1931, was able to cite only eight localities. At about this time it was advertised widely for the phenomenally high production for silage or forage with three or four cuttings a year; no doubt many farmers and smallholders gave it a trial and found it difficult to eradicate. It persisted wherever thrown out and is now the commonest comfrey in the county. It is known to backcross with *S. officinale* when the two grow in proximity and A. E. Wade has named the following as this cross: 35A1, Bletchingley, near Pendell, 1939, CEB, Hb. Wallace; Pendell Court, 1959, BAK; 45A1, S of Limpsfield towards Itchingwood Common, 1937, CEB, Hb. Wallace.

S. orientale L.		White Comfrey.		Established alien.
Salmon, 1931.		Rare.
Roadsides and copses, an outcast from gardens. 94D3, Charterhouse Hill, roadside and copse, 1948, OP det. NYS (*Watsonia*, **2**, 37, 1951); 04A3, Bramley, Chinthurst Lane, 1966, JFL; 04B1, nr Scrubbins Pond, Palmerscross, 1962, JCG; 06C4, Chertsey, Willow Walk, 1969, RMB; 17E4, Kew, bombed site, Sandycombe Road,

1950, KEB; 17E2, Ham Common, 1958, LJJ; 26C2, Sutton, cutting on Epsom Downs line, 1955, LJJ; 26E4, Mitcham Common, nr Beddington Lane, 1960, JEL; 35C2, old tip between Caterham and Godstone, 1957, BMCM, 1961, RARC; 35C5, Warlingham, laneside, 1957, RARC; 36C3, Croydon, edge of golf course near Oaks Farm, 1952, LJJ; 37C3, old railway, 1965, RARC.

S. caucasicum Bieb.   Established alien.
Only record: 35B1, Bletchingley, coppice nr cottages, Coldharbour, probably a garden outcast, 1960, DPY, det. AEW.

S. tuberosum L.   Tuberous Comfrey.   Established alien.
Brewer, 1863. 16D3, Upper Long Ditton opposite church, well established on both sides of hedgerow, 1962, JES; 25B2, open wood between Buckland and Colley Hills, 1942, IAW & JSLG, Hb. Kew; 26D4, Mitcham, small coppice, naturalised, 1957, RCW; 35C1, Godstone Marsh, Church Lane, roadside and field nr house, 1957, RARC. The Mitcham and Godstone localities are wet, with conditions similar to those in which the species thrives best in other counties.

S. grandiflorum DC.   Creeping Comfrey.   Established alien.
1929, Biddiscombe, *Rep. Watson BEC*, **3**, 481.   Rare.
94C3, Hurtmore Bottom, open bushy ground, well-established colony 2–3 sq yd in area, 1948, OVP (*Watsonia*, **2**, 47, 1951), a grand patch and vigorous, 1959, OVP; 04A5, by track through Chantries, 1957, JCG, 1972, SFC; 05B3 (?), Send, by pond at Woodhill, 1928, Biddiscombe (as above); 05B3, Hort ex nr Sendhurst Grange, Send, 1928, etc, Hb. Mus Brit; 35A5, Kenley, 'Pondfield Road', long naturalised on grass verge, 1950, AC; 35B1, Bletchingley, naturalised in a sunken lane, Miss M. Bryan, det. BV (*Watsonia* **1**, 254, 1950), refound in 1960, BAK.

S. bulbosum Schimp.   Established alien.
04A4, by overgrown and shaded pond by road to Great Tangley Manor, 1971, AL, conf. JEL; well established but pond recently partly filled in, 1972, JEL & JES. For an account of this species see Lousley (*Proc. BSBI*, **4**, 43–44, 1960).

Borago L.
B. officinalis L.   Borage.   Alien.   1836.   Rare.
Formerly much grown by bee-keepers and frequently established

for a time as a garden escape. Now occasionally on refuse tips. 06B5, Thorpe, Mill Lane in hedgerow, 1970, GHG; 35D4, Nore Hill, in old chalkpit near beehives, 1957, RARC.

Trachystemon D. Don
T. orientalis (L.) G. Don
A park-type garden alien sometimes established. 93A3, Grayswood, nr cottages, 1973, HM-P; 05A1, Guildford, Stoke Park, a naturalised patch 10ft x 5ft in wood at side of by-pass, 1969, JFL; 13C4, nr Vann House, 1965, RARC; 17E2, Petersham Common, behind Star & Garter, 1961, BW, 1966, BWu.

Pentaglottis Tausch
P. sempervirens (L.) Tausch (*Anchusa sempervirens* L.)
Established garden escape.   1762.   Frequent, 89.   Map 250.
Roadsides, hedges, waste ground. Increasing, but usually near houses and often barely out of a garden.

Lycopsis L.
L. arvensis L. (*Anchusa arvensis* (L.) Bieb).   Small Bugloss.
Native.   Locally common, 154.   Map 251.
Cultivated ground on sand or chalk soils. The distribution follows closely the occurrence of sands of the Bagshot and Greensand series; on chalk the plant is rare and records from chalkpits are more numerous than from arable. It is decreasing in common with other agricultural weeds.

Pulmonaria L.
P. officinalis L.   Garden Lungwort.   Established alien.
1931.   Rare.
Roadsides and copses. 94B3, ditch by road from Somerset Farm to Gatwick, a large patch, 1943, Mrs D Turner, only a little left, 1963, AB; 94C3, Hurtmore Bottom, 1948, OVP; 05E1, Sheepleas, by footpath, 1968, JFL; 35C2, Marden Park, naturalised, 1957, EMCI; 35E3, Titsey Plantation, roadside opposite, 1971, SFC.

Brunnera Stev.
B. macrophylla (Bieb.) I. M. Johnston   Garden alien.
94E1, nr Munstead, hedgerow opposite old cottage, 1962, OVP, obvious outcast, 1964, DPY; 25E5, Coulsdon, S end of Downs Road, 1962, EMCI, BW; 35C4, Warlingham, chalk scrub below garden, established, 1958, RARC.

Myosotis L.
M. scorpioides L. (*M. palustris* (L.) Hill)
Water Forget-me-Not.   Native.   1827.
Common throughout the county.
Wet places by rivers, streams and ponds and marshy meadows.

M. secunda A. Murr. (*M. repens* auct.)
Creeping Forget-me-not.   Native.   1850.   Rare, 29.
Boggy, usually peaty, places on heaths, and commons. Mainly on the peaty heathlands of the SW and Mr Clarke finds it commonly in the SE.

M. caespitosa K. F. Schultz   Tufted Forget-me-not.
Native.   1827.   Common, 110.   Map 252.
Ditches and marshy places, by ponds and streams.

M. sylvatica Hoffm.   Wood Forget-me-not.   Native.
1850.   Locally frequent, 58 (also 25 as garden escape).   Map 253.
Woods and copses. In two areas, on the chalk round Titsey, and on the Weald Clay about Capel and Ockley, the native status of Wood Forget-me-not can hardly be in doubt. It is in abundance in woods in areas with few habitations. It also occurs freely from seeded garden plants and these we have endeavoured to distinguish on the map. There are many border-line cases and much has depended on the individual judgement of our recorders.

M. arvensis (L.) Hill   Field Forget-me-not.   Native.
1640.   Very common.
Arable fields and gardens, edges and clearings in woods, and hedgebanks.

M. discolor Pers. (*M. versicolor* Sm.)
Changing Forget-me-not. Native. 1827. Frequent, 107. Map 254.
Open communities on fairly light soils, sandy tracks, chalky field borders, ant-hills on chalk grassland.

M. ramosissima Rochel (*M. hispida* Schlecht, *M. collina* auct.).
Early Forget-me-not. Native. 1838. Frequent, 86. Map 255.
Sandy heaths, banks, field borders, ant-hills on chalk. The distribution is very similar to that of the last species, but this is a more common constituent of short turf on sandy heath.

Lithospermum L.

L. purpurocaeruleum L.   Blue Gromwell.
Established garden alien.
94A5, Great Down, naturalised in a wood, a patch of many square yards, 1966, OVP.

L. officinale L.   Common Gromwell.   Native.   1838.
Rather rare, 23.   Map 256.
Rides and borders of woods, hedgerows and copses on the chalk. Usually in very small numbers.

L. arvense L.   Corn Gromwell.   Native.   1827.
Very rare, 13.   Map 257.
Cultivated fields, waste ground, rubbish tips, chalkpits. Brewer, 1863, found this 'common'; Salmon, 1931, 'uncommon', and we have found it 'very rare'. Some of the records on the map are based on small numbers of plants, and others on casuals. At the present rate this species will soon be extinct.

Echium L.
E. vulgare L.   Viper's-bugloss.   Native.   1831.
Locally common, 48.   Map 258.
At its best in rather bare places on the chalk where it is often plentiful, but even there rather patchy; elsewhere in sandy fields, roadsides and banks. On Banstead Downs, 26C1, it was collected by ECW in 1935; not found from 1950 to 1960 in spite of special search by five botanists, and a single plant found in 1963, DPY.

CONVOLVULACEAE

Convolvulus L.
C. arvensis L.   Field Bindweed.   Native.   1793.
Abundant throughout the county.
Cultivated land, hedgerows, waste ground, railings by railways and elsewhere. An agricultural pest which is also quite common in towns.

Calystegia R.Br.
C. sepium (L.) R.Br. (*Convolvulus sepium* L.)
Great Bindweed.   Native.   1666.   Frequent.
Ditches, wet alder and willow spinneys, and roadsides, also in gardens. Corolla nearly always white, rarely pink.

x *C. silvatica* = *C. x lucana* (Ten.) G. Don.
Frequent near London, distribution less well known elsewhere, 21.
The data were supplied mainly by E. B. Bangerter based on the
statistical approach of the LNHS *Calystegia* survey, see Bangerter
(*Lond. Nat.*, **46,** 15–23, 1967).

C. pulchra Brummitt & Heywood (*C. dahurica* auct.)
Established garden alien.   Frequent, 44.   Map 259.
Garden fences, roadsides near houses, by canals and railways.
Mainly near London, Redhill and Dorking, with scattered records
elsewhere.
x *C. sepium.*   Apparently very rare.
17D2, Ham, in alley S of Ham House grounds, 1967, JLG,
teste RKB.

C. silvatica (Kit.) Griseb. (*C. sylvestris* (Willd.) Roem. & Schult.,
Volvulus inflatus Druce *pro parte, Calystegia inflata* auct.)
Naturalised garden alien.
Common throughout the county, 295.   Map 260.
Hedges near gardens, waste places, railway fences, canal towpaths.
This S European species is known to have been in Britain before
1835, presumably as a garden plant, and appears to have owed its
spread to the sale to gardeners as the American Great Bindweed.
It was not until 1921 that it attracted attention as a wild plant in
Britain (Lousley, 1948). It was not known to Salmon, 1931, and
yet it is now common throughout the county, in rural and urban
areas alike, as too many gardeners know to their regret.

Cuscuta L.
C. europaea L.   Greater Dodder.   Native.   1793.
Rare, 27.   Map 261.
Banks of rivers and streams on nettles. (Plate 19). Mainly along the
Mole where it is frequent and occurs in 17 tetrads, now less frequent
along the Thames, and a few localities by the Wey. By far the most
common host is *Urtica dioica*, but recorders have also noted it on
Ulmus sp, *Solidago* sp, *Malva sylvestris*, *Chrysanthemum vulgare*,
Lactuca serriola and *Galeopsis tetrahit*. The records do not state
specifically that these were all parasitised (ie penetrated by haustoria)
and some may have been used only for support.

C. epilinum Weihe   Flax Dodder. Once a pest on crops of flax, not recorded in the present century.

C. epithymum (L.) L.   Common Dodder.   Native.
1835.   Frequent, 82.   Map 262.
Parasitic on ling, gorse, clovers, thyme and many other species. The distribution falls roughly into two parts: the Dodder parasitic mainly on ling, *Calluna vulgaris*, characteristic of the acid heathland in the NW and W of the county, where it is often locally abundant; and the Dodder mainly parasitic on Leguminosae, which occurs on the chalk. The latter was included by Salmon, 1931, as a separate species, *C. trifolii* Bab., based on a plant which was a serious pest of clover crops in the first half of the nineteenth century. This was extended to cover parasites on chalk grassland legumes, such as the Dodder which still occurs at 05E1, Horsley Sheepleas.

C. campestris Yuncker   Alien.
17E4, Kew, established in Herbarium grounds, apparently parasitic on *Bromus carinatus*, 1950–6, EM-R, Kent & Lousley (*Handlist*, 1957, 354); 26B1, Ewell, Mizen's market garden, on basil, 1958, AEE, det. Kew, 1959, DPY.

SOLANACEAE

Lycium L.
L. barbarum L. (*L. halimifolium* Mill.)
Duke of Argyll's Tea-Tree.   Established alien from gardens.
Planted in hedges, persistent and never far from houses. 94E4, Shalford Common in two places, 1955, OVP; 16C3, Esher Railway Station by lay-by, 1958, BW; 17E4, Kew, beside railway between Mortlake and the Thames, 1958, BW, 1961, WTS; 17E4, between Mortlake and Kew, by towpath, 1948, EBBa; Upper Shirley, in spinney near cottages, 1957, CNHSS.

L. chinense Mill.   Established alien.
27D2, Wandsworth, nr Prison, 1952, JAC, Hb. Mus. Brit, det. N. Feinbrun; 27D3, Wandsworth, several around Neal's nursery at above locality, also on opposite side of road near railway bridge (*L. barbarum* also present), 1966, Hb. DPY. This species has been greatly misunderstood and is evidently very rare in Britain. See Feinbrun & Stearn, 1964 (*Israel J. Bot.*, **12**, 114–23).

Atropa L.
A. bella-donna L.   Deadly Nightshade.   Native.   1640.
Locally frequent, 53.   Map 263.

Scrub, wood borders on the chalk, elsewhere bird-sown in gardens
and waste places.

Hyoscyamus L.
H. niger L.   Henbane.   Casual, perhaps formerly native.
1793.   Rare, 16.   Map 264.
Waste places, allotments and farm lands. The appearance of Hen-
bane is often erratic, and it can reappear after long intervals. Two
interesting records are: 15D3, Leatherhead, allotments, 1946–51,
JESD; 35C4, Halliloo Farm, site of grubbed-up hedge, where A.
Beadell had found it many years earlier, 1968, RARC.

Physalis L.
P. alkekengi L.   Japanese Lantern.
A garden plant which can be very persistent. 94E5, Guildford, in
hedge on site of demolished farm buildings, 1969, JFL. More often
persistent on sites of refuse tips, as at 26E4, Beddington Lane, site
of rubbish dump, 1957, RCW.

Solanum L.
S. dulcamara L.   Bittersweet.   Native.   1597.
Common, even in towns.
Hedges, woods, waste ground, in towns in damp alleys and gardens.

S. nigrum L.   Black Nightshade.   Colonist.
1832.   Locally frequent.
Cultivated and waste places.

S. sarrachoides Sendtn.   Alien.   Rare, but increasing.
04A4, Shalford Common, abundant in field, 1972, AL; 05A1,
Guildford, refuse tip, 1972, AL; 26B1, Ewell, North Looe, many
plants in a garden field, 1962, DPY. In other counties this species,
easily recognised by the green berries with strongly accrescent
calices, and rather yellowish usually hairy leaves, is firmly established. It may become so in Surrey.

S. sisymbriifolium Lam.   Alien.
16B1, Oxshott, near Fairmile Park, 1961, comm. RHS Wisley.
Probably introduced with building materials and seen by JES, JCG,
JEL, Hb. Lousley.

Datura L.

D. stramonium L.   Thorn-apple.   Alien.

1793.   Rather rare, 35.

Gardens, allotments, waste places. This species is known to be introduced with wool and skins, and this may explain its appearance at Gomshall Tannery, 04E4, but we are unable to suggest a reasonable explanation of how it arrived at most of its localities. The seeds can remain dormant in the soil for a lengthy period in my garden, and some of the appearances on allotments and gardens are almost certainly due to digging up old seed. It was grown in the Whitgift Grounds, Haling Park, 36B3, for the Oxford Medicinal Plants scheme during the 1939–45 war (var. *inermis*) and did not disappear until 1968 after disturbance of the ground, CTP.

SCROPHULARIACEAE

Verbascum L.

V. thapsus L.   Great Mullein.   Native.   1793.   Common.
Banks, roadsides, downs, open woods and railway cuttings. An erratic species, which often appears suddenly after road building or other activities have provided suitable open habitats for the seedlings. Usually in rather small numbers.

V. phlomoides L.   Garden Mullein.
Naturalised garden alien.   1861.   Uncommon but increasing.
Roadsides, waste places and refuse tips. Persistent for over 35 years at 94D5, Guildford By-pass, cutting on the Hog's Back, 1936, R. B. Ullman, Hb. Lousley, 1972, JEL.
x *V. thapsus* 05A1, Guildford Refuse Tip, 1972, JLP.

V. lychnitis L.   White Mullein.   Native.   1770.
Rare, 10.   Map 265.
Chalkpits, chalky banks and roadsides, gravel pits. Restricted to the eastern end of the chalk, from Chipstead Valley, Coulsdon and Warlingham in the S to Wallington and Croham Hurst in the N. We have also a record from 36C5, Anerley, which is off the chalk. At 35C4 and 35D4, Nore Hill Gravel Pits, chalk adjoins the gravel and in the 1930s it grew only on a chalky slope above the chalkpit, though it has since been found on gravel. The Surrey localities must be regarded as the westerly extension from the large area of north Kent where the plant is more plentiful.
x *V. nigrum* and *V. lychnitis* x *thapsus* have occurred in the past but have not been found during the period of our survey.

V. nigrum L.   Dark Mullein.   Native.   1640.
Locally common, 124.   Map 266.
Roadsides and open habitats on hedgebanks, field borders, chalk-pits and chalk grassland. *V. nigrum* is generally regarded as characteristic of calcareous soils but, as the map shows, this is far from being the case in Surrey—less than half our records are on the chalk, and the plant is absent from considerable areas of the outcrop. There are more localities on the Lower Greensand, and others scattered over the remaining geological deposits.

x V. thapsus = *V. x semialbum* Chaub.
Probably occurs freely but the only records we can give are: 15D1, Box Hill, chalky field above R. Mole and also slope above Stepping Stones, 1942, JEL, Hb. Lousley; 15E2, chalky valley near High Ashurst, 1944, JEL, Hb. Lousley; 26E4, Beddington Lane Dump, 2 plants with parents, 1958, RCW.

V. blattaria L.   Moth Mullein.   Garden alien.
1724.   Very rare, 5.
94B1, rough ground, Witley Common, probably introduced by army during war, 1955, OVP, 1970, shown to JEL; 16B3, Esher Place Park, on new embankment, 1956, DPY (*Proc. BSBI*, **2**, 258, 1957); 17D4, Queen's Cottage grounds, 1965, BWu; 25D4, Chipstead Valley, fallow field near Well Copse, 1955, BMCM, JEL, Hb. Lousley; 33B5, Copthorne, waste ground beside the 'Hunter's Moon', 1968, BMCM.

V. virgatum Stokes.   Twiggy Mullein.
Garden alien, sometimes established for a few years.   1829.
94B1, Witley Common, old camp site, 1955, OVP sp *non vidi*; 25B5, Epsom, Great Burgh, persisted on a grassy bank, 1956, DPY; 35A2, Pendell tip, 1964, RARC; 35C4, Warlingham, top of Bug Hill, 1960, RARC; 35E3, Oxted, rough field above Oxted chalkpit, plentiful for several years, 1943, JEL.

Misopates Raf.
M. orontium (L.) Raf.   Lesser Snapdragon.   Colonist.
1781.   Rather rare, 53.   Map 267.
Sandy arable fields, gardens, roadsides, waste places. A decreasing species still locally frequent in the W of the county on light soils.

Antirrhinum L.
A. majus L.   Snapdragon.   Garden alien.   1836.
Occasionally naturalised on old walls and railway cuttings.

Linaria Mill.

L. purpurea (L.) Mill.   Purple Toadflax.
Garden escape established on old walls and in waste places.
1862.   Not common.

L. repens (L.) Mill.   Pale Toadflax.   Native.
Locally on chalk.   1913.   Very rare.
Downs and field borders on the chalk, also in quarries, roadsides, railways and waste places as an introduction. This is accepted as native on the chalk escarpment at Buckland, where it was recorded by A. J. Crosfield (*Proc. Holmesdale*, **1910–13**, 90) and where it has been seen at intervals by ECW and JEL since 1928, and we have the following recent records: 25B2, below Buckland Hill, 1959, SFC, 1965, CTP and DH, 1972, RARC, 1973, JEL. The following records are believed to refer to garden escapes, but some of those on the chalk may possibly be native: 94E5, Guildford, Great Quarry, 1965, RARC; 25D5, Park Downs, 1948, DPY, unable to refind, 1961, DPY; large patch, 1973, BRR; 25E1, Redhill, wall of railway embankment, 1966, BMCM; 26D5, Merton, Bunce's Meadow, 1956, RCW; 27E4, Nine Elms rail goods yard, RARC, DPY, AB, EJC; 34E2, Lingfield, persistent on garden wall, 1963, RARC; 35C4, W of Halliloo Farm, 1960; two localities on roadsides, 1972, RARC & JEL; 35D4, Chelsham, by Nore Hill dewpond, a new arrival, 1965 JRP & RARC.

L. vulgaris Mill.   Yellow Toadflax.   Native.
1762.   Very common.
Roadsides, hedgebanks, round cultivated fields, on exposed chalk, commons and waste places.

Chaenorhinum (DC.) Reichb.
C. minus (L.) Lange (*Linaria minor* (L.) Desf.)   Small Toadflax.
Native or colonist.   1783.   Locally common, 142.   Map 268.
Cultivated fields, waste places, railway tracks and yards. Until very recently this species was regarded as a cornfield weed especially common on chalk and sand. Lately it has extended its range rapidly along railway tracks where it thrives on the cindery permanent way. In Surrey there is little evidence of this; it grows sometimes on railway premises, but the main habitat is still cornfields, most common on the chalk, and almost equally so on light soils on the Weald Clay.

Plate 17   *Early Gentian,* Gentianella anglica, *from Banstead Downs (1973)*

Plate 18   *Green Hound's-tongue,* Cynoglossum germanicum, *in Norbury Park in 1973; it was first recorded there in 1666*

Plate 19   *Greater Dodder,* Cuscuta europaea, *on nettle, Wonham Mill*

Plate 20   *Greater Yellow Rattle*, Rhinanthus serotinus, *Coulsdon*

Kickxia Dumort.
K. *spuria* (L.) Dumort. (*Linaria spuria* (L.) Mill.)
Round-leaved Fluellen.   Native.   1597.
Locally frequent, 86.   Map 269.
Cornfields and other arable fields, mainly on the chalk and Weald
Clay.

K. *elatine* (L.) Dumort. (*Linaria elatine* (L.) Mill.)
Sharp-leaved Fluellen.   Native.   1597.
Common, 168.   Map 270.
Arable fields, mainly on the chalk and Weald Clay, but much more
common than the last species.

Cymbalaria Hill.
C. *muralis* Gaertn., Mey & Scherb. (*Linaria cymbalaria* (L.) Mill.)
Ivy-leaved Toadflax.   Naturalised alien.   1732.
Frequent throughout the county. Old walls and rubble.

Scrophularia L.
S. *nodosa* L.   Common Figwort.   Native.   1836.
Very common.
Hedgebanks, rides and margins of damp woods and moist fields,
sometimes in drier situations in waste places.

S. *auriculata* L. (S. *aquatica* auct.)   Water Figwort.   Native.
1832.   Frequent throughout the county.
Mainly by ditches, streams and rivers and in damp woods and
marshy places, but also on the chalk in woods or round their
edges in places which are superficially dry. The following are
examples of chalk habitats: 15D1, Box Hill, on the higher ground,
JEL; 25B2, Buckland Hills, 1972, RARC; 35C4, Caterham, Birch
Wood reserve, 1972, RARC; 35E3 & 45A3, Titsey Plantation East,
large colonies after felling of woods, 1972, RARC. The occurrence
in such places of a marsh plant, and failure to mention it in literature,
has puzzled me since I was faced with a difficult identification
problem in boyhood; I think the explanation may be that the
junction of Clay-with-Flints and chalk provides a seepage area
which is always wet just below the surface.

S. *umbrosa* Dumort. (S. *alata* Gilib.)   Green Figwort.
Native.   1913.   Extinct.
Not seen during our present survey. 15D2(?), E bank of R. Mole

nr Norbury Park, 1934, RWR, Kent & Lousley, *Handlist*, 1954, 205, and 1920, T. J. Foggitt, Hb. Mus. Brit; 36A3, Mill Pond, Waddon, 1863, ABe (*J. Bot*, **51,** 61, 1913).

S. vernalis L.   Yellow Figwort.   Persistent garden alien.
1762.   05E5, Cobham, Painshill Park, naturalised in old grotto, 1965, EJC; 27B1, Wimbledon Common, waste ground under bushes, 1950, AWJ.

Calceolaria Fabr.
C. chelidonioides H. B. K.
Alien, a persistent weed in gardens.   1964 (see below).
17E4, Kew, waste ground near Herbarium, 1922, WBT, Lousley, *Proc. BSBI*, **5**, 338–41, 1964; 25B1, Buckland, persistent weed, 1963, BMCM, Lousley *loc. cit.;* 27D4, Battersea Park, weed in flower garden, 1965, DPY—and doubtless elsewhere.

Mimulus L.
M. guttatus DC. (*M. langsdorffii* Donn ex Greene).
Monkey Flower.   Naturalised garden plant.   1863.   Rare, 34.
Map 271.
By streams, canals and ponds. Surprisingly, this alien has not increased its range since Salmon, 1931. It is still in some of the old places in the Tillingbourne valley, and in the system of the Wey, and in one place by the Wandle, but has not spread along these waters.

M. moschatus Dougl. ex Lindl.   Musk.
Naturalised garden alien.   1866.   Rare.
By streams, ditches, ponds and in swampy fields. 93B2, ponds nr Imbham's Farm, 1961, JCG; 96A2, Bagshot, in great quantity in ditch in Waterer's nursery, 1968, MRM; 96E5, Egham, Ulverscroft, tiny patch in pond, 1965, EJC; 04A2, Bramley, Birtley House grounds, 1972, BMCM; 14A4, by Tillingbourne at bottom of Townhurst Wood, 1962, JCG; 14B4, two extensive colonies in swampy fields to E of Tillingbourne brook, 1965, RARC—perhaps the same as Wotton, Mrs Wedgwood (*Rep. BEC*, **6,** 391, 1922); 17E1, Isabella Plantation, 1957, LJJ; 34E1, Dormers, along stream and adjoining marsh, 1963, RARC.

Limosella L.
[*L. aquatica* L. Mudwort. Native. 1805. Extinct. A species which thrives on rich mud exposed in dry seasons round ponds, especially ponds used

by ducks and geese. Salmon, 1931, recorded 26 localities, including four in which he had seen the plant himself. Since then the only records have been from 94E, Peas Pond, Peasemarsh, 1930, J. E. Little, R. J. Burdon and G. M. Ash, and pond on Shalford Common, 1934, JEL, Hb. Lousley. The two localities in Surrey with the longest histories were in 16C4, a pond on Weston Green, where Mudwort was collected in 1838, 1851, 1852, 1853, 1867, 1897, 1902 and 1916 (Rake in Hb. Lousley); and in 94A3, Cut Mill Pond where it was collected in 1810?, 1830, 1857, 1874, 1881, 1882, 1883, 1893 and 1895. In both cases there is a sequence of dates suggesting runs of summers with low water tables but, although these may be repeated, geese and ducks are also needed to keep down the vegetation and enrich the mud for the reappearance of Mudwort.]

Erinus L.
E. alpinus L. Garden alien. 14B4, Wotton, old wall W of Wotton House, 1955–7, LJJ; 25A3, Headley Park, old brick wall by road, 1964, DCK.

Digitalis L.
D. purpurea L.   Foxglove.   Native.   1763.   Frequent.
Open places in woods, heaths, commons, hedgebanks, railway banks, with a preference for acid soils, and avoiding the chalk.

D. lutea L.   Alien.
15E1, Brockham Hill, bomb-crater, with many other aliens, 1948, JEL & RG; also Boxhill, edge of wood, 1947, Brenda E. Hains, Hb. Kew, probably refers to same locality—Lousley, 1949 (*London Nat.*, **28**, 29, 1949). It persisted for a good many years. 15D3, Leatherhead, chalk cutting on by-pass, 1959, I. Thornley, 1958, BMCM, 1965, AGH, 1974, JEL; 15D3, Leatherhead, chalk railway cutting, 2 clumps, 1960, AWJ; 25A1 (and 15E1), Brockham disused chalkpits, nr ponds, 1958, BMCM, 1960, BMCM & JEL, 1974, BRR, JEL; 25A1, Betchworth, quarried chalk on downs, numerous plants, 1967, BWu. There have also been other scattered records from the chalk.

D. ferruginea L., *D. grandiflora* Miller, and *D. lanata* Ehrh. were all established for a few years at 15E1, Brockham Hill, bomb-crater, 1948, JEL & RG. All three species were also persistent at 25A1 (& 15E1), Brockham chalkpits, 1958, BMCM, 1960, BMCM & JEL.

Veronica L.
V. beccabunga L.   Brooklime.   Native.   1836.   Common.
Streams, ponds, ditches and wet places in meadows.

V. anagallis-aquatica L.   Blue Water-speedwell.   Native.
1793 (for the collective species with the next).
Rather rare, 28.   Map 272.
By streams and ponds. Occasionally by the Mole and Wey, but
mainly by ponds.

V. catenata Pennell. (*V. aquatica* Bernh., non Gray).
Pink Water-speedwell.   Native.
1928, C. E. Britton (*J. Bot.*, **64,** 43.)   Rare, 14.   Map 273.
By R. Thames, ponds, gravel pits and small streams.

V. scutellata L.   Marsh Speedwell.   Native.   1762.
Frequent, 47.   Map 274.
Pond margins, boggy meadows and ditches, usually under acid
conditions.

V. officinalis L.   Heath Speedwell.   Native.
1836.   Common.
Dry places on heaths, commons, open places in woods, pastures and
hedgebanks.

V. montana L.   Wood Speedwell.   Native.   1805.
Common, 281.   Map 275.
Damp woods, copses and hedgebanks. Thrives especially in the wet
woods in the S of the county, but almost equally common on the
Lower Greensand, Chalk and parts of the London Clay.

V. chamaedrys L.   Germander Speedwell.   Native.
1793.   Common.
Grassy places on commons and pastures, hedgebanks and wood
borders.

V. longifolia L.
A garden plant which persists and is sometimes recorded as *V.
spicata.* 06E3, Walton Common, persistent on ground formerly
cultivated but well established and known to RAB for over 10 years,
1965, RAB & JES; 25A2, Headley Heath, in ditch near car park,
1964, EJC; 26E4, Mitcham Common, on roadside bank on heathland, 1957, DPY.

V. serpyllifolia L.   Thyme-leaved Speedwell.   Native.
1836.   Common.
In grassland on commons, pastures, roadsides, damp woods and
lawns.

V. peregrina L.   American Speedwell.
Alien, spread from nurseries to gardens.
05A2, Guildford, sewage works, 1959, Mrs E. Hodgson; 17E4,
Kew Gardens, weed of cultivated ground, 1918, W. C. Worsdell,
Hb. Kew; 34D3, Blindley Heath, plentiful in vegetable garden,
1964, GIC; 34C4, Blindley Heath, 1969, GIC. See E. B. Bangerter
(*Proc. BSBI*, **5**, 303–13, 1964 & **6**, 215–20, 1966).

V. arvensis L.   Wall Speedwell.   Native.   1836.
Common on dry soils throughout the county.
Heaths and commons, dry banks, wall-tops, and weeds in gardens
and arable fields.

V. acinifolia L. Alien. Persisted for at least 14 years in cornfields. 93D3,
found by Mrs C. H. Wilde & E. B. Bishop in great quantity 'near
Milford, Surrey', 1920 (*Rep. BEC*, **6**, 730, 1920); Chiddingfold, in
enormous quantities in one field, less in another, 1920, G. C. Druce
(*Rep. BEC*, **6**, 34–5, 1921); arable field, Chiddingfold, 1920, Mrs Wilde
(*Rep. Watson BEC*, **3**, 177, 1922); in original field where Woking Field
Club found it 4–5 years ago, but has increased, 1923, W. Biddiscombe
(*Rep. BEC*, **7**, 398, 1924); Chiddingfold, abundant in fallow, 1934, JEL &
GMA. It persisted after this and is a likely species to become established,
but recent search has failed to refind it.

V. triphyllos L.   Fingered Speedwell.   Doubtfully native.
1871.
Weed in sandy fields. 06D1, Byfleet, nr Sanway, in Salmon, 1931;
scarce 1932, plentiful 1934, JEL; abundant 1958, JEL, BW, JES;
4 plants, 1970, EJC. Not seen in any of the other localities given
by Salmon.

V. hederifolia L.   Ivy-leaved Speedwell.   Native.
1836.   Very common.
Cultivated and waste ground, churchyards, hedgebanks. Attempts
have been made to divide this species into subsp. *hederifolia*, with
somewhat fleshy leaves with rather acute teeth and corolla 6–9mm
bright blue streaked with violet, and subsp. *lucorum* (Klett &

Richter) Hartl. (*V. sublobata* M. Fischer, *V. trilobata* auct.), a smaller plant with leaves thin in texture with small lobes and corolla 4–6mm pale lilac to white. Both occur in Surrey though subsp. *lucorum* is rare, but insufficient work has been done to say if the characters can be applied generally or to give the distribution.

V. persica Poir. (*V. buxbaumii* Ten,. non Schmidt, *V. tournefortii* auct.)   Common Field Speedwell.   Alien.   1837.
Abundant throughout the county.
Cultivated fields and waste ground. First recorded from Britain in 1825, it was frequent in Surrey by 1863, and Salmon, 1931, found it unnecessary to give localities.

V. polita Fr.   Grey Field Speedwell.   Native.   1835.
Frequent, 138.   Map 276.
Cultivated fields and gardens, and rarely on hedgebanks. Perhaps more frequent on the chalk.

V. agrestis L.   Green Field Speedwell.   Native.   1838.
Frequent, 99.   Map 277.
Rich cultivated soil, nearly always in gardens, nurseries, allotments and churchyards.

V. filiformis Sm.   Slender Speedwell.
Garden alien, thoroughly established.   1934.
Very common, 192.   Map 278.
Banks of rivers and streams, roadsides, commons, churchyards and cemeteries. *V. filiformis* is a native of the Caucasus and Asia Minor, grown in British gardens since 1808, which became a popular rock garden plant about 45 years ago and soon spread to lawns. The first record for Surrey was from 05D4, Ockham, on a roadside bank where A. H. Carter showed it to me in 1933, Lousley (*Rep. Watson BEC*, **4**, 233–4, 1934). The early spread in Surrey is given by Bangerter & Kent (*Proc. BSBI*, **2**, 197–217, 1957, *op cit.* **4**, 384–97, 1962, *op. cit.* **6**, 113–18, 1966). Fruits have been found in the county, but it spreads mainly by the extensively creeping stems.

Pedicularis L.
P. palustris L.   Marsh Lousewort.   Native.   1666.
Rare, 9.   Map 279.
Marshy meadows and peat bogs. A decreasing species. Formerly

widespread over the county, but now practically confined to wet places on the peaty commons in the extreme W.

P. sylvatica L.   Lousewort.   Native.   1601.
Frequent, 81.   Map 280.
Damp heathland and commons. Avoids the chalk, most common in the W.

Rhinanthus L.
R. angustifolius C. C. Gmelin subsp. *ansgustifolius* (*R. serotinus* (Schönh.) Oborny, *R. major* Ehrh., non L.).
Greater Yellow Rattle.   Native.   1968 (see below).   Very rare.
Chalk grassland. (Plate 20). 25D4, Chipstead Valley, meadow on hillside above Dene Farm, 1966, REG, 1968, EJC & SBS, conf. DJH, 1969, SBS & DPY; 35A4, Coulsdon, 'Happy Valley', abundant amongst grass and scrub on chalk hillside, BWu conf. DJH & DPY (*Proc. BSBI*, **1**, 197, 1968). The abundance of Yellow Rattles at this spot had been known for many years, but *R. angustifolius* is not easily distinguished from *R. minor* on the Surrey chalk.

R. minor L.   Yellow Rattle.   Native.   1836.
Uncommon, 56.   Map 281.
Grassland on downs, meadows, pastures and roadsides. Appears to have decreased since Salmon, 1931. Autumnal ecotypes with narrow intercalary leaves have been distinguished as *R. stenophyllus* Schur., 35D4, near Warlingham, on the chalk downs, Salmon, 1931, and nr Warren Barn, 1943, JEL, Hb. Lousley; and *R. calcareus* Wilmott, 15D2, Box Hill, above Juniper Hall, 1957, FR; 35A4, Devilsden, sloping chalkland pasture, 1960, HAS, Hb. Sandford, det. DPY.

Melampyrum L.
M. pratense L.   Common Cow-wheat.   Native.   1836.
Locally frequent, 106.   Map 282.
Woods, heaths, and heathy roadsides. Absent from large areas. A variable species which was studied in great detail by C. E. Britton who cited many gatherings from Surrey in his papers. These culminated in 'The Genus Melampyrum in Britain', *Trans. Proc. Bot. Soc. Edinb.*, **33**, 356–79, 1943. Here he includes subsp. *meridionale* C. E. Britton from 14A2, Holmbury Hill; 14A3, Felday;

and 14B3, Friday St, where it grows on the Hythe Beds of the Lower Greensand with, and probably parasitic on, *Vaccinium myrtillus*. Also subsp. *ericetorum* (D. Oliver) C. E. Britton, which he found, inter alia, widespread in the Leith Hill and Hurt Wood areas, such as 14A3, Felday; 14B3, Friday St and Abinger Common; 14B4, Wotton. This grows with woody heathland species, such as *Calluna vulgaris*, on which it may be parasitic. Reference should be made to this and earlier papers for other Surrey records. More recently A. J. E. Smith has divided British *M. pratense* into subspecies related to soils (*Watsonia*, **5**, 336–67, 1963): subsp. *pratense* of acid soils in woodlands, hedgerows and bogs. This is the common Surrey plant and includes the two 'subspecies' mentioned above, and subsp. *commutatum* (Tausch ex Kerner) C. E. Britton, of calcareous or base-rich habitats, with broader leaves which Britton cites from 15C1(?), Ranmore. This seems to provide a sound basis for future work since basic soils will produce different woody species as hosts and it is probable that the Cowwheat 'splits' are related to the plants on which they grow.

Euphrasia L.
E. *officinalis* L. *sensu lato*.   Eyebright.   Native.
1836.   Common.
Grassy places—meadows, downs, heaths, pastures and open woods. The Eyebrights of Surrey have received a great deal of attention and especially from H. W. Pugsley whose revision (*J. Linn. Soc. (Bot.*), **48**, 467–544, 1930) appeared after the genus had been printed for Salmon's *Flora*, 1931. Many Surrey specimens were named for E. C. Wallace and myself by Pugsley, and lately for D. P. Young and others by P. F. Yeo. Considerable effort has been made to bring the account up to date, but we have found that Eyebrights in Surrey have become much less common than they were 40 years ago. *E. nemorosa* and *E. pseudokerneri* are still quite common, but we have been unable to refind *E. anglica* in a number of places where we knew it earlier, and *E. micrantha* appears to be extinct. The following are the segregates now recognised:

[*E. micrantha* Reichb. (*E. gracilis* (Fr.) Drej.)   Native.   1897.
Heaths on peaty sandy or gravelly soils, usually not under especially wet conditions. Extinct. Salmon, 1931, gave 31 localities, many of which are supported by herbarium specimens, and most of the habitats are still suitable, such as the commons in the W, yet members of the committee familiar with the species in Scotland and also Dr Yeo have repeatedly

failed to refind it. The nearest is a specimen from 06A1, Horsell Common, 1966, WEW, determined by Dr Yeo as *E. micrantha x nemorosa?*.]

[*E. curta* (Fr.) Wettst.
The two records given in Salmon, 1931, are unlikely to be this species.]

E. nemorosa (Pers.) Wallr.   Native.
By far the most common segregate occurring in a wide range of habitats. Var. *calcarea* Pugsley was described from Box Hill, and var. *transiens* Pugsley was based on a type from Chobham Common, but these varieties are no longer regarded as of value.
x E. pseudokerneri is common where the parents are abundant on the chalk. Pugsley described it from 15D1, Box Hill, Pugsley, 1930, 537; and named it for us from 25D4, Chipstead railway bank, ECW; 26C1, Banstead Downs, ECW, 1952, PFY; 35D2, South Hawke, JEL; 35D4, Warlingham, below Nore Hill, JEL; and P. F. Yeo named it from 25C1, Reigate Hill, ECW. The records are before the present survey and grid references only approximate.

E. pseudokerneri Pugsl. (*E. kerneri* auct., *E. stricta* auct.)
Native.   1896.   Locally abundant, 13.
Chalk grassland on shallow soils. It may still be at 94B5, Hog's Back above Puttenham, 1902, CES, Hb. Mus. Brit.; 94C5, Hog's Back, Guildford, 1941, N. D. Simpson; and 05E1, Horsley Sheepleas, 1943, JEL, but there are no later records from the W of the county.

[*E. brevipila* Burnat & Gremli. Given in Salmon, 1931, but regarded as an error.]

E. anglica Pugsl. (*E. rostkoviana* of Salmon.)   Native.
1897, Townsend, *J. Bot. (Lond).*, **35**, 466.
Rather rare, 40.   Map 283.
Commons, heaths and golf courses, usually on rather acid soils, on Clay-with-Flints, but rarely in chalk grassland.
x E. pseudokerneri—15D2/15E2, Box Hill, 1952, PFY.

For an account of some Surrey Euphrasias see P. F. Yeo (*Proc. BSBI*, **2**, 421–3, 1957).

Odontites Ludw.
  O. verna (Bellardi) Dumort. (*Bartsia odontites* (L.) Huds.)
  Red Bartsia.   Native.   1805.   Common.
  Pastures, roadsides, woodland rides, and waste ground. The common Surrey plant is subsp. *serotina* Corb. Map 284. Subsp. *verna* occurs rarely.

Parentucellia Viv.
  P. viscosa (L.) Caruel (*Bartsia viscosa* L.)   Yellow Bartsia.
  Alien.   1931, Salmon.
  94B1, Rodborough (Witley) Common, nr Borough Farm, colony of nearly 100 plants on disturbed open ground near site of military camp, 1958, GSE, 1958, c300 plants, JEL, Hb. Lousley, about 30 plants, 1965, DCK. The colony was gradually reduced and has not been seen recently. The appearances of this species on the eastern side of England are difficult to explain; it is a Mediterranean plant which thrives on the western side of Britain but appears to be near the limit of its range.

OROBANCHACEAE

Lathraea L.
  L. squamaria L.   Toothwort.   Native.   1597.
  Locally frequent, 45.   Map 285.
  Parasitic on hazel and elm, reported also on ash and Common Laurel. Toothwort is very closly associated with the chalk although the hosts occur freely elsewhere. It is most common on the chalk in the E (an extension of the area just across the boundary into Kent where it is common) and then extends at close intervals along the chalk outcrop almost to Newlands Corner, to reappear in 84A4 on chalk nr the Hampshire border. Other isolated localities are in 94E1, where it may be on Bargate Stone, and 13E4 on Weald Clay.

  L. clandestina L.   Purple Toothwort.   Alien—from gardens.
  83D2, side of Hindhead road N of Shottermill. Introduced by former owner of Frensham Hall, DWF, 1965, EM-R & EJC.

Orobanche L.
  O. rapum-genistae Thuill. (*O. major* auct.)
  Greater Broomrape.   Native.   1562.   Very rare, 1.
  Parasitic on the roots of Broom. 84B3, Farnham, 8 plants in clump amongst *Sarothamnus scoparius* on sandy verge of by-pass, 1945,

WEW, 1946, WEW, not seen since; 03A3?, Alfold, Sidney Wood, twice in recent years, REG, comm. ECW, 1973. A decreasing species nearly extinct. Salmon, 1931, gave about 30 reliable records extending over nearly 370 years, but it has always been an erratic plant with appearances coinciding with an abundance of Broom.

O. elatior Sutton (*O. major* auct.)   Knapweed Broomrape.
Native.   1746.   Rare, 11.   Map 286.
Parasitic on Greater Knapweed, *Centaurea scabiosa*. Like most Broomrapes this species varies widely in numbers from year to year even in the regular localities. It is restricted to the chalk and, although the localities are not always the same, it is probably no rarer than it was 40 years ago.

O. minor Sm.   Common Broomrape.   Native.   1832.
Locally frequent, 56.   Map 287.
Parasitic on Leguminosae and many other plants. Mainly on the chalk but also on other soils. Increasingly frequent in gardens where *Senecio greyi* has been the host in 27B2 and 16E1, and Mr A. J. Bellett of Little Bookham achieved something of a record in his garden, 15B2, where the parasite was found on 'petunias, pelagoniums, garden carrots, strawberries and caraway' and was particularly vigorous on the caraway (*Proc. R. Hort. Soc.*, **97**, 53, 1972.)

O. hederae Duby   Ivy Broomrape.   Native.   1899.
Very rare.   Parasitic on Ivy.   .
17E4, Kew Gardens, long known from various places in the Gardens on ivy, as, for example, nr the Order Beds, 1933–53, DHK. In 1948 two plants were found on the towpath outside the Gardens, WHS.

[*O. picridis* F. W. Schultz ex Koch. Salmon, 1931, gives three records. These are probably all based on large forms of *O. minor*.]

LENTIBULARIACEAE

Pinguicula L.
[*P. vulgaris* L.   Common Butterwort.
Doubtful and in any case now extinct. 14C2, 'On ye common below Cold-harbour Church in 1859, disappeared since', MS note probably by J. L. Jardine, in a copy of Brewer's *Flora*, 1863, in possession of R. Palmer;

25B1, Reigate Heath, reported recently planted, 1950, did not persist, BMCM.]

Utricularia L.
U. vulgaris L.   Greater Bladderwort.   Native.   1837.
Exceedingly rare.
07A1–2, Langham Pond, 1953, AWW; 25A5, Epsom, Baron's Pond, 1973, BRR, in smaller quantity, 1974, JEL.

U. neglecta Lehm. (*U. major* auct.)   Native.   1879.   Very rare.
85E3, Mytchett Lake (as *U. vulgaris*), 1947, RAB, 1965, JES & RAB.
This species can only be distinguished from the last when in flower.

U. minor L.   Lesser Bladderwort.   Native.   1850.   Rare, 4.
Peaty pools in acid bogs. 94A1, Thursley Common, known here for over a century and seen regularly by many observers, by Pudmore, 1972, RF & JEL; 95B2, Pirbright, nr Henley Park Lake, 1965, DCK; 95B3, Pirbright Common, bog N of Chair Hill, 1956, RAB; 06B1, Sheerwater Bog, 1949, BSBI exc., since destroyed by drainage and building.

VERBENACEAE

Verbena L.
V. officinalis L.   Vervein.   Native.   1793.
Locally frequent, 64.   Map 288.
Chalk downs, quarries, waste places and roadsides, in dry places on chalk or gravel soils. Most common on the chalk.

LABIATAE

Mentha L.
M. requienii Benth.   Corsican Mint.   Alien.
This delightful little garden plant has been found: 14B2, nr Leith Hill, on cindery track, c1953, ECW, path verge grassed over, 2 small plants, 1963, ECW & JES. Thoroughly established in woodland rides in Kent, this may yet be found in quantity in the Leith Hill area.

M. pulegium L.   Pennyroyal.   Native.   1746.
Very rare, 3, and perhaps extinct.
Margins of small ponds and wet commons. 05D5, in car park, introduced from RHS Gardens, Wisley, 1974, JFL; 25B4, Burgh

Heath, by pond, now in small quantity and much trampled, 1958, JEL, ECW, DPY, 1962, DPY, 1966, BWu failed to refind; 26E4, Mitcham Common, wet hollow, 1957, RCW, JEL, 1960, still fairly plentiful, JEL; 34E2, Lingfield, pond at 'The Garth', 1957, BAK, 1959, DPY, soon after destroyed by tipping, DPY. Burgh Heath was an old locality, where Mr Wallace had known it since 1928, and I first saw it on Mitcham Common about 1930; we both knew it on several other Surrey commons. Salmon, 1931, gave 50 localities where the plants had been seen. We have been able to refind it in only three since 1950—an indication of the great change that has taken place in commons and their ponds, often associated with the reduction in geese and ducks.

M. arvensis L.   Corn Mint.   Native.   1838.
Common, 177.   Map 289.
Cultivated fields, rather bare ground in field gateways, rides in woods, waste ground and by streams. Variable, but most of the forms are related to the nature of the habitat.

M. x verticillata L. (*M. sativa* L., *M. aquatica x arvensis*).
Native.   Frequent.
By ditches, ponds, lakes and on roadsides.

M. x gentilis L. (*M. arvensis x spicata*), including *M. cardiaca* (Gray) Bak.   Alien, and established garden mint. 1863. Rare, 7.
Ponds and streams and in marshes. 04A4, Shalford, village green pond, 1965, EJC, 1970, GWJ, det. RMH; 14B3, Wotton, Broadmoor (with *cardiaca*), BSBI exc., det. RAG; 16B2, Esher, West End Common, two patches, 1962, RAB, JES (known here earlier); 17E2, Petersham, NW of church, 1955, LJT; 26E3, Hackbridge, by Wandle, 1957, RCW, det. RAG; 27C3, weedy ground nr Wandle mouth, 1965, RARC, det. DPY; 36B2, Sanderstead, on building plot, 1951, DPY (*BSBI Year Book*, **1952,** 108).

M. smithiana R. A. Graham (*M. rubra* Sm. = *M. aquatica x arvensis x spicata*).   Alien, an established garden mint.
1724.   Rare, 7.
Roadsides, ponds, ditches and by streams. 04D4, Albury, Brook, on roadside, 1950, BSBI exc., det. RAG; 14D3, Holmwood, ditch near Folly Farm, 1951, LJJ; 14D4, N. Holmwood, by pond nr church, 1951-7, LJJ; 15D5, nr Ashtead Station, by stream, 1958,

JEL; 26E5, Mitcham Common, on refuse tip, 1956, DPY, conf. RAG; 35A1, Nutfield Marsh, by the stream, 1947–58, LJJ; 35C4, Bughill Farm, 1963, RARC, det. RMH; 36B1, Riddlesdown, disturbed ground nr houses, 1953–9, LJJ. The Nutfield Marsh locality has been known since Luxford, 1838.

M. aquatica L.   Water Mint.   Native.   1763.
Very common throughout the county.
Rivers, streams, ditches and pond sides and marshes.

M. x piperita L. (*M. aquatica x spicata*).   Peppermint.
Alien, an established cultivated mint.   1756.   Rare.
By ponds and streams. 14B3, S of Friday St, 1965, RARC, and 14B4, N of Friday St, on SE side of lake, RARC—both det. DPY. Salmon, 1931, gives records from Friday St and Wotton. The northomorph *citrata* (Ehrh.) Boivin formerly grew in a marshy place at 93D3, Chiddingfold, Vann's Lane (*Rep. BEC*, **9**, 521, 1931), and was seen there at intervals by JEL until the habitat became too overgrown. Our only reports of Bergamot Mint since 1950 are from 26B2, Nonsuch Park, 1963, DCK, det. RMH, and 26D4, Mitcham, bank of Wandle, 1956, RCW, det. RAG.

M. suaveolens Ehrh. (*M. rotundifolia* auct.).
Round-leaved Mint.   Escaped garden mint.   1805.   Very rare.
15D5, small colony of about 6 stems on waste ground between road and footpath on edge of Ashtead Common, 1965, JES, det. DPY.

M. x villosa Huds. (*M. spicata x suaveolens*) Escaped garden mint. 94C3, Hurtmore Quarry, 1970, GWJ, det. RMH; 04B3, Morley Common, 1970, GWJ, det. RMH. The northomorph *alopecuroides* (Hull)=*M. alopecuroides* Hull is much more frequent, and the distribution, including the records above, is shown in Map 290 as 36 tetrads. It is well established on commons, by roadsides and other places where an excellent tall garden mint is thrown out as excessively rampant in cultivation.

M. spicata L. (*M. viridis* L.)   Spearmint.
Established garden mint.   1756.   Frequent.
The familiar garden Spearmint, with bright green, glabrous and sharply serrate leaves, is frequent in waste places where it is thrown

out from gardens, and occasionally established in fields. The hairy plant, recorded by our observers as *M. longifolia*, is mainly the hairy form of *M. spicata*, but includes also some gatherings of *M. villosa* which have not been distinguished.

M. villosonervata auct. (*M. longifolia x spicata*).   Very rare.
94E3, Catteshall, Unstead Park Farm, 1970, GWJ, det. RMH.

The mints of Surrey have been studied by several specialists and especially by J. Fraser, A. L. Still and R. A. Graham. In recent years R. M. Harley has been revising the genus, using modern techniques. Unfortunately much of our work was based on the old system and it is not possible to fit all our records into the new arrangement. *Mentha arvensis, M. x verticillata*, and *M. aquatica* are native, and *M. pulegium* is probably so. But the remainder depended for their distribution on the exchange of roots between housewives and sale by nurseries, to provide a constant supply of surplus roots to be thrown out on commons and roadsides. Since 1930 there has been a considerable reduction in well-established ex-garden mints in wild places, and an increase in those noted on refuse tips. This is a reflection of the substitution of organised refuse collection for the former practice of dumping household and garden rubbish on the commons.

Lycopus L.
L. europaeus L.   Gipsywort.   Native.   1836.   Common.
Riverbanks, by streams, ditches and ponds, and in marshes. The seeds are apparently water-carried as it soon appears on any bit of embankment or staging by the Thames in central London when sufficient soil has collected.

Origanum L.
O. vulgare L.   Marjoram.   Native.   1814.
Locally common, 137.   Map 291.
Chalk grassland, chalkpits, hedgebanks and scrub, railway embankments, bridges, walls and other dry sunny places. Very common throughout the area of the chalk, but also many scattered localities, mainly on Lower Greensand. Away from the chalk, bridges, walls and railway embankments are favoured habitats, no doubt because they are sunny and warm and provide calcium carbonate from mortar or the materials used to make up the embankments.

Thymus L.

T. pulegioides L. (*T. ovatus* Mill, *T. glaber* Mill, *T. chamaedrys* Fries). Large Thyme. Native. 1856.
Locally common, 99. Map 292.
Chalk and other dry grassland on downs, commons and pastures, especially on molehills. The commonest thyme of chalk grassland, and also common in the Bargate Beds area of the Lower Greensand, scattered localities elsewhere.

T. drucei Ronn. (*T. serpyllum* auct., *T. pycnotrichus* auct., *T. neglectus* Ronn., *T. britannicus* Ronn.) Wild Thyme.
Native. 1718. Local and rather rare, 60. Map 293.
Dry sunny banks, usually in rather bare places, mainly on the chalk. This species is much less common than the last in Surrey, and especially in the W.

Calamintha Mill.

C. ascendens Jord. (*C officinalis* auct.) Common Calamint.
Native. 1835. Rare, 14. Map 294.
Dry hedgebanks, roadsides, walls and a railway embankment. 93A1, Lythe Hill, in fallow, 1963, EMH, since planted with conifers; 94D1, OVP; 03A5, Hascombe, by wall, 1955, JCG; 04A3, nr Snowdenham House, roadside, 1957, JCG; 05A1, Guildford, by London Road Station, 1957, EN; 05B5, C5, roadsides and church-yard about Pyrford Church, Salmon, 1931, 1948, WEW, BW, 1966, DPY, etc.; 15B1, B2, C1, C2, D1, D2, numerous old records and still about Headley Lane, Ranmore, Polesden Lacy, etc; 26C2, Sutton, chalk railway cutting, known to ECW for many years.

[*C. nepeta* (L.) Savi. Extinct. Salmon, 1931, gave several records from round Mickleham but the only reports of this species received by us have turned out to be errors for *C. ascendens*.]

Acinos Mill.

A. arvensis (Lam.) Dandy (*Calamintha acinos* (L.) Clairv., *C. arvensis* Lam.). Basil-thyme. Native. 1835.
Locally common, 56. Map 295.
Open habitats in chalk grassland, chalkpits, arable fields and roadsides. Almost completely restricted to the chalk—the records from 94B1, Witley Common, and 16D2, Chessington, derelict railway embankment, were from places where chalk had been imported.

Clinopodium L.
C. vulgare L. (*C. clinopodium* Benth.)   Wild Basil.   Native.
1836.   Common, 155.   Map 296.
Hedges, wood borders and scrub, occasionally in chalk grassland.
Common throughout the chalk range, and locally frequent on
Lower Greensand, a few records elsewhere.

Melissa L.
M. officinalis L.   Balm.
Alien, sometimes established for a time from bee-keeping activities.
1763.   Probably rarer than formerly, we have the following
records:
94D1, Tuesley, 1955, OVP; 15B4, Bookham Common, E side, AWJ
(*London Nat.*, **33**, 41, 1954); 17D3, Richmond, towpath near Old
Deer Park, 1959, BW; 26D5, Bunce's Meadow, 1957, RCW. For
further records see Kent & Lousley (*Handbook*, 222).

Salvia L.
S. verticillata L.   Whorled Clary.   Established alien.
1885.   Rare, 5.
94D3, Godalming, railway goods yard, 1932, ECW, at intervals
JEL and others, 1967, EJC; 35A2, Bletchingley, 100 yards W of
Rockshaw, 1950, 1954, since destroyed by roadwork, LJJ (this is
near the locality 'Side of road from Merstham towards Rocks
Shaw', where Dr F. Bossey found it in 1885 (see Salmon, 1931, 522);
35B2, Caterham, Gravelly Hill near Red House, 1951, not seen
since, LJJ; 35D4, Nore Hill, old chalkpit, 1921, 1922, 1937, 1942,
JEL, 1950, RARC, but gone by 1957; 37C5, Surrey Commercial
Docks, nr Redriff Road, 1957, LMPS.

S. pratensis L.   Meadow Clary.
Native on the downs, alien elsewhere.   1787.   Very rare.
Chalk grassland. 04B4, Blackheath, on grassy bank, planted, 1969,
IBA; 14C5 (?), below Ranmore nr Denbies, edge of cornfield, 1950,
MBG, Hb. Mus. Brit; 25A3, Headley Park, in meadow, 1943, JEL,
destroyed by war-time agriculture; Smith-Pearse, 1917, knew it as
well established at Headley; 25C2, Colley Hill, several places, 1951,
RWD, 1961, DP, 1962, BMCM, etc., 2 plants in one place and
1 in another, 1970, DPY, perhaps also in C1. It has been on the
downs near Reigate since at least 1902, C. E. Salmon (*J. Bot.*, **40**,

411), but does not flower regularly and the flowers are often picked or slashed when they appear. 34C4, South Godstone, Hookstile, planted, 1960, RARC.

S. horminoides Pourr. (*S. verbenaca* auct.)   Wild Clary.
Native.   1597.   Rare, 8.
Our few records show an interesting history of persistence, so the earlier records are quoted. Dry banks and pastures, towpaths by the Thames. 94C3, Hurtmore, A. W. Bennett (1833–1902), c1955, OVP; 94E4, OVP; 94E5, bottom of St Catherine's Hill, Milne & Gordon (*Indigenous Botany*, 1792–3); 1950, LJJ; 07A1, Egham, by the road to Staines, J. S. Mill in Brewer, 1863; Egham by-pass, 1948, AWW; 15B3, lane by School, and roadside verge nr Grove-side, 1963, both colonies known for over 30 years, AME, by Bookham railway station, 1952, AWJ; 17D1, towpath nr Tedding-ton Lock, Salmon, 1931, 1946, LJJ; 25C1, Reigate Castle grounds, Luxford, 1838, 1952, LJJ; 26B2, Ewell, abundant, Beeby in Salmon, 1931, footpath by Ewell East Station, 1952, LJJ. A decreasing species.

Prunella L.
P. vulgaris L.   Selfheal.   Native.   1784.   Common.
Grasslands, heaths, commons, woodland rides and waste places.

P. laciniata L.   Cut-leaved Selfheal.   Established alien.
1906.   Very rare, 1.
Chalk grassland. The only present locality is 25A2, Pebblecombe valley, 1954–7, BMCM, nearly smothered by *Bromus erectus* in 1958, but 2 plants still there in 1962, BMCM, 1 plant with three hybrids, 1965, BWu, still there 1972, JEL. The following records refer to stations known to Salmon, 1931: 05D1, above Clandon, 1906, CEB (*J. Bot.*, **44**, 428); chalk grassland nr Fuller's Farm, 1944, JEL, ploughed soon after and plant destroyed; 06B1, golf links near Pyrford, 1912, Lady Davy (*J. Bot.*, **50**, 287–8); West Byfleet Golf Course, 1932, 1937, JEL, not seen since; 36, Addington, A. Beadell, Salmon, 1931, Addington, 1936, ALS, Hb. Lousley.
x P. vulgaris = *P. x hybrida* Knaf (*P. x intermedia* Link)
06B1, W. Byfleet Golf Course, 1932, JEL; 25A2, Pebblecombe valley, 1954, 1964, BMCM, 1965, BWu, 1972, JEL, now more numerous than *P. laciniata*.

Betonica L.
B. officinalis L. (*Stachys officinalis* Trev., *S. betonica* Benth.)
Betony.   Native.   1763.   Common.
Open woods, copses, heaths, grasslands and field borders.

Stachys L.
S. annua (L.) L.   Alien, perhaps only casual.
25E3, Merstham, clover field near Greystone Limeworks, 1943,
JEL, Herb. Lousley; 34B3, nr Outwood Mill, 1960, JAC, conf.
DPY.

S. arvensis (L.) L.   Field Woundwort.   Native.   1782.
Locally frequent, 67.   Map 297.
Weed in cultivated fields especially on sandy soils.

S. palustris L.   Marsh Woundwort.   Native.   1597.
Common, 203.   Map 298.
By rivers, streams, ditches and wet fields, also in cultivated fields in
surprisingly dry places. Well distributed through most of the county
but most common on the Weald Clay.
x *S. sylvatica* = *S. x ambigua* Sm.   Rare.
15B4, Bookham Common, at edge of wood (*Lond. Nat.*, **33**, 41,
1954); 25C5, Banstead, weed in allotments, 1962, DPY; 26D1,
Banstead, footpath from Mellow Close to Mental Hospital, 1957,
EM.

S. sylvatica L.   Hedge Woundwort.   Native.   1836.   Common.
Wood borders, hedgebanks, roadsides, and waste places. Equally
common in shady passages and gardens in towns.

Ballota L.
B. nigra L.   Black Horehound.   Native.
1827.   Common.
Roadsides, waste ground and hedgebanks, often round houses
in villages.

Lamiastrum Heister ex Fabr.
L. galeobdolon (L.) Ehrend. & Polatschek (*Lamium galeobdolon*
(L.) L., *Galeobdolon luteum* Huds.)   Yellow Archangel.
Native.   1830.   Locally common, 211.   Map 299.
Woods, hedges, shady lanes. Common on chalk and the heavy soils
of the Weald Clay and part of the Lower Greensand, and on

Bagshot Beds about Englefield Green, absent from most of the remainder of the county.

The subspecies in Surrey is said to be subsp. *montanum* (Pers.) Ehrend. & Polatschek, the diploid, which has a more southern distribution than subsp. *galeobdolon* which is known only from N Lincolnshire, Wegmüller (*Watsonia*, **8**, 277–8, 1971).

Lamium L.

L. amplexicaule L.   Henbit.   Native.   1777.   Common.
Cultivated ground and gardens, roadsides, mainly on light soils.

L. moluccellifolium Fr. 05D4, Ockham, 1 plant in potato field on S side of Portsmouth Road, 1968, JFL, named at BM (Nat. Hist.). Casual.

L. hybridum Vill.   Cut-leaved Deadnettle.   Native.
1827.   Rare, 50.   Map 300.
Cultivated fields, gardens, nurseries and waste ground. Most frequent on sandy soils but also on chalk and elsewhere.

L. purpureum L.   Red Deadnettle.   Native.   1827.   Abundant.
Cultivated fields, gardens, and waste places.

L. album L.   White Deadnettle.   Denizen.
1836.   Very common.
Waste ground, hedgebanks and roadsides near houses.

L. maculatum L.   Spotted Deadnettle.   Garden alien.   Rare.
Commons, roadsides, waste places, usually near houses, seldom persistent for long. 94C3, Hurtmore Bottom, 1948, OVP; 24D5, Earlswood, two places, 1951, BMCM; 25A5, Epsom Downs, 1944, JEL; 27B1, Wimbledon Common, several places, 1952, AWJ; 35, Caterham, 1957, RARC; 35D5, Chelsham, Mill Common, 1944, JEL. More frequent than these records suggest.

Leonurus L.

L. cardiaca L.   Motherwort.   Garden alien.   1795.
17D1, Ham pits, 1954, RAB, BW; 17D4, towpath outside Kew Gardens, 1950, BW (by coincidence recorded from here in Brewer, 1863), gone by 1966, DPY; 36A3, Beddington, grass verge of bridle path nr Salcott Road, 1949, 1950, LJJ.

Galeopsis L.

G. angustifolia Ehrh. ex Hoffm. (*G. ladanum* auct.)
Red Hempnettle.   Native.   1836.   Rare, 29.   Map 301.
Weed in cornfields and other cultivated fields. Confined to the
chalk. Has decreased rapidly since 1950 and now exceedingly rare.

G. tetrahit L. *sensu lato*.   Common Hempnettle.   Native.
1640.   Common.
Open woodland, hedgebanks, and cultivated ground. Now treated
as two species as follows:
    G. tetrahit L. *sensu stricto*   The commoner of the two.
    G. bifida Boenn.   Recorded from the following tetrads:
    95A1, 04C1, 04E5, 06C3, 14D2, 16C1, 26E4, and 34A5.
Under-recorded. These are not always easily distinguished.

G. speciosa Mill.   Large-flowered Hempnettle.   Colonist.
1878.   Very rare.
Weed in arable fields, disturbed ground. 14E2, Newdigate, kale
field at Kingsland, 1960, BMCM, BW; 15C4, Fetcham, field by
footpath between Canon's Court and River Lane, 1948, LJJ;
25A5, Epsom Downs, by Langley Bottom Road, 1968 in abundance,
less in 1969 & 1970, none in 1971, BRR; 25C4, field nr Kingswood
Station, 1950, JEL, CTP; 35D3, Marden Park in potato field,
1964, RARC.

Nepeta L.

N. cataria L.   Catmint.   Native.   1763.   Very rare, 7.
Lanes and field borders, on chalk. 94C5, OVP; 94D5, Compton,
nr Hog's Back, 1 plant on edge of cornfield, 1964, GSE; 04A5,
Pewley Down, 1957, JCG; 04C5, between Newlands Corner and
Silent Pool, 1957, JCG; 16E3, Surbiton, on dumped rubble, 1964,
EJC; 25A4, Walton Downs, 1958, LJJ; 26C1, Banstead Downs, nr
the Golf House, 1957, EM; 35C4, Warlingham, Tippet's Piece,
1960, RARC; 36D2, Addington, 1946, now under new road,
CTP. A decreasing species.

Glechoma L.

G. hederacea L. (*Nepeta hederacea* Trev., *N. glechoma* Benth.).
Native.   1836.   Common throughout the county.
Hedgebanks, woods and fields especially on damp and heavy soils.

Marrubium L.

M. vulgare L.   White Horehound.   Native.   1718.   Very rare, 2.
04C5, between Newlands Corner and Silent Pool, 1 plant, 1952,
PDO; 16C3, Ditton Common, nr cottages, 1 plant, 1956, 1958,
JES. Accepted as a native because one record is in a place where
introduction is not very likely, and the other from an area with
very old records. The decrease in this species in the last 50 years is
remarkable and the reason not known. It is apparently extinct.

Scutellaria L.

S. galericulata L.   Skullcap.   Native.   1827.
Frequent, 106.   Map 302.
By rivers, streams, lakes and ponds. Very rarely as a garden weed,
as at 06A4, nr Lyne, 1965, EJC. Skullcap is common along the
Thames, Wey, Basingstoke Canal and Mole, especially by cal-
careous waters.

x S. minor = S. x hybrida Strail (*S. nicholsoni* Taub.)
This hybrid was first found in Surrey (and Berkshire) at Virginia
Water in 1883 by G. Nicholson (*Rep. BEC*, **1**, 93, 1885) and was
described by Taubert as *S. nicholsoni*. In 1931 it was abundant,
near Wick Pond, Virginia Water, 96E5, Lousley (*Rep. BEC*, **9**,
838, 1932). Fresh material was brought by Mrs L. M. P. Small
about 1967, but on a visit in 1972 I failed to refind it. It is likely to be
still at Virginia Water and also at Holmwood Common and woods
N of Newdigate where it was found by J. Fraser (Salmon, 1931, 525).
In Kent and Sussex this hybrid occurs in abundance in rides in
woods on clay, where it increases vegetatively, and it should be
searched for in woods in the S of Surrey.

S. minor Huds.   Lesser Skullcap.   Native.   1666.
Frequent, 92.   Map 303.
Ditches, marshes and bogs on commons and heaths, in woods and
by ponds, on acid soils.

Teucrium L.

[*T. chamaedrys* L.   Wall Germander.   Planted alien.
15D2, planted from Camber Castle, Sussex, on a wall at Juniper Hall,
1965, JPHS.]

T. botrys L.   Cut-leaved Germander.   Native.
1844.   Very rare, 3.
Open broken ground on chalk, chalky fallow fields. (Plate 21).

[04B3, Wonersh, Norley Common, 1907, CES, Herb. Mus. Brit. The specimen is correctly named, and Salmon is known to have been on Norley Common on that date, but has been re-labelled by R. S. Standen, and the locality mislabelled, ECW.] 15D1, E1, E2, Box Hill—the plant has appeared in several parts of the hill in addition to the original locality 'In a wild stony locality . . . at the back of Box-hill', and the habitats change when the broken chalky ground becomes overgrown or shaded. 25D4, Chipstead Valley— found by C. E. Britton in 1909 in an arable field where it persisted until about 1940. The present locality is the adjacent field, which was ploughed in both World Wars and, after a couple of poor crops, allowed to revert, and the plant appeared in abundance. It was in great quantity up to 1950, by 1960 was confined to the E end of the field. After further decline it reappeared after strips of the field had been ploughed as a conservation measure in 1966, and was down to 43 plants in 1970, and very few in 1972. Plentiful in 1974 after ploughing. Its persistence here depends on conservation work by the owners on the advice of Surrey Naturalists' Trust. 36D2, Addington, Long Bottom—apparently known here as early as 1853, and the locality variously described as near Sanderstead and near Selsdon. Our last record is 1954, RARC, and it could not be found in 1966 when the field was to be built over.

T. scorodonia L.   Wood Sage.   Native.
1763.   Very common.
Open woods, commons and heaths, avoids exposed chalk.

Ajuga L.
A. chamaepitys (L.) Schreb.   Ground Pine.   Native.
1746.   Very rare, 16.   Map 304.
(Plate 22). Open chalky habitats on banks, by chalkpits, round the edges of arable fields and tracks. In such places it has appeared where rabbits have been active, where chalk has been dug for pipe-laying, and in gardens on the chalk. Always dependent on temporary habitats, this species has decreased rapidly in the last 20 years. In 1973 it appeared in abundance in part of a field in the Chipstead Valley which was ploughed during the previous January.

A. reptans L.   Bugle.   Native.   1836.   Common.
Woods, copses, roadsides, meadows and pastures, especially when damp.

PLANTAGINACEAE

Plantago L.

P. major L.   Greater Plantain.   Native.   1836.   Very common.
Roadsides, waste ground, lawns and pastures. A plant of rather
open habitats, but frequent in grassland where competition is
reduced by grazing or mowing. In lawns it thrives as an ecad with
small, prostrate leaves, and short peduncles which remain just out
of reach of the blades of the lawn mower.
A closely allied taxon, *P. intermedia* Gilib., the rank of which is
uncertain, occurs round the margins of ponds in Surrey. For a
discussion of this see Lousley (*Proc. BSBI*, **3,** 33–6, 1958).

P. media L.   Hoary Plantain.   Native.   1793.
Locally common, 120.   Map 305.
Abundant and characteristic of chalk grassland, pastures, roadsides
and lawns. Mainly on the chalk and by the Thames, where calcareous material has been brought down. Most of the other
occurrences are on roadsides, where chalk may have been used to
build up the foundations of the road, or in lawns, where turf may
have been brought from chalky places.

P. lanceolata L.   Ribwort Plantain.   Native.
1836.   Very common.
Grassland, especially on the chalk, pastures, roadsides, waste
ground, and a weed in cultivated ground.

P. maritima L.   Sea Plantain.   Introduced.
36B4, Norwood, Davidson Road, several large clumps with
P. coronopus in derelict turfing contractor's yard, evidently introduced with seaside turf, 1962, DPY.

P. coronopus L.   Buckshorn Plantain.   Native.   1548.
Locally common, 158.   Map 306.
Dry rather open places on sand and gravel on heaths, commons
and pastures. Most common on the Lower Greensand and Bagshot
Beds.

P. indica L. (*P. arenaria* Waldst. & Kit., *P. ramosa* Aschers.)
An occasional alien.
16C2, Claygate, garden weed in Telegraph Lane, 1961, JES; 27C2,
Earlsfield, tip nr railway, 1958, RCW. This has become naturalised
in other counties.

Littorella Berg.
L. uniflora (L.) Aschers.   Shoreweed.   Native.   1814.
Rare, 11.   Map 307.
Canal, lakes and ponds. 84C1, Frensham Little Pond; 85E2, E3, E4, Basingstoke Canal; 86D1, Camberley, Lower Lake, abundant, 1952, RAB, rare, 1958, WEW; 95A4, B4, Basingstoke Canal; 96B1, Lightwater, lake, 1965, WEW; 05D5, Boldermere, recorded most years, abundant, 1972, JEL, JES; 25B1, Reigate Heath, 1956, RAB; 34C1, Hedgecourt Pond, still locally abundant, 1960, RARC.

CAMPANULACEAE

Wahlenbergia Schrad.
(*W. hederacea* (L.) Reichb. (*Campanula hederacea* L.)
Ivy-leaved Bellflower.   Native.   1836.   Very rare, 1.
93A3, Hurthill Copse, 1957, FR, 1958, ALJ, 1966, JES, ECW.
A decreasing species, but it was already very rare 40 years ago.

Campanula L.
C. latifolia L.   Great Bellflower.   Established garden alien.
1838.
Parks and hedges. 25C1, Reigate, Priory Grounds, almost certainly planted, 1960, BMCM, DPY; 34E1, Dormans, hedge, no evidence as to status, 1963, RARC.

C. trachelium L.   Nettle-leaved Bellflower.   Native.
1793.   Locally common, 115.   Map 308.
Hedgebanks, open woods and copses. Mainly on the chalk, more local on the Lower Greensand, and a few scattered localities elsewhere but avoids the clays.

C. rapunculoides L.   Creeping Bellflower.
Established garden alien.   1858.   Rather rare, 32.
Parks, hedgebanks, roadsides, commons and cornfields.

C. persicifolia L.   Peach-leaved Bellflower.
Established garden alien.   1931 (Salmon, *Flora*).   Rare, 6.
Woods, roadsides, commons, field border, downland scrub. 84E3, Crooksbury Hill, well established on roadside in pine woods, 1961, VML; 93B5, Witley Park, in shrubbery, 1959, SFC; 94A3, Putten-

ham Common, in bracken far from houses, 1967, JEL, WEW; 13A5, Ewhurst, Buildings Wood, naturalised in dense woodland with *Lysimachia ciliata*, 1965, RARC; 35C2, Gravelly Hill, 'normal race which is native in Europe', plentiful in 1954–5 after myxomatosis but decreased gradually as scrub closed over area, CDP; 36E1, Addington, Featherbed Lane, shady field border, 1952–9, HB.

C. glomerata L.   Clustered Bellflower.   Native.   1763.
Locally common, 45.   Map 309.
Chalk grassland, also in alluvial meadows by the Thames.

C. rotundifolia L.   Harebell.   Native.   1696.
Common, 222.   Map 310.
Dry grassy places on heaths, commons, pastures and banks. Very widely distributed but almost absent from clay soils.

C. patula L.   Spreading Bellflower.   Native?
1792.   Very rare, 3.
Gravel pits, fields, railway embankments and on a common. 84A3, Wrecclesham, gravel pits and railway embankment, known in district since 1850, and recently seen annually, 1960, DPY, 1961, VML, etc.; 84C1, Frensham, Lane End, in field, 1958, VML, 1961, DPY; 93A2, Almshouse Common, bank nr garden, 1966, EMH; 95B4, Pirbright, on railway embankment, c1930, Lady Davy, 1933, JEL, about 6 plants, 1965, WEW; the embankment is mown and it is difficult to see how many plants appear.

C. rapunculus L.   Rampion Bellflower.
Established garden alien.   1762.
25A1, railway bank nr Betchworth Station, 1950, BMCM, 1965, RARC, 1971, JEL. This was collected from here by Arthur Bennett in 1873, C. E. Salmon in 1890, E. S. Salmon in 1896, and there is a specimen labelled 'Betchworth', 1866, collected by H. E. Fox (Hb. Mus. Brit.). It grew in pastures and hedgebanks about Hersham, 16A2, for about 30 years from 1838, but seems to have disappeared from there and the many other places listed in Salmon, 1931. Rampion was formerly a very popular vegetable widely grown for the use of its roots in winter salads, but is now less commonly grown.

C. alliariifolia Willd.   Established garden alien.
05A1, Guildford, railway bank E of London Road Station, 1956, MN, 1967, JFL; 25A1, old quarry on Downs near Betchworth, 1 plant, 1967, BWu; 25C1, Colley Hill, steep chalky hillside under trees, 1962, DPY.

C. medium L.   Canterbury Bells.
Alien from gardens, thoroughly established on chalk railway banks. 25D4, Chipstead, 1950, DPY; 25E4, Hooley, 1953-7, LJJ; 26C2, Sutton, 1956, RCW; 35B5, between Riddlesdown and Upper Warlingham Stations, 1953, KEB; 35C4, nr Succomb's Bridge, 1952, LJJ; 35C5, S of Upper Warlingham Station, 1957, RARC; 36A1, on the Cliff at Purley, 1960, DPY.

Legousia Durande
L. hybrida (L.) Delarb. (*Specularia hybrida* (L.) A. DC.)
Venus's-looking-glass.   Native.   1763.
Rare, 29.   Map 311.
Cornfield weed, now almost restricted to the chalk. Decreasing, no longer in many of the localities reported since 1950.

Phyteuma L.
P. tenerum R. Schulz (*P. orbiculare* auct.).
Round-headed Rampion.   Native.   1695.   Rare, 30.   Map 312.
Chalk grassland. Still abundant in some localities though reduced by the increase in scrub in others.

Jasione L.
J. montana L.   Sheepsbit.   Native.   1666.
Rather rare, 53.   Map 313.
Heaths, commons, roadside banks, sand-pits, sandy fields, golf courses, racecourse on sandy and gravelly soils. In the W, and especially the SW of the county. The most easterly localities reported are 16C5, Hurst Park Racecourse, 1962, BW, JES; 25B1, Buckland, Shagbrook, on sandy ground under pines, 1962, BMCM.

RUBIACEAE

Sherardia L.
S. arvensis L.   Field Madder.   Native.   1763.   Common.
Cultivated ground and waste places. Rare in NW Surrey.

Asperula L.
A. cynanchica L.   Squinancywort.   Native.   1746.
Locally common, 43.   Map 314.
Chalk grassland, especially on slopes where the soil is shallow.

Galium L.
G. cruciata (L.) Scop. (*Cruciata chersonensis* (Willd.) Ehrend.,
C. laevipes Opiz).   Crosswort.   Native.   1832.
Locally common, 96.   Map 315.
Wood borders, hedges and scrub, roadsides and chalk grassland.
Crosswort is common along most of the chalk, but otherwise the
distribution is patchy and difficult to explain.

G. odoratum (L.) Scop. (*Asperula odorata* L.).   Woodruff.
Native.   1834.   Frequent, 106.   Map 316.
Woods on base-rich soils. A characteristic plant of beechwoods on
the chalk but also on base-rich soils in damp woods on the London
Clay and elsewhere. Fairly commonly planted under trees in wooded
gardens and six records (not shown on map) have been received
from such places.

G. mollugo L.   Hedge Bedstraw.   Native.   1666.
Common, especially on the chalk.
Hedgebanks, wood-borders, scrub and grassland.
subsp. *erectum* Syme (*G. erectum* Huds. 1778 *pro parte*) is a weed of
agricultural seed mixtures found in arable fields and sometimes
established on banks on the chalk. Rather rare, 40. Most of our
records come from roadsides and in recent years introduction in
seed mixtures used to cover areas bared in road construction is
probably more common than the traditional introduction by
farmers.

G. verum L.   Lady's Bedstraw.   Native.   1793.
Common, 182.   Map 317.
Dry grassland of commons, chalk downs, pastures and roadsides.
This species avoids acid soils but its rarity on Weald Clay is
probably due to the scarcity of well-drained habitats.

G. pumilum Murr. (*G. sylvestre* Poll. non Scop.).   Native.
1899.   Very rare, 5.   Rather bare slopes on the chalk.
05D1, E. Clandon, Salmon, 1931, Fuller's Farm, 1945, JEL, NYS,
destroyed by ploughing; 15C1, Denbies Drive, 1927, ECW, 1935,

1948, 1958, JEL, 1960, SFC; 15E3, Headley Warren Reserve, in field which has not been ploughed for at least 100 years, small colony, 1965, EM-R, JEL, JES, WEW; 25C2, Reigate Hill, 1894, HWP, Colley Hill, 1917, EBB, 1942, JEL, ECW; 35D4, Nore Hill, 1929, CEB (*Rep. BEC*, **9**, 229, 1930), 1925, ABead, JEL, and regularly until habitat destroyed by extension of pit.

G. saxatile L. (*G. hercynicum* auct.)   Heath Bedstraw.
Native.   1793.   Common.
Heaths, commons and woods on acid soils.

G. palustre L.   Marsh Bedstraw.   Native. 1763. Common.
Marshes, wet woods, by ponds and streams.

G. uliginosum L.   Fen Bedstraw.   Native.   1838.
Frequent, 51.
Boggy heaths, peaty meadows, marshes and swamps.

G. tricornutum Dandy (*G. tricorne* Stokes *pro parte*).
Corn Bedstraw.   Naturalised alien.   1778.
Very rare, probably extinct.
A native of Mediterranean countries, formerly established in cornfields from repeated introductions. Our only records are from 24A3, Leigh, weedy cornfield nr Shellwood Cross, 1955, DPY; 24D4, Redhill refuse tip, a casual, 1958, BMCM.

G. aparine L.   Cleavers.   Native.   1836.   Common.
Hedges, copses, waste places and cultivated ground.

G. spurium L. (*G. vaillantii* DC)   Alien.
Our only record: 25B5, Epsom College, in cultivated field, 1939, AEE, Hb. Lousley.

G. parisiense L. (*G. anglicum* Huds.)   Wall Bedstraw.
Native (and alien).   1793.   Very rare, 2.
26B1, Ewell, Downs Farm, casual with other aliens, 1959, DPY; 34D3, Crowhurst, Moat Farm on old wall, 1960, RARC. This species has long been known on old walls in the adjacent parts of Kent but only just extends into Surrey.

CAPRIFOLIACEAE

Sambucus L.

S. ebulus L.   Danewort.   Denizen.   1718.   Very rare.
13E5, Capel, in the park of Lyne House, 1957, BH, BMCM, 1967, SFC. The name 'Danewort' was explained by John Aubrey in connection with Gatton, where there is a tradition that there was a great slaughter of the Danes by the women of the neighbourhood at Battlebridge Field. He says that the Danewort was shown as having sprung from the Danish blood (*Nat. Hist. Antiq. Surrey*, **4**, 217, 1718), and has a similar story about Slaughterford in Wiltshire where he changes the name to 'Danesblood'.

S. nigra L.   Common Elder.   Native.   1836.   Very common.
Woods, scrub, roadsides and waste places. The fruits are distributed by birds to waste ground and gardens in towns, as well as throughout the rural areas. Var. *laciniata* Mill. occurs occasionally, perhaps bird-sown from gardens, as, for example: 24C1, wood near Gatwick Airport, 1956, BMCM, destroyed by extension of airport, 1958; 25A2, Headley Heath, in a dry chalk valley, CEB (*Rep. BEC*, **5**, 822, 1920), still there, 1944, JEL.

Viburnum L.

V. lantana L.   Wayfaring-tree.   Native.   1793.   Locally common.
Scrub, woods, and hedges, very common on the chalk, also occasionally on Bargate Beds and by the Thames.

V. tinus L.   Laurustinus.   Naturalised shrub from gardens.
Apparently self-sown on chalk escarpment. 14B5, above Landbarn Farm, 1962, JCG; 15C1, about Denbies Drive, 1958, JEL, RFC, 1960, BW, 1966, BWu; 15E1, Box Hill, E of Salomon's Memorial, 1958, KAB, JEL.

V. opulus L.   Guelder-rose.   Native.   1793.
Frequent, 321.   Map 318.
Hedges, woods, and scrub, preferring moist situations but also in well-drained habitats. Distributed throughout the county, but seldom abundant. Yellow-fruited plants were found near Gomshall, 04E5, Sandwith (*NW Nat*, **20**, 274, 1946).

Symphoricarpos Duham.

S. rivularis Suksd. (*S. racemosus* auct.)
Garden shrub which persists. Not rare, but usually on the site of

cottage gardens, or on railway banks, or in estates where there has been extensive planting of shrubs.

Lonicera L.
L. xylosteum L.   Fly Honeysuckle.   Garden alien.   1836.
93B4, hedge by main road between Brook and Upper Birtley, 1960, GSE—perhaps the same place as 'hedge near Brook', Miss C. Perry (*Fl. Met.*, 40, 1836); 93E4, Lower Vann, hedge by road, 1960, GSE; 94E4, Artington, pit between Littleton and Braboeuf, 1967, EJC.

L. periclymenum L.   Common Honeysuckle.   Native.
1836.   Common.
Hedges, woods and copses.

L. caprifolium L.   Perfoliate Honeysuckle.
Established garden alien.
94B5, Wanborough, quarry below the Hog's Back, 1966, OVP; 05E2, persisted in the copse behind W. Horsley Church until at least 1935, JEL; 15E2, Headley Lane, N side near Warren Farm, 1923, JEL, 1949, BW; 24D1, Leigh, hedge S of Shellwood Farm, 1959, BMCM, BW, JEL; 26D1, hedge by track leading to Oaks Park, 1963, LMPS.

L. henryi Hemsley
A garden plant thoroughly established at 14E3, Holmwood Common, in woodland nr Fourwents Pond, 1950, BMCM, BW, 1961, BMCM, JEL, 1963, DPY—still there. This vigorous climber covers an area of about 20 x 20m, twining up trees to produce flowers 3m from the ground—see Lousley (*Proc. BSBI*, **4,** 416, 1962).

L. nitida Wils.
Very commonly planted as a hedge on the chalk and occasionally reported as more or less naturalised, as at 35D2, Caterham, below Hill Top by grass verge, 1955, GFL, 1960, RARC.

Leycesteria Wall.
L. formosa Wall. A garden shrub, sometimes spreading from seed. 25D1, Redhill, 1964, EMCI; 35B2, Caterham, nr North Park Farm. in three places, 1955, GFL.

ADOXACEAE

Adoxa L.
A. moschatellina L.   Moschatel.   Native.   1657.
Common, 215.   Map 319.
Woods, copses and hedgebanks. Rare or absent over much of the
N of the county.

VALERIANACEAE

Valerianella Mill.
V. locusta (L.) Betcke (*V. olitoria* (L.) Poll.)
Common Cornsalad, Lamb's Lettuce.   Native.   1763.
Rather rare, 52.   Map 320.
Cultivated fields and gardens, hedgebanks and old walls, mainly
on light soils.

V. carinata Lois.   Keel-fruited Cornsalad.   Native.
1849.   Very rare, 8.
Railway tracks and embankments, gardens, nursery, sandy ground
by roadside. Recorded from 95D5, 14B3, 15D2, 25A1, 25B1,
34A5, 34E2 and 35A5. Still at 25A1, Betchworth Station, as
recorded in Salmon, 1931.

[*V. rimosa* Bast. Broad-fruited Cornsalad. Native, 1847. Extinct. Salmon,
1931, gave 13 records; we have none.]

V. dentata Poll.   Narrow-fruited Cornsalad.   Native.
1805.   Rare, 45.   Map 321.
Cornfields, mainly on the chalk.   A rapidly decreasing species.

Valeriana L.
V. officinalis L. (*V. mikanii* Syme. *V. sambucifolia* Mikan f.)
Common Valerian. Native. 1732. Frequent, 74. Map 322.
Streamsides, hedgerows, and meadows in wet places, and also
wood borders and hedgebanks in dry places on the chalk.

V. dioica L.   Marsh Valerian.   Native.   1722.
Rare, 17.   Map 323.
Marshy meadows and copses. Mainly in the SW.

Plate 22   *Ground Pine,* Ajuga chamaepitys, *with leaves of* Teucrium botrys *in the background*

Plate 21   *Cut-leaved Germander,* Teucrium botrys, *Chipstead (1973)*

Plate 23 *Chinese Mugwort,* Artemisia verlotorum, *Wimbledon Common (1973)*

Plate 24 *Starfruit,* Damasonium alisma, *in Surrey (1973)*

Centranthus DC.
  C. ruber (L.) DC.   Red Valerian.
  Alien from gardens established on walls, chalkpits, railway banks
  and road cuttings. Widespread but shows no rapid increase.

DIPSACACEAE

Dipsacus L.
  D. fullonum L. (*D. sylvestris* Huds.)   Teasel.   Native.
  1832.   Common, 198.   Map 324.
  Field borders, damp woods, ditches and rough pastures. Most
  common on Weald and London Clay, avoids sandy or acid soils.

  D. sativus (L.) Honck (*D. fullonum* sensu Huds. non L.).   Fuller's Teasel,
  Occurs rarely on refuse tips, mainly from seed for cage birds. Persisted
  for 4–5 years on a rough patch at 05A1, Guildford Tip, 1972, AL.

  D. pilosus L.   Small Teasel.   Native.   1633.
  Rare, 11.   Map 325.
  Wet woods by the Wey and Mole, and woods on the chalk.

Knautia L.
  K. arvensis (L.) Coult. (*Scabiosa arvensis* L.)
  Field Scabious.   Native.   1836.   Common.
  Abundant in chalk grassland and round the edges of woods on the
  chalk; dry grassland and hedgebanks elsewhere.

Scabiosa L.
  S. columbaria L.   Small Scabious.   Native.   1762.
  Locally common, 73.   Map 236.
  Dry pastures and banks on the chalk, also on alluvial soils likely
  to be rich in calcium carbonate by the Thames.

Succisa Haller
  S. pratensis Moench (*Scabiosa succisa* L.)
  Devil's-bit Scabious.   Native.   1836.   Common.
  Heaths, commons, rough pastures, and wet woods, often on
  rather acid soils. Avoids the chalk but does occur, very rarely,
  in chalk grassland where leached, as at 25D4, Chipstead Valley.

COMPOSITAE

Helianthus L.

H. annuus L. is increasingly common as a casual on refuse tips, now introduced mainly with cage-bird foods.

H. tuberosus L. occasionally established for a time from Jerusalem Artichokes on refuse tips and waste places, but often overlooked as it seldom flowers.

H. rigidus (Cass.) Desf. Grown in gardens as Perennial Sunflower, persists on waste ground and in rough fields, and may eventually become established.

Bidens L.

B. cernua L.   Nodding Bur-marigold.   Native.   1666. Frequent, 107.   Map 327.

By ponds, ditches and streams.

B. tripartita L.   Common Bur-marigold.   Native. 1827.   Common.

By ponds, ditches and streams, more common than the last.

B. vulgata Greene.

This alien from N. America was found in 36B4, Croydon, St James' Rd at edge of lorry park, 1956, DPY (*Proc. BSBI*, **2**, 240, 1957). Dr Young's manuscript note says 'in great quantity', but his published account says 'several plants'; the site is now built over.

Galinsoga Ruiz. & Pav.

G. parviflora Cav.   Gallant Soldier.   Naturalised alien. 1861.   Locally common, 144.   Map 328.

Weed of arable land, gardens, nurseries, churchyards and roadsides. This South American plant is known to have been grown in Kew Gardens in 1796 and escaped from there to Kew Bridge, 1861, to asparagus grounds between Richmond and Sheen, 1860, and to other places in the vicinity. Salmon, 1931, knew it as plentiful round Kew, Richmond, Ham, Barnes and Putney, and also at Merton and Mitcham. It was recorded also from Milford, 1893, and from Guildford, 1918. It is now widely distributed, but is still most abundant in the Greater London area and in the gardens of the 'stockbroker belt'. The spread has been from nurseries. In 1926 it was abundant in nurseries at 17D2, Ham, and in 1915 in nurseries at 26D5 (Lousley, *Rep. BEC*, **8**, 263, 1926); both these (and others) served as centres for distribution with plants, and especially potted plants, sold for gardens.

G. ciliata (Raf.) Blake (*G. quadriradiata* auct.)
Shaggy Soldier.   Naturalised alien.   First record: Lousley, 1946 (see below).   Locally frequent, 68.   Map 329.
Weed in cultivated fields, gardens, nurseries and roadsides in towns. This species seems to have been first found in Britain at Acton, Middlesex, in 1909 but did not appear in Surrey until 1943 when it was found in fields at 16D2, fields E of Claygate, and 37A2, John Peed's Nursery at Tulse Hill, Lousley (*Rep. BEC.*, **12,** 680–1, 1946). The spread in the next few years was rapid and, like *G. parviflora*, seems to have been based on nurseries from which it was distributed with garden plants. Recently the extension has been less rapid and the distribution is still patchy. The rapid spread in Surrey coincided with the period when it was spreading rapidly in Britain generally, and also in several European countries where from approximately 1940 to 1960 it spread twice as fast as *G. parviflora* already present.

Gaillardia Fougeroux
G. aristata Pursh   Naturalised alien.
94B1, Witley Common, on heathland away from roads, 1959, JEL, Hb. Lousley, several patches, 1961, RWD, large patch naturalised, 1965, DCK. (The last two were determined by Dr Young as *G. pulchella* Foug., but this is an annual species unlikely to persist in a closed habitat.)

Senecio L.
S. jacobaea L. Common Ragwort. Native. 1836. Very common. Pastures, chalk downs, roadsides and waste ground.

S. aquaticus Hill   Marsh Ragwort.   Native.   1836.
Common throughout the county.
Marshes, wet meadows, and commons, by ponds and streams.
x *S. jacobaea*=*S. ostenfeldii* Druce This hybrid has been recorded by Prof. Ostenfeld from 94B2, Elstead, bank of R. Wey at Somerset Farm (*Rep. BEC*, **7,** 39, 1924), where I found it again in 1935. From experience in other counties it might be expected to occur rather frequently where the two parents occur in close proximity, as they often do.

S. erucifolius L.   Hoary Ragwort.   Native.   1836.
Common.
Commons, pastures, rough roadsides, hedgebanks, wood borders.
Most abundant on clay, especially the Weald Clay in the S, and
also on chalk grassland, less common elsewhere.

S. squalidus L.   Oxford Ragwort.   Naturalised alien.
1913.   Very common, 274.   Map 330.
Railway embankments, old walls, bombed sites, waste ground,
quarries. The spread from Oxford Botanic Garden, where it has
been cultivated since at least 1690, to London and southern
England has been well described by Kent (*Proc. BSBI*, **2**, 115–18,
1955; **3**, 375–9, 1960; **5**, 210–13, 1964). The early spread was
mainly along railways, and the first evidence for Surrey is a specimen collected by C. E. Britton from a 'partly made road by railway
between Tadworth and Epsom Downs' in 1904, Herb. Kew. At
the outbreak of war in 1939 Oxford Ragwort was still quite rare in
Surrey and restricted to the NE, except for Reigate and Thorpe.
From 1940 onwards the rate of spread has been more rapid and,
as the map shows, it has extended from London towards the SW
along the railways, and also in between them.
x *S. viscosus* = *S.* x *londinensis* Lousley
London Ragwort.   Rare.
06C1, Byfleet, Dartnell Park, building plot, 1966, DPY, det. JEL;
17D2, by River Ham, 1950, BW; 35A1, Bletchingley, by Pendell
Court, 1955, BMCM, det. JEL; 36B3, Croydon, Cranmer Road,
waste land, 1963, DPY. This hybrid was described as new to
science from a plant found at 37A1, Streatham, bombed site in
Baldry Gardens, 1944, Lousley (*Rep. BEC*, **12**, 869–74, 1946 &
Proc. Linn. Soc., **158**, 21–2, 1947). The hybrid arises rather freely
when the two parents grow together, and was frequent on bombed
sites for some years after the war. It is now much rarer owing to
the decrease in *S. viscosus*.
x *S. vulgaris.* 15D1, Dorking Tip, 1964, BMCM, det. JEL; 25D1,
Redhill, edge of car park, 1965, EMCI, det. JEL & DPY. No doubt
much more frequent than these records suggest but, owing to the
variation in the parents, considerable caution is necessary in
naming this hybrid from morphological characters alone.

S. sylvaticus L.   Heath Groundsel.   Native.   1836.
Common, 211.   Map 331.
On sand or gravel in open vegetation on heaths and commons.

x *S. viscosus* = *S.* x *viscidulus* Scheele. 84C1, Frensham Little Pond, 1947, JEL & CW (Lousley, *Proc. BSBI*, **1**, 37–9, 1954). Likely to be rare rather than unrecognised as the two parents are not commonly in association.

S. viscosus L.   Sticky Groundsel.   Alien.   1763.
Rather rare, 168.   Map 332.
Railway yards, waste places and roadsides. Possibly native on shingle near the sea, but in Surrey an alien usually associated with railways or building activities. Much rarer than the map suggests as few of the localities persist for more than a year of two.

S. vulgaris L.   Common Groundsel.   Native.
1836.   Abundant.
Weed in gardens and arable fields, and in waste places. Plants with conspicuous ray florets, var. *hibernicus* Syme (var. *radiatus* auct.), occur rarely: 34E2, Lingfield Station yard, 1959, EMCI; 36A1, Purley, railway goods yard, many plants, 1950, DPY, not seen since, DPY.

S. tanguticus Maxim.
Garden alien established at 15E1, Brockham Hills Quarry, 1967, JHPS, 1968, DPY.

[*S. integrifolius* (L.) Clairv. (*S. campestris* DC.). Native. 1849. Extinct. Not seen for many years at the old locality on the Hog's Back nr Puttenham, or elsewhere in the county.]

Doronicum L.
D. pardalianches L.   Leopard's-bane.
Established garden alien.   1832.   Rare.
94A5, Great Down, OVP, colonising wood for several acres, 1967, DPY; 94C4, Compton, laneside hedge E of Watts' Gallery, 1965, DPY; 94D3, Charterhouse, several sq yd in copse nr farm buildings, 1950, OVP; 14A3, Abinger, by Woodhouse Farm, 1960, FDSR; 35C2, Winder's Hill, wood nr ruins of Marden Castle, 1946–58, LJJ, 1959, JPSR; 35C4, top of Bug Hill, two colonies, 1954, RARC, DPY, 1958, JPSR, 1960, RARC, 1963, DPY; 45A2, Limpsfield, West Heath, abundant, H. S. Salt in Salmon, 1931; West Heath, a 10yd patch, 1951, RWD; Paines Hill Corner, 1965, BAK.

D. plantagineum L.
There is a specimen from 15A1, small copse opposite Bethlehem Lodge between Effingham and Dog Kennel Green, 1949, MBG, Hb. Mus. Brit, but Dr Young failed to refind in 1958.

Tussilago L.
T. farfara L.   Coltsfoot.   Native.   1836.   Abundant.
Railway banks, waste ground, roadsides and arable fields, especially on clay.

Petasites Mill.
P. hybridus (L.) Gaertn., Mey. & Scherb. (*P. ovatus* Hill., *P. vulgaris* Desf.)   Common Butterbur.   Native.
1746.   Rare, 30.   Map 333.
River and streamsides and wet meadows. By the Thames, Wey and Mole. Only male plants now known, the female has only been recorded once, from Egham over 125 years ago.

P. albus (L.) Gaertn.   White Butterbur.
Established garden alien.
14B2, Leith Hill Place, on both sides of the road, first shown to me about 1930 by A. H. Carter (*Rep. BEC*, **12**, 492, 1944), and seen many times since, 1972, JEL & JES.

P. japonicus (Sieb. & Zucc.) F. Schmidt   Great Butterbur.
Established garden alien.
95D4, nr Hook Heath, SW of Woking, extensively naturalised on both sides of stream, 1965, EJC; 16D4, Long Ditton, naturalised along stream bank, 1962, JES.

P. fragrans (Vill.) C. Presl   Winter Heliotrope.
Established garden alien.   1838.   Frequent, 103.   Map 334.
Roadsides, railway banks and cuttings, waste places.

Inula L.
I. helenium L.   Elecampane.   Denizen.   1863.   Rare, 4.
By ditches near old farms and as a recent introduction. 05C1, Clandon Park, a large colony in clearing by stream, 1963, JES & ECW; 15E1, Brockham Hill, bomb crater, 1949, JEL, 1967, BWu; 15E5, Ashtead, by the larger pond NW of the parish church, 1964, EJC; 16D2, in a hedge by a farmhouse between Chessington and Horton, H. C. Watson, Brewer, 1863; beside a stream S of Chessing-

ton, 1946, CLC (*Lond. Nat.* **26,** 75, 1947); Park Farm, Chessington, 1958, JES & BW; 25A1, ditch nr Kemps Farm, Betchworth, Salmon, 1931, 1945, 1949, 1969, 1974, JEL, reported also by LJJ, BMCM and others.

I. conyza DC.   Ploughman's Spikenard.   Native.   1836.
Locally common, 102.   Map 335.
Grassy slopes, chalk pits and open scrubby woodland. Mainly on the chalk, and limestone exposures on the Lower Greensand, rarely on gravel and sand elsewhere.

[*Telekia* Baumg.
T. speciosa (Schreb.) Baumg. 83E2, pond by Hindhead-Shottermill road 1965, EJC. Scarcely naturalised but included here because this species has been recorded in several other counties as *Inula helenium* to which it bears some resemblance.]

Pulicaria Gaertn.
P. dysenterica (L.) Bernh.   Common Fleabane.   Native.
1832.   Common.
Pondsides, marshes, wet commons and roadsides, occasionally in apparently dry places on the chalk.

P. vulgaris Gaertn. (*Inula pulicaria* L.)   Small Fleabane.
Native.   1763.   Very rare, perhaps extinct.
Gravelly pond margins, especially where ducks and geese are kept. 95C1, Worplesdon, Wood St., 1934, WEW, 1 plant, 1962, BW & WEW; 95E2, Stringer's Common, Salmon, 1931, 1934–48, WEW, 5 plants, 1950, BW, few plants, 1952, J. Codrington. 16B2, Esher, West End Common, numerous old records back to Brewer, 1863, 4 plants after pond margin disturbed, 1961, RAB & JES, 6 plants, 1962, JES. Salmon, 1931, deplored the reduction this species had suffered, which he attributed to improved land drainage, but few people then would have expected the dramatic decrease since. The reason is partly drainage, but even more the falling off in the keeping of geese and ducks which kept down the grass and enriched village greens with their droppings.

Filago L.
F. vulgaris Lam. (*F. germanica* L. non Huds.)
Common Cudweed.   Native.   1836.   Rare, 27.   Map 336.
Cultivated fields, commons and heaths, usually on acid sandy soils. A decreasing species.

F. lutescens Jord. (*F. apiculata* G. E. Sm.)   Red-tipped Cudweed.
Native.   1848.   Rare, 11.   Map 337.
Sandy fields and sides of tracks. A rapidly decreasing species
restricted to the Lower Greensand. Since 1970 only in 84C3 and
04B3.

F. pyramidata L. (*F. spathulata* C. Presl)
Broad-leaved Cudweed.   Native.   1713.   Very rare, 2.
Cultivated field and bare ground on chalk. 25D4, Chipstead,
abundant in field with *Teucrium botrys*, 1950, DPY, 1951, ECW,
1963, DCK; 26C1, Banstead Downs, old chalk heaps, 1957, RCW,
1968, BWu, 1973, JEL.

F. gallica L. Narrow-leaved Cudweed. Alien. Was seen in 04B5, cornfield
between Chilworth Woods and St. Martha's Hill, 1867, J. C. Melvill,
Hb. Mus. Brit., but no later record.

F. minima (Sm.) Pers.   Small Cudweed.   Native.   1836.
Locally frequent, 72.   Map 338.
Dry places on sandy or gravelly commons, heaths, paths and fields.
Mainly in the SW.

Gnaphalium L.
 G. sylvaticum L.   Wood Cudweed.   Native.   1805.
Rather rare, 57.   Map 339.
Rides in woods, heaths and commons. Has decreased in the E of
the county, still locally frequent in the SW.

 G. uliginosum L.   Marsh Cudweed.   Native.   1836.
Common.
Damp places round arable fields, roadsides, gardens and pond
margins.

Antennaria Gaertn.
 A. dioica (L.) Gaertn. Mountain Everlasting. Native. There are several
old records for 26C1, Banstead Downs, but it has not been reported
for over a century.

Solidago L.
 S. virgaurea L.   Golden Rod.   Native.   1763.
Locally common, 209.   Map 340.
Dry woods, commons and heaths. Perhaps most frequent on acid
soils, rare on chalk.

S. canadensis L. (*S. altissima* auct.)   Naturalised alien.
1926, (see below).   Very common, 211.
Waste ground, roadsides, railway banks, commons, building land,
and streamsides. Introduced into British gardens from North
America as long ago as 1648, the rapid spread to the present
abundance commenced only about 20 years ago. Salmon, 1931,
did not mention the species. The earliest evidence we have traced
is from 25C3, Kingswood, 1926, CEB, Hb. Kew.

S. gigantea Ait. (*S. serotina* Ait., non Retz.)   Naturalised alien.
1916 (see below).   Frequent, 86.
Waste places, roadsides and commons. Introduced into Britain
from North America in 1758 and a more attractive, though less
aggressive, garden plant than the last. It seems to have been established a little earlier and our earliest evidence is from 16C2?,
Claygate, 1916, CEB, Hb. Kew, and 26E5, waste ground, Mitcham
Common, 1928, ECW, Hb. Lousley.

Aster L.
A. tripolium L.   Sea Aster.   Native.   1882.   Very rare, 1.
Embankment of tidal Thames. 27B5, above Hammersmith Bridge,
CEB (Salmon, 1931), river wall 150yd upstream from Hammersmith Bridge, 1956, BW. Probably common by the tidal Thames
before the growth of London and construction of embankments
destroyed the habitats.

The Michaelmas Daisies naturalised in Britain belong to a critical
group of which the nomenclature and synonymy adds to the
difficulties. Our plants cannot be matched with those native in
North America, and the key to the problem is probably to be found
by identifying the early introductions first. Working from this on
the combinations of characters shown in our plants it may be
possible to demonstrate how these characters are likely to have arisen
in Europe. The more aggressive naturalised plants are relatively
unattractive as garden plants and in fact soon replace modern
hortal introductions if allowed to do so. C. E. Britton paid considerable attention to naturalised asters of the Thamesbank in
Surrey (*Rep. BEC*, **6**, 384, 1922 & **9**, 710–18, 1932).

A. novae-angliae L., a rather distinct plant with rough, hairy,
glandular stems and reddish rays is found occasionally on refuse

tips, waste ground, lanesides and commons, but shows little tendency to spread.

A. novi-belgii L.   Michaelmas Daisy.
The most common naturalised plant with wishy-washy blue rays in rather small flowers. 1870. Common in the east, 77.
Riverbanks, waste ground, commons and railway banks. Includes subsp. *floribundus* (Willd.) Thell. and subsp. *laevigatus* (Lam.) Thell.

A. lanceolatus Willd. (*A. paniculatus* auct.)
A tall plant with linear-lanceolate leaves not broadened at the base and slightly amplexicaul. 1865. Common in the north-east, 47.
Riverbanks, roadsides and waste ground. Probably the species to which the name 'Michaelmas Daisy' was first applied in the mid-seventeenth century and likely to be the '*A. leucanthemus* Desf.' which H. C. Watson found from 1860 to 1865 'near Thimble Bridge, Thames Ditton' (*Rep. Thirsk Nat. Hist. Soc.*, **1864**, 12, 1865).

Other names which have been applied to Surrey plants include *A. tradescanti*, L., *A. salignus* Willd., and *A. x versicolor* Willd.

Erigeron L.
E. acer L.   Blue Fleabane.   Native.   1814.
Locally common, 127.   Map 341.
Dry grassland, banks and walls, mainly on the chalk where it is common.

E. philadelphicus L.   Alien.   Rare.
94D1, side of path leading to Milford Chest Hospital, 1960, KFA; 94E3, Farncombe, well naturalised near lake and car park, 1958, BMCM: 15D1, Boxhill, numerous plants on 'The Whites', 1966, RCW.

Conyza Less.
C. canadensis (L.) Cronq.   Canadian Fleabane.   Alien.
1827.   Common.
Waste and cultivated ground and roadsides, especially on sand or gravel, less common on chalk.
x C. Erigeron acer = Erigeron x huelsenii Vatke. This easily recognised hybrid was found as a single plant growing with the parents

in 1884 in 84, Tilford, by E. S. Marshall (*J. Bot.*, **45**, 164, 1907). Since then it has been found in various other counties, sometimes in considerable quantity. It is easiest to find in the autumn when it continues to flower after the parents pass into fruit.

Bellis L.
B. perennis L.   Daisy.   Native.   1836.   Very common.
Short grassland of meadows, lawns and roadsides. The daisy thrives where ground is compacted by trampling and the grass short owing to grazing or mowing. Hence it often picks out the pattern of tracks and paths over grassland and is abundant in lawns which are not treated with selective weedkillers.

Eupatorium L.
E. cannabinum L.   Hemp Agrimony.   Native.   1827.
Frequent, 115.   Map 342.
River and streamsides, ditches and moist woods. Mainly by the Thames, Wey, Basingstoke Canal and Wandle, but also in many other places away from these. Sometimes in woods on the chalk which seem quite dry, such as 35D4, Woldingham, in beech wood on chalk, 1963, RARC.

Anthemis L.
A. tinctoria L.   Yellow Chamomile.   Alien.   1864.   Rare.
94E4, Shalford, on tipped rubble on building site, 1969, JFL; 94D3, Godalming railway goods yard, var. *discoideus* Willd., Bishop in Salmon, 1931, 1935, 1944, 1949, etc, JEL, 1967, DPY.

A. cotula L.   Stinking Chamomile, Poison Magweed.
Native.   c1700 (see below).   Locally abundant, 152.   Map 343.
Cultivated land and waste places. Seldom seen in the N of the county, but abundant in cornfields and other arable on the heavier soils in the S.E.S. Marshall said that in W Surrey it was called 'Poison Magweed' (Mayweed?) by the harvesters, whose hands were often much inflamed by its acrid juice (*J. Bot.*, **27**, 220, 1889). The earliest evidence for the species in the county is a specimen collected by Stonestreet from Peckham Field, c1700, in Herb. Du Bois at Oxford (*Rep. BEC.*, **9**, 688, 1932).

A. arvensis L.   Corn Chamomile.   Native.   1713.
Rare, 21.   Map 344.
Arable fields, mainly on the chalk. A decreasing species which is

rarer than the records suggest. Most of the occurrences off the chalk were of a few plants which did not persist, and the most reliable area on the chalk is between Headley and Ashtead: 15E3, Headley, in large cornfield, frequent, 1964, RARC; 15E4, nr Thirty Acre Barn, 1929, ECW, 1932, JEL, common, 1944, JEL, 1958, DPY, 1964, 1 plant, RARC; 25A4, Ashtead, arable field along Headley Road, 1958, DPY.

Chamaemelum Mill.
C. nobile (L.) All. (*Anthemis nobilis* L.)   Common Chamomile.
Native.   1548.   Rare, 32.   Map 345.
Damp short grass on commons, roadsides and especially near ponds used by geese and ducks. A decreasing species but persistent in some old localities. It was first recorded from Britain by William Turner in 1548 from 'Rychmund grene' and, if no longer there, it is still plentiful on lawns in 17E4, Kew Gardens. John Aubrey in 1718 wrote 'Camomil grows here (Purbright) very common and wild' (*Nat. Hist. & Antiq. Surrey*, **3**, 215), and it is still on Pirbright Common, 95A3. It is sometimes distributed in garden turf, as, for example, 95E5, Woking, Horsell Rise, several square yards in 'grovely' turf bought for lawn, 1965, WEW, but will only persist when the lawn conditions are suitable.

Achillea L.
A. millefolium L.   Yarrow.   Native.   1836.   Very common.
Meadows and pastures, roadsides, field borders, downs and lawns. Equally common in suburban London.

A. ptarmica L.   Sneezewort.   Native.   1763.
Common, 190.   Map 346.
Damp commons, meadows, heaths and marshes. A commonly cultivated plant in gardens, which is sometimes established on commons and roadsides, often as the double-flowered form 'The Pearl'. These records have been excluded from the map.

Tripleurospermum Schultz Bip.
T. maritimum (L.) Koch subsp. *inodorum* (L.) Hyland. ex Vaarama (*Matricaria inodora* L.)   Scentless Mayweed.   Native.   1724.
Common.
Weed of arable land, waste places and roadsides.

Matricaria L.
M. recutita L. (*M. chamomilla* auct.)   Scented Mayweed.
Native.   1806.   Uncommon, but easily overlooked.
Cultivated and waste ground.

M. matricarioides (Less.) Porter (*M. discoidea* DC., *M. suaveolens*
(Pursch) Buchen.)   Pineapple Weed.   Alien.
1901.   Very common.
Roadsides, waste places, field borders. G. C. Druce claimed that the
first evidence for this species in Britain was a plant gathered
between Richmond and Kew in 1871 (*Gard. Chron.*, **69** (Ser. 3),
187, 192) but later gave the same details in error for a specimen
from Aber, Caernarvonshire (*Comital Flora*, 164, 1932). Our first
reliable record is from Kew Green in 1900, when S. T. Dunn
remarked on the rapid spread (*Rep. Watson BEC*, **1**, 20, 1901).
Salmon found it too common to need detailed localities (Salmon,
1931, 390); it is now found throughout the county. The spread has
been mainly through the transport of seeds in mud on boots and
on the tyres of vehicles, and for this reason it is often abundant
round gateways giving access to fields or on tracks across commons,
etc.

[*M. tschihatchewii* (Boiss.) Voss. (*Chamaemelum tschihatchewii* Boiss.)
There is a specimen in Herb. Kew labelled 'Wild in Surrey', comm.
Dr. G. C. Druce, 14.5.21', and this is no doubt the plant which Lady
Davy found growing 'near *Arabis glabra* and *Ulex*' on a sandy common
near Byfleet (*Rep. BEC*, **6**, 291, 1922). It was grown for a time in Britain
as 'Lawn Pyrethrum', in view of its resistance to drought, but nothing
further is known of it at Byfleet.]

Leucanthemum Mill.
L. vulgare Lam. (*Chrysanthemum leucanthemum* L.).
Ox-eye Daisy.   Native.   1836.   Common throughout the county.
Meadows, grassy banks, downs and roadsides. A characteristic
plant of railway embankments.

L. maximum (Ramond.) DC.   Shasta Daisy.
Occasionally established for a time on roadsides, commons and
waste places as a garden throwout.

Chrysanthemum L.
C. segetum L.   Corn Marigold.   Colonist.   1836.
Locally frequent, 112.   Map 347.
Arable fields on acid soils, mainly on sand. The abundance from

year to year in individual fields depends on agricultural management, but occasionally fields are coloured a golden yellow from
end to end with the flowers.

C. serotinum L.
Well-established garden outcast at 05E4, Ockham, Mays Green,
1957, JEL, 1958, BW, 1970, DPY.

Tanacetum L.
T. parthenium (L.) Schultz Bip. (*Chrysanthemum parthenium* (L.)
Bernh., *Matricaria parthenium* L.)   Denizen.   1836.
Frequent in walls and hedgebanks about houses.
Usually rather obviously a garden escape.

T. vulgare L. (*Chrysanthemum vulgare* (L.) Bernh.).   Tansy.
Native.   1666.   Frequent, 163.   Map 348.
River and streambanks, hedges and roadsides. Mainly by the rivers
Thames, Wey, Tillingbourne and Mole where it is clearly native.
Tansy was formerly much cultivated for medicinal and pot-herb
use and, although we have removed the obvious garden plants from
the map, some of the other localities away from rivers are likely to
have originated from garden escapes.

Cotula L.
Two alien species are established on lawns:
C. dioica Hook. 16B3, Esher, female plants on lawn of 'Little
Lammas', Clive Road, known since 1945, 1959, JES, Hb. Lousley.
C. squalida Hook.f. 05A5, Woking, Maybury, lawn of 'Yarrow-
field', Mrs Kenyon, 1959, Hb. Lousley.

Artemisia L.
A. vulgaris L. Common Mugwort. Native. 1836. Common.
Roadsides, hedgerows, waste ground and railway yards.

A. verlotorum Lamotte.   Chinese Mugwort.
Naturalised alien.   1946 (see below).
Locally frequent, 85.   Map 349.
Roadsides, commons, gravel pits, sewage works, and clearing
in a wood. (Plate 23). First recorded for Surrey (and Britain) from
towpath between Kew and Chiswick bridges, Ham and Ripley in

1946, Lousley (*Lond. Nat.*, **25**, 13, with further details in *Rep. BEC*, **13**, 161–2). From a full account by J. P. M. Brenan (*Watsonia*, **1**, 209–23, 1950) it seems that it had been known to Iolo A. Williams and Francis Druce at 17D1, Ham Pits since 1938–9 and to G. M. Ash at 94C3, Godalming 'for some years'. Brenan was able to cite ten Surrey localities. It is now known to be frequent along the Thames and in the NE of the county with scattered records elsewhere. The means by which *A. verlotorum* spreads is not fully understood but it is evident that the production of ripe seed on such a late-flowering species in Britain must be rare. It is also evident that the species requires open communities in which to become established, and probably also fairly rich organic soils. There is little evidence of short-distance extension other than by rhizomes— for example, all our records at Godalming refer to the same clump which has been known since before 1946. The most likely explanation is that it is spread with builders' rubble and it is frequently associated with places where this has been tipped.

A. absinthium L.   Common Wormwood.
Naturalised alien.   1718.   Rare, 22.
Waste ground, roadsides, railway sidings. Perhaps most frequent about Mitcham and Wimbledon Common. The present distribution is based mainly on recent introductions and, with the possible exception of Mitcham, none of the present occurrences seem connected with the records given in Salmon, 1931.

Echinops L.
Garden Globe Thistles are occasionally well established on slopes and several species have been recorded from Britain as *E. sphaerocephalus* L. For a useful key to the species as now understood see *Proc. BSBI*, **7**, 243–4, 1968.

E. sphaerocephalus L. 36B1, railway bank between Purley Oaks and Riddlesdown Station, very plentiful, 1942 to 1959, HB. (sp. *non vidi*.)

E. exaltatus Schrad. (*E. commutatus* Juratzka). 94B5, growing naturally in a small private chalk quarry, Hog's Back, Puttenham, 1968, EJC; 15E1, Brockham Hill bomb crater and adjacent down, 1949 onwards, JEL.

Carlina L.

C. vulgaris L.   Carline Thistle.   Native.   1832.
Locally common, 63.   Map 350.
Mainly on chalk grassland and there common, but also on well-drained places on sand or other light soils.

Arctium L.

Very little attention has been paid to this genus by our recorders who have had difficulty in separating the segregates by the supposed characters.

A. lappa L. (*A. majus* Bernh.).   Greater Burdock.   Native.
1836.   Uncommon.
Damp places near rivers, streams and canals, wood borders, hedgebanks and waste places.

A. minus Bernh. subsp. *minus*   Lesser Burdock.   Native.
1838.   Very common throughout the county.
Roadsides, waste places, field borders and wood borders.
subsp. *nemorosum* (Lejeune) Syme (*A. nemorosum* Lejeune, *A. vulgare* auct.).   Apparently rather rare.

Carduus L.

C. tenuiflorus Curt.   Slender Thistle.   Alien.   1827.   Very rare.
By roads, tracks and arable fields. 25C1, Banstead Downs, by asylum, 1926, JEL, Herb. Lousley; 25D4, Chipstead, lucerne field near Dene Farm, 1954, JEL, DMcC, HB; 25E4, field by Starrock Road and Starrock Green, 1954, LJJ; 35C2, Godstone, Flower Lane, 1962, RARC; 45B2, Limpsfield, W of Moorhouse, by track through arable field, 1957, CNHSS. Elsewhere in England this species is widespread as a native on the coast, and it occurs as a wool introduction where shoddy is used as a manure. The way in which it is introduced in Surrey is not known.

C. nutans L.   Musk Thistle.   Native.   1837.
Locally frequent, 34.   Map 351.
Characteristic of rather open places on the chalk, but also on sandy soils, especially on the Lower Greensand, on alluvial soils by the Thames, and sometimes on transported soils.

C. acanthoides L. (*C. crispus* auct.)   Welted Thistle.
Native.   1709.   Common, 147.   Map 352.
Roadsides, edges of arable fields, wood borders, and river banks.

Avoids clays and very acid soils. Occurs also on transported soils in localities excluded from the map.

x *C. nutans=C. x orthocephalus* Wallr. This hybrid has been reported rather frequently from the chalk in the past and no doubt still occurs.

Cirsium Mill.

C. eriophorum (L.) Scop. (*Cnicus eriophorus* (L.) Roth).
Woolly Thistle. Native. 1952 (see below). Very rare.
Chalk grassland. 35A4, Old Coulsdon, W of Caterham Barracks on steep chalky scrubby bank, 1951, HB, in Lousley (*Lond. Nat.*, **31**, 11, 1952); in fine flower, 1952, JEL (*Lond. Nat.*, **32**, 80, 1953); 1954, BMCM, EMCI, BW, annually but often not flowering in alternate years, spreading down to more level ground where in July 1972 there were at least 18 fine plants just coming into flower, which were cut down by Croydon Corporation workmen a fortnight later; 5 plants in flower, 1974, JEL.

C. vulgare (Savi) Ten. (*Carduus lanceolatus* L., *Cnicus lanceolatus* (L.) Willd.) Spear Thistle. Native. 1836.
Too common throughout the county.
Field, roadsides, open woods and waste ground.

C. palustre (L.) Scop. (*C. palustris* L.) Willd.) Marsh Thistle.
Native. 1838. Very common throughout the county.
Damp meadows, hedgerows, damp woods, and roadside ditches.

C. oleraceum (L.) Scop. 25D2, Gatton Park, relic of old water-garden, 1955, JEL, BW, BMCM, EMCI, 1966, CNHSS.

C. acaule Scop. (*Cnicus acaulos* (L.) Willd.) Dwarf Thistle.
Native. 1746. Locally abundant, 97. Map 353.
Abundant on grazed chalk grassland, grazed alluvial meadows by the Thames, and on dry commons. Blackstone recorded it from Box Hill in 1746 and it is still too plentiful there for the comfort of picnickers.

C. heterophyllum (L.) Hill 25D2, Gatton Park, relic of old garden, 1955, JEL, BW, BMCM, EMCI, gone by 1966, RARC, AGH.

C. dissectum (L.) Hill (*Cnicus pratensis* (Huds.) Willd.)
Meadow Thistle. Native. Rare, 47. Map 354.
Bogs, peaty commons, and heaths. Common on the acid commons of NW Surrey.

x *C. palustre* = *C.* x *forsteri* (Sm.) Loud. (*Cnicus forsteri* Sm.).
A rare but distinctive hybrid. 34C1, Hedgecourt Mill-pond,
W. H. Beeby, in Salmon, 1931, 402, other records from here 1849–
1910, swamp SW of Mill-pond, 1933, JEL, 1956, BAK, 1968,
JEL & MEY, apparently destroyed by building of new pool,
1971, BMCM.

C. arvense (L.) Scop. (*Cnicus arvensis* (L.) Roth).
Creeping Thistle.   Native.   1724.   Abundant.
Pastures and arable fields, roadsides, waste places; a noxious weed.
A very variable species. Plants with flat, not undulate, leaves,
almost without prickles have been distinguished as var. *setosum*
C. A. Mey, and are regarded as aliens. A very remarkable tall form
of this with entire leaves (f. *integrifolium*) occurred for some years
at 04C4, by Postford Ponds, 1932, ECW (*Rep. BEC*, **10**, 437–8,
1933), but is no longer there. Var. *setosum* is less frequent now than
it was formerly.
x *C. palustre* = *C. celakovskianum* Knaf 06B1, near Basingstoke
Canal, Sheerwater, 1949, DHK, Hb. Lousley, det. W. A. Sledge
(*BSBI Year Book*, **1951**, 64).

Silybum Adans.
S. marianum (L.) Gaertn.   Milk Thistle.   Alien.
1719.   Very rare.
Waste places and roadsides. Reported from 16E5, 25D5, 45B2, but
not persistent.

Onopordum L.
O. acanthium L.   Cotton Thistle.   Alien.   1827.   Rare.
Roadsides and refuse tips, probably always a garden escape.
Reported from 06B4, 15D2, 16A1, 26B2, but not persistent.

Centaurea L.
C. scabiosa L.   Greater Knapweed.   Native.
1836.   Locally common.
Chalk grassland, hedgebanks and roadsides. Rare off the chalk and
then usually an obvious introduction, or on chalky railway banks,
as at 94C1, Enton, on a poultry farm; 27D5, Wandsworth, factory
road and on a railway bank; 27E1, Lower Streatham, on railway
bank; 38B1, Southwark, derelict garden, St Thomas St.

C. cyanus L.   Cornflower.   Colonist in cornfields.   1836.
Very rare.
Cornfields and other arable fields, also as a casual garden escape
in waste places. The following are all from arable fields: 95D2,
Worplesdon, 1963, WEW; 06B5, Thorpe, three arable fields blue
with cornflowers, 1965, CPP, WEW, JES, ECW, 1966, DPY;
less plentiful, 1972, JEL; 06D2, Addlestone, Wey Manor, in
barley, 1968, GHG; 15E3, Headley, 1970, RARC, HWM-P; 25A2,
Pebblecombe Hill, 1950, BMCM; cabbage field above Betchworth
station, 1950, JEL; 34A5, S Nutfield, field adjoining Bowerhill
Wood, 1946, 1947, 1956, BAK; 34B5, Bletchingley, in grass ley,
1959, RARC, DPY; 35A1, Nutfield, E of The Priory, 1943, JEL;
35C2, cornfield, 1959, LJJ. Decreasing, but never as common in
Surrey as in some other counties.

C. jacea L.   Brown Knapweed.   Alien.   1832. Very rare.
Disturbed grassland, golf links and waste places. In the past
C. jacea has been rather frequently introduced, probably with
grass mixtures and perhaps from central Europe. It hybridises
freely with native *C. nigra* L. (aggregate) to form hybrid swarms
in which *jacea* becomes rare and disappears after a few years, as in
the populations shown to me by Lady Davy on the golf links at
Byfleet and by A. Beadell near Nore Hill, Warlingham. No convincing evidence of *C. jacea* has been reported in recent years,
though a colony of *Centaurea* in 35D3, Lunghurst Valley, Woldingham, found in 1972 by RARC, deserves further study.
Studies of British *Centaureas* by C. E. Britton contain many references to Surrey populations (*Rep. BEC*, **6**, 163–73, 1921; **6**, 406–17,
1922, **8**, 149–52, 1927) and reference should also be made to
E. M. Marsden-Jones and W. B. Turrill's *British Knapweeds* (1954)
which is based on experimental work and includes a list of Surrey
records (58–61).
x C. nigra. See above, also 25A5, Epsom Downs, 1952, Marsden-
Jones & Turrill (*British Knapweeds*, 59).

C. nigra L.   Common Knapweed.   Native.   1836.
Very common.
Grassland, downs, and roadsides. There are two subspecies:
subsp. *nigra* (*C. obscura* Jord.) is rare. 06D4, Chertsey Mead, 1962,
JEL, and doubtless elsewhere. Subsp. *nemoralis* (Jord.) Gugler is
the common plant. Many intermediates between the two subspecies
occur.

C. calcitrapa L.   Red Star-thistle.   Alien.   1827.
Now only casual.
17E4, towpath by Kew Gardens, 1950, BHSR, BW.

C. solstitialis L.   Yellow Star-thistle.   Alien.   1805.
Introduced with lucerne seed and now very rare. 84A2, Elstead, in lucerne, 1972, RF; 36D2, Addington, 1 plant in lucerne field, 1950, CNHSS, DPY.

C. diluta Aiton. An alien now frequently introduced with bird-seed and reported from lanes and gardens as well as refuse tips, but always casual.

Serratula L.
S. tinctoria L.   Saw-wort.   Native.   1763.
Locally frequent, 73.   Map 355.
Rides and margins of woods, rough pastures, roadsides and commons. Most plentiful on the clays in the SW.

Cichorium L.
C. intybus L.   Chicory.   Native.   1782.
Frequent, 110.   Map 356.
Fallow fields, roadsides, waste places, especially on chalk. Also occasionally as an obvious introduction on refuse tips, building sites and docks, or sometimes planted as a garden plant.

Lapsana L.
L. communis L.   Nipplewort.   Native.   1836.   Common.
Hedgebanks, wood margins, waste and cultivated ground and under walls.

Arnoseris Gaertn.
A. minima (L.) Schweigg. & Koerte.   Lamb's succory.
Native.   1746.   Very rare, perhaps extinct.
Sandy cultivated fields, usually associated with *Hypochoeris glabra*. Formerly rather widespread, but very rare since 1950: 85E4, edge of cornfield near Frimley Green, abundant, 1953, JEL, and at intervals to 1966, scarce, BWu; 04E4, Gomshall, in three fields by sandy track, abundant, 1944, 1945, JEL, still widespread, 1950, RWD, limited to corner of one field, 1958, RWD, not found, 1970, JES, WEW, nor 1972, JEL, JES.

Hypochoeris L.
H. radicata L.   Common Cat's-ear.   Native.
1836.   Common.
Meadows and pastures and other short grassland.

H. glabra L.   Smooth Cat's-ear.   Native.   1696.
Rare, 26.   Map 357.
Sandy and gravelly places in arable fields, by tracks on heaths, and in sandpits. Decreasing and now restricted to the W of the county; now extinct in 25B1, Reigate Heath and not reported from any of the old localities round Weybridge, Walton, Esher, Claygate, etc.
x *H. radicata* A very rare hybrid. 84B3, Tilford, in sandy field, July 1961, and grown on, AB & BMCM, Hb. Lousley; 04B3, below Derry's Wood, 1953, ECW & NYS, Hb. Lousley, etc. (*Proc. BSBI*, **1**, 99, 1954).

Leontodon L.
L. autumnalis L.   Autumn Hawkbit.   Native.   1836.   Frequent.
Meadows, pastures and roadsides.

L. hispidus L.   Rough Hawkbit.   Native.   1836.   Common.
Meadows, pastures, roadsides, especially on the chalk.

L. taraxacoides (Vill.) Mérat (*L. hirtus* auct., *Thrincia hirta* Roth, *T. leysseri* Wallr.).   Native.   1792.   Frequent.
Dry grassland on chalk downs, commons and roadsides.

Picris L.
P. echioides L. (*Helminthia echioides* Gaertn.)   Bristly Oxtongue.
Native.   1633.   Frequent, 106.   Map 358.
Roadsides, hedgebanks and field borders. Mainly on clay soils, occurring on the Weald Clay, Gault, London Clay and also on Clay-with-Flints and chalk.

P. hieracioides L.   Hawkweed Oxtongue.   Native.
1832.   Rather rare, 52.   Map 359.
Dry grassland, mainly on downs and roadsides on the chalk, alluvial soils near the Thames, and railway embankments and roads made up with chalk elsewhere.

Tragopogon L.

T. pratensis L. subsp. *minor* (Mill.) Wahlenb. (*T. minor* Mill.)
Goat's-beard.   Native.   Common.
Meadows, pastures, railway banks, roadsides and waste places.
Salmon, 1931, gave a very good many records for *T. pratense* subsp.
pratensis, but this has not been recognised by our recorders.

T. porrifolius L.   Salsify.   Alien.
Occurs occasionally as a casual but never becomes a permanent
part of the flora as it does round the Thames estuary in Kent and
Essex. Our best record is from 04A5, Guildford, nr St Luke's
Hospital, 1960, BMCM.
x *T. pratensis* subsp. *minor*. 34E5, Gibbs Brook water-meadows SW
of Foyle Farm, flowers orange-purple, 1963, RARC, det. DPY.
This hybrid was found in quantity with the parents between
Putney and Barnes in 1910 and there is an interesting account by
Britton and Todd (*J. Bot.*, **48**, 203–4).

Lactuca L.
L. serriola L. (*L. scariola* L.).   Prickly Lettuce.
Naturalised alien.   1832.   Locally common, 147.   Map 360.
Waste places, roadsides, especially on disturbed or dumped soils.
First recorded for Surrey from between Mitcham and Croydon
in 1832 and yet Salmon, 1931, was only able to give two records in
addition to the first. The spread started about 1934, accelerated
during the war, and has been rapid since 1950. In other counties
round London there has been a rather similar timetable. The
spread has been by wind dispersal from central London outwards;
in part it has been along railways and main roads, and in part it has
been helped by distribution by contractors' lorries from sand and
gravel pits. One can only guess why it took more than a century to
get really aggressive, but no doubt some of the earlier occurrences
were recorded in error as *L. virosa*.

L. virosa L.   Great Lettuce.   Native?   1763.
Rare, 26.   Map 361.
Hedgebanks, ditches, roadsides and gravel pits. A fine plant sometimes about 2m tall, with purplish-black fruits, and biennial, which
is sometimes confused with *L. serriola*.

Mycelis Cass.
M. muralis (L.) Dumort. (*Lactuca muralis* (L.) Gaertn.)

Wall Lettuce. Native. 1763. Common.
Shady lanes, beech and other woods on chalk and on walls.

Sonchus L.
[*S. palustris* L. Salmon, 1931, accepted a record from the banks of the
Croydon Canal between the Dartmouth Arms and Norwood, but it has
not been reported for over a century.]

S. arvensis L. Corn Sowthistle. Native. 1836. Very common.
A creeping weed of cultivated land, marshes and riverbanks.
Plants differing in being glabrous and in other characters (*S.
uliginosus* Bieb.) which occur in Surrey are discussed by Lousley
(*Rep. BEC*, **12**, 471, 1944, and *Proc. BSBI*, **7**, 151–7, 1968).

S. oleraceus L. Smooth Sowthistle. Native.
1836. Common.
Cultivated ground, waste places, and roadsides.

S. asper (L.) Hill Prickly Sowthistle. Native.
1863. Common.
Cultivated and waste ground.

Cicerbita Wallr.
C. bourgaei (Boiss.) Beauverd
35C5, Warlingham, top of Bug Hill, a long-established garden
escape, 1958, S. Fletcher, Hb. Mus. Brit.; 1963, DPY.

C. macrophylla (Willd.) Wallr. (*Lactuca macrophylla* (Willd.)
A. Gray). Blue Sow-thistle. Established alien. Rare, 18.
Roadsides, riverbanks, waste ground. A persistent garden alien
which is increasing.

Hieracium L.
Hawkweeds are a difficult critical group in which minor variations
are perpetuated by apomixis, and a very large number of these have
been given names. The publication in 1948 of H. W. Pugsley's
A Prodromus of the British Hieracia brought together the earlier
work in orderly form and provided a sound basis for the recent
studies of P. D. Sell and Dr Cyril West and others. The present
account is based on specimens named by Pugsley, indicated by one
asterisk, and by Sell and/or West, indicated by two. A few records
prior to 1950 are included but no attempt has been made to relate

all the 26 species listed in Salmon, 1931, to modern concepts of the taxa represented.

Sect. *Amplexicaulis* Fr.

    H. speluncarum Arv.-Touv.   Alien.

    17D2, Ham House, two clumps established for many years by an old wall, 1960, RAB**, abundant on walls, DPY**. Recorded in *Lond. Nat.* **46**, 11, 1967 as *H. pulmonarioides* in error.

Sect. *Vulgata* Fr.

    H. scotostictum Hyland. (*H. praecox sensu* Pugsley).
    Alien, probably introduced during 1914–18 war.
    Rare, 13.   Map 362.
    By railways, roadsides, wood borders. First noticed in 1920 by A. H. Wolley-Dod at 35A5, Smitham Bottom in a railway cutting, and for the first 20 years all our localities were extensions of this by railway or the Brighton Road. It has since spread more widely.

    H. exotericum Jord. *sensu lato.*   Native.   Local.
    83C5, nr Churt, 1958, JEL**; 94E1, Winkworth Arboretum, 1967, JES, ECW**; 05E1, roadside by Mountain Wood, 1965, CPP**; 15C1, Denbies Drive, 1962, EAB**; 15D2, Mickleham, nr Fridley, ERD**; 35C2, Caterham, War Coppice Road, 1963, DCK**. Dr West has also seen *H. exotericum sensu stricto* in Surrey at Redhill, 1940, CW** and Limpsfield, 1949, CW**.

    H. grandidens Dahlst. Very close to the previous species, but rare. 14B5, Wotton, 1924, CEB**; 15D2, Box Hill, 1967, JRP**; 35D3, nr Woldingham Station, 1930, CEB**, 1961, RSRF; 35B2, Pilgrims Way, Caterham, 1932, CEB*.

    H. sublepistoides (Zahn) Druce.   Native.   Frequent.
    13B5, Abinger, Walliswood, 1966, DPY**; 14C5, Westcott, Bury Hill, 1961, BMCM**; 15, frequent about Box Hill, Bookham, Leatherhead, Ranmore and Polesden Lacey; 25D3, Chipstead, by High Lane, 1963, DCK**; 35B2, Caterham, under War Coppice Camp, 1970, RARC**; 35C4, Warlingham, Stuarts Road, 1954, JEL**; 45B1, Limpsfield Chart, 1965, DPY**.

H. surrejanum F. J. Hanb. Described by F. J. Hanbury in 1894 as new to science from material found near Witley, 93B4, by E. S. Marshall in 1889. It grew on the banks of a road cutting (= Brook Rocks) nr Brook, and I collected it there in 1930, and it is also recorded from Hindhead.

H. lepidulum (Stenstr.) Omang was added to the British Flora by Pugsley from 15D1, Boxhill, railway bank nr railway station, 1922 onwards, HWP, 1944, JEL*. It has also been found at 93, Thursley, 1953, JER**; 14B5, Westcott, 1928, CEB**; 25E3, Merstham, main roadside, 1956, FRB**; 35, Coulsdon, CW** and Dr West tells me that he has seen it at 96, Bagshot.

H. pollichiae C. H. Schultz (*H. roffeyanum* Pugsl.). 44A2 (?), Dormansland, 1957, FRB & CW**. It is interesting that this grew in 1937 in the garden of Lady Davy, Churchfield, West Byfleet, 06D1, det. Sell & West, as she formerly lived at Cuckfield in Sussex where it has grown in the lanes for nearly 300 years.

H. maculatum Sm.   Alien.   Rare.
05D1, Hatchlands, on verge of drive, 1962, BW**; 14B4, Wotton House, in grounds and on walls abundant, 1967, JES, BMCM (also Brewer, 1863); 15E1, Brockham Hill, 1950, JHPS**; 25B4, Tadworth Station, 1957, EAB; 35C4, Warlingham, railway bank, Stuarts Road, 1954, JEL**; 25E5, Brighton Road nr Coulsdon, 1958, CW**. Sometimes confused with *H. scotostictum* which also has spotted leaves.

H. diaphanum Fr. (including *H. anglorum* (A. Ley) Pugsl.) 37C3, Nunhead Cemetery, 1969, RMB det. CEAA. Old records from 95, nr Brookwood, 1929, JEL*; 27, Putney Heath, HWP* and 36, Addington, W. F. Miller*.

H. cheriense Jord. ex Boreau (*H. tunbridgense* Pugsl). 34E1, Felcourt, Chartham Park, 1960, JEL**; 35D4, Worms Heath, by Fairchildes Road, 1943, JEL*; 38A1, Lambeth, by Royal Festival Hall, 1966, JEL**.

H. strumosum (W. R. Linton) A. Ley (*H. chlorophyllum* auct.; *H. acuminatum* Jord. *H. lachenalii* auct.)   Native.

Frequent, 25.   Map 363.

Roadside, railway and other banks, most frequent on the chalk.

Sect. *Tridentata* Fr.

H. cantianum F. J. Hanb. 84, Farnham, Tilford Road, 1951, E. J. Gibbons**; 06B1, Sheerwater, by Basingstoke Canal, 1949, JEL; 36C3, Addington & Shirley Hills, 1953, FRB**. We also have old records for 93, Thursley; 93B4, Witley, nr Brook; 04B3, Wonersh; and 34A1, Burstow.

H. trichocaulon (Dahlst.) Johans. (*H. tridentatum sensu* Pugsl. *pro parte*).   Native.   Frequent.

93A2, Haste Hill, 1950, Haslemere Museum**; 94A3, Puttenham, Cutt Mill, 1968, DPY**; 95A2, Pirbright Common, 1966, DPY**; 96A1, Bagshot Heath, Hangmoor, 1957, JEL**; 96D4, Chobham Common, 1957, FA**; 96E5, Virginia Water churchyard, 1965, EJC**; 03A5, Hascombe, 1958, JEL**; 06B1, Woking, New Zealand Golf Course, 1962, BW**; 15D4, Leatherhead, railway embankment, 1967, DPY**; 36C3, Shirley, FRB**; 36C1, Sanderstead, Kings Wood, 1960, DPY**; 37C5, Surrey Commercial Docks, 1972, JEL, LMPS**. Probably under-recorded in the W of the county.

H. acamptum Sell & West. 93C5, Witley, 1896, ESM**. This very rare species is based on material collected by Marshall at Church Lane, Witley.

H. eboracense Pugsl. 36C3, Addington Hills, 1953, FRB**. There are old records from 04, Chilworth; 93, Witley; 84, Frensham Heights, and 83, Valley Rd*, Churt.

H. calcaricola (F. J. Hanb.) Roffey   95C4, Brookwood, railway bank, 1957, MW**; 35A4, Coulsdon Common, 1968, A. Silverside**.

Sect. *Umbellata* Fr.

H. umbellatum L.   Native.   Frequent, 32.   Map 364.

Commons and roadsides. More frequent than the map suggests and especially so on heavy soils.

Sect. *Sabauda* Fr.
H. perpropinquum (Zahn) Druce (*H. bladonii* Pugsl.).
Native. Common, 93. Map 365.
Wood borders, and tracksides in rather heathy places.

H. virgultorum Jord. Only record 95, Worplesdon, railway
bank, 1912, J. Comber** but not seen recently.

H. rigens Jord. 93B5, 'Brook nr Haslemere', 1949, RAG**;
25B4?, Burgh Heath, 1946, ECW**. Pugsley reported this
from Wimbledon, Leith Hill, Pirbright and Cobham.

H. salticola (Sudre) Sell & West 14, Leith Hill, 1923, J.
Roffey**; 27B3?, Barnes Common, 1954, CW**; 27B2,
Putney Heath, 1956, CW**.

Pilosella Hill
P. officinarum C. H. & F. W. Schultz (*Hieracium pilosella* L.)
Mouse-ear Hawkweed. Native. 1836. Common.
Dry short grassland in pastures, commons, roadsides and chalk-
downs.

P. aurantiaca (L.) C. H. & F. W. Schultz subsp. *brunneocrocea*
Pugsl.) Sell & West (*Hieracium brunneocroceum* Pugsl.).
Alien, from gardens. Rare.
Roadsides. 83E5, 1964, SFC; 93D5, Witley Common, in a very
wild-looking situation, 1960, GE; 24D5, Redhill, 1963, EMCI;
25B3, Walton Heath, plentiful, 1966, CWW; 26E2, Croydon
aerodrome, 1960, DPY**; 34D5, Godstone, by station approach,
1956, BAK.

Crepis L.
C. vesicaria L. subsp. *taraxacifolia* (Thuill.) Thell. (*C. taraxacifolia*
Thuill.) Beaked Hawksbeard. Colonist.
1849. Common.
Tracksides and waste ground, cultivated fields, meadows, with a
preference for chalk. Although not recorded until 1849, this species
was in Surrey by 1809 (Salmon, 1931, 416) and, as in other counties,
was at first confused with *C. foetida*. The rapid extension took place
in the nineteenth century, but it is probably still spreading slowly.

C. setosa Haller F.   Bristly Hawksbeard.   Alien.   1848.   Rare.
Roadsides and refuse tips, formerly in clover fields, probably
introduced with the seed. 15A5, Cobham, Downside, on roadside,
1958, DT, about 20 plants seen later by BW, JEL and others;
25A1, Brockham, refuse tip, 1968, JEL, BMCM, EMCI; 26D4,
Mitcham, beside Watermeads, 1958, RCW.

C. biennis L.   Rough Hawksbeard.   Colonist.   1805.
Rare, 19.   Map 366.
Roadsides, arable fields and waste places. Where it occurs on the
chalk it is usually abundant, in the more numerous localities on
other soils often less so.

C. capillaris (L.) Wallr. (*C. virens* L.)   Smooth Hawksbeard.
Native.   1836.   Common.
Grassland of commons, heaths and meadows, and in arable fields
and on waste ground. A polymorphic species with the two extremes
of the luxuriant plant of damp meadows and rich arable (var.
anglica Druce & Thellung) at one end of the series, and the small,
diffuse plants with few cauline leaves of rabbit-infested places (var.
diffusa DC.) at the other.

Taraxacum Weber
The study of dandelions in Britain has usually been on the basis
of division into four or five species as represented by the treatment
in Clapham, Tutin & Warburg's *Flora of the British Isles*. This has
been the standard for most of our field recording and the results
are as follows:

T. officinale Weber (*T. vulgare* Schrank).
Common Dandelion.   Native.   1836.
Abundant throughout the county in grassland and waste places.

T. palustre (Lyons) DC.   Marsh Dandelion.   Native.
Very rare and not recorded by our field workers, but *T. anglicum*
Dahlst. (see below) occurs in hay-meadows liable to seasonal
flooding.

T. spectabile Dahlst.   Red-veined Dandelion.   Native.
Local, in marshes and by streams and ponds, as, for example,
96D5, Virginia Water, 1949, AHGA, Hb. Mus. Brit; 25B1, Reigate

Heath in damp grass and nr Dungates Farm in damp cow pasture, 1956, CNHSS, det. DEA.

T. laevigatum (Willd.) DC. (*T. erythrospermum* Andrz, ex Bess.) Lesser Dandelion.   Native.   Common, 114.
Dry grassland, mainly on chalk and sand.

Some botanists prefer to treat them as a large number of mainly agamospermous microspecies grouped into sections which approximate to the species of the lumpers. This treatment was introduced into Britain by G. C. Druce about 1919 when he enlisted the help of H. Dahlstedt, a Scandinavian botanist, in naming plants collected by himself or his correspondents. Many sheets from Surrey were collected by W. A. Todd, and Dahlstedt's determinations were published in the *Report of the Botanical Society and Exchange Club of the British Isles*, and in Salmon, 1931. These are not repeated here.
A. J. Richards has recently revised the British species of *Taraxacum* using modern methods and he recognises 132 microspecies which are described, and mostly illustrated, in *The Taraxacum Flora of the British Isles*, 1972. This did not appear in time to be used in our field work, but the list below includes all species cited by Richards for V-c 17, to which has been added determinations by Richards or J. L. van Soest of the Netherlands of material we have collected or which is in the herbarium of the Natural History Museum.
The following microspecies are accepted by Richards for Surrey, V-c 17, or have been named by J. L. van Soest:

Sect. *Erythrospermum* (H. Lindb. f.) Dahlst.
    T. brachyglossum (Dahlst.) Dahlst. 27B3?, Barnes Common, 1953, AHGA, Hb. Mus. Brit., det. JLvS.
    T. argutum Dahlst.
    T. lacistophyllum (Dahlst.) Raunk.   Probably common.
94D2, Godalming, lane above Station, 1934, JEL, Hb. Lousley det. AJR; 25, Epsom Downs, 1947, AJW, Herb. Mus. Brit. det JLvS; 25B5, Epsom Downs, Great Burgh, 1955, DPY, det. DEA; 25E4, Farthing Downs, 1953, AHGA, Hb. Mus. Brit. det. JLvS; 27B3?, Barnes Common, 1953, AHGA, Herb. Mus. Brit., det. JLvS.
    T. rubicundum (Dahlst.) Dahlst. 25, Epsom Downs, 1947, AJW, Herb. Mus. Brit., det. JLvS.
    T. silesiacum Dahlst. ex Hagl.

T. disseminatum Hagl.

T. fulvum Raunk.

T. fulviforme Dahlst.

T. oxoniense Dahlst. 04D5, chalkpit behind Silent Pool, 1970, AB, det. AJR; 25Cl, Reigate Hill, 1949, AHGA, Hb. Mus. Brit, det. JLvS; 25E5, Coulsdon, Marlpit Lane, 1953, AHGA, Hb. Mus. Brit, det. JLvS.

T. glauciniforme Dahlst.

T. simile Raunk.

[Sect. *Obliqua* Dahlst.—not represented in Surrey].

Sect. *Palustria* Dahlst.

T. anglicum Dahlst.

Sect. *Spectabilia* Dahlst.

T. faroense (Dahlst.) Dahlst.

T. spectabile Dahlst.

T. laetifrons Dahlst.

T. nordstedtii Dahlst. 96D5, Virginia Water, 1949, AHGA, Hb. Mus. Brit., det. JLvS.

T. adamii Claire. 96D5, Virginia Water, 1949, AHGA, Hb. Mus. Brit, det. JLvS.

Sect. *Vulgaria* Dahlst.

T. cylanolepis Dahlst.

T. sellandii Dahlst.

T. ancistrobolobum Dahlst.

T. sublaciniosum Dahlst. & H. Lindb. f.

T. stenacrum Dahlst.

T. alatum H. Lindb.f. 04C5, Weston Wood, 1970, AB, det. AJR.

T. lingulatum Markl

T. croceiflorum Dahlst. 35A5, Coulsdon, Bradmore Green, 1953, AHGA, Hb. Mus. Brit, det. JLvS.

T. lacerabile Dahlst. 04C5, roadside verge nr Silent Pool, 1970, AB, det. AJR.

T. expallidiforme Dahlst.

T. insigne Ekman ex Raunk.

T. sublaeticolor Dahlst.

T. valdedentatum Dahlst.

T. laciniosum Dahlst.

T. cherwellense A. J. Richards

T. aequilobum Dahlst.

T. erkmanii Dahlst.

T. porrectidens Dahlst.

T. aurosulum H. Lindb. f.

T. xanthostigma H. Lindb. f.

T. mucronatum H. Lindb. f.

T. cordatum Palmg.

T. longisquameum H. Lindb.f. 04C5, Newlands Corner, 1970, AB, det. AJR.

T. dahlstedtii H. Lindb. f.

T. duplidens H. Lindb. f.

T. bracteatum Dahlst.

T. hamatum Dahlst. 36C2, Ballard's Way, Croydon, 1928, JEL, Hb. Lousley, det. AJR.

T. hamatiforme Dahlst.

T. marklundii Palmg.

T. maculatum Jordan

T. duplidentifrons Dahlst. 04B5, Colyer's Hanger, chalk hill, 1970, AB, det. AJR.; 04D5, chalkpit behind Silent Pool, 1970, AB, det. AJR.

T. hemipolyodon Dahlst. 25E4, Farthing Downs, 1923, JEL, Hb. Lousley, det. AJR.

T. polyodon Dahlst. 04B5, Colyer's Hanger, chalk hill, 1970, AB, det. AJR.

T. crispifolium H. Lindb. f.

T. obliquilobum Dahlst.

Monocotyledones
ALISMATACEAE

Baldellia Parl.

B. ranunculoides (L.) Parl. (*Alisma ranunculoides* L.)
Lesser Water-plantain. Native. c1670. Rare, 8. Map 367.
By canal, heathy pools, and lakes. A decreasing species owing to drainage and park maintenance (Richmond Park), but never as frequent as might be supposed from the statement in Salmon, 1931.

Alisma L.

A. plantago-aquatica L.   Common Water-plantain.   Native.
1666.   Common, 281.   Map 368.
By ponds, canals, streams and ditches. Distributed over almost

the whole county with the exception of the chalk and built-up areas.

A. lanceolatum With.   Narrow-leaved Water-plantain.
Native.   1713.   Uncommon, 44.   Map 369.
By canals, ponds, lakes and small streams. A very characteristic plant of almost the whole length of the Basingstoke Canal, and also common by the ponds on the Weald Clay in the SE of the county.

Damasonium Mill.
D. alisma Mill.   Starfruit.   Native.   1666.
Very rare. (Plate 24).
Small, gravelly ponds. Very erratic in appearance being apparently dependent on water level falling in early summer. Formerly in many widespread ponds but greatly decreased by Salmon's time. I have seen it in the following localities: 95E2, Stringer's Common, last seen c1965; 16B2, Esher, West End Common, last seen c1966; 25A3, Headley Heath, in two ponds some distance apart, 1944, JEL, 1958, CNHSS, 1964, RARC; 1965, JRP, 1973, 1974, JEL; 26E4, Mitcham Common, in two ponds, last seen 1940.

Sagittaria L.
S. sagittifolia L.   Arrowhead.   Native.   1597.
Rather rare, 35.   Map 370.
In shallow water in ponds, canals, ditches and rivers. Mainly by the rivers Wey and Mole and in the Basingstoke Canal.

S. latifolia Willd.   Alien.
A garden plant which persists in streams, ponds and ditches when introduced. 05C4, Ockham Mill, 1965, JRP, JM, BWu, 1974, JFL; 16E1, Lower Stew Pond, Epsom Common, 1941, JEL, 1952, F. Gregory, pond filled in and plant destroyed, 1956, DPY; 24E2, nr Langshott Manor, Horley, in roadside ditch, BMCM, 1961, FDSR, 1967, DPY.

BUTOMACEAE

Butomus L.
B. umbellatus L.   Flowering Rush.   Native.   1597.
Rare, 23.   Map 371.
Rivers and canals and ponds. Most plentiful in the Basingstoke

Plate 25 *Water Soldier*, Stratiotes aloides, *choking the waterway, Basing-stoke Canal, Ash Vale (1973)*

Plate 26 *Martagon Lily*, Lilium martagon, *Mickleham (1973), in a copse where it has been known since 1826*

Plate 27   *Wild Daffodil,* Narcissus pseudo-narcissus, *Glovers' Wood (1973)*

Plate 28   *Narrow-leaved Helleborine,* Cephalanthera longifolia, *in Surrey (1973)*

Plate 29   *Marsh Helleborine,* Epipactis palustris, *in chalk grass-land, Box Hill (1973)*

Canal, Wey Navigation, and rivers Mole and Thames. Decreasing, especially in the Thames where it was formerly plentiful.

HYDROCHARITACEAE

Hydrocharis L.
H. morsus-ranae L.   Frogbit.   Native.   1597.
Rare, 24.   Map 372.
Canal, backwaters by the Thames, ponds. Frogbit is still plentiful in parts of the Basingstoke Canal and was abundant throughout almost the full length before parts became dry. Less common than formerly by the Thames, but now known from various ponds in the S of the county, in some of which it may have been introduced as an ornamental plant.

Stratiotes L.
S. aloides L.   Water Soldier.   Alien.   1827.
Locally abundant.
Choking the Basingstoke canal, and in ponds. (Plate 25). 85E1, Ash, canal, 1963, WEW; 85E2, Ash, canal, 1956, Miss Lloyd, dominant over vast areas, 1963, WEW; 85E3, canal at Mychett, 1959, JEL; 85E4, canal, Frimley Green, 1963, WEW; 96D4, pond on Wentworth Golf Course, 1965, JES, RAB; 07A2, Runnymede, most northerly pond, very recent introduction, 1969, JES, WEW; 34C4, field pond S of South Farm, 1964, RARC. For many years Water Soldier grew in ponds at Wandsworth Common, where it was introduced from Chelsea Physic Garden, and disappeared before the end of the nineteenth century. In the Basingstoke Canal I first saw two plants in the early 1950s and it increased vegetatively so rapidly that it soon choked the canal and prevented boating. The owner of the boating business told Mr Warren that it was brought there in the turnups of Canadian soldiers' trousers— it was far more likely to be introduced by an aquarist. The other localities are certainly due to aquarists or gardeners.

Elodea Michx.
E. canadensis Michx.   Canadian Waterweed.   Alien.   1852.
Common, 88.
Slow-moving streams, ditches and ponds. The spread of this species introduced from North America commenced from Market Harborough, the centre of the English canal system, in 1847. It appeared in Surrey in ditches at Wandsworth Common in 1852,

and in ditches by the Thames between Kew and Richmond the following year, and spread rapidly. The present distribution cannot be explained solely by water transport; it still grows in canals and rivers but is far more common in ponds. In Salmon, 1931, it was said to be decreasing. It is certainly less aggressive than it was during the period of rapid spread, but during the last 40 years the population in Surrey has remained about the same.

E. ernstiae St John (*E. callitrichoides* auct. angl.).   Alien.
1957 (B. Welch in *Proc. BSBI*, **2**, 261).   Rare.
Ditches near Thames, and ponds. 06E4, boat pool beside Walton Bridge, 1967, EJC; 14C5, Bury Hill Lake, 1972, JES, JCG, ECW; 17E1, Richmond Park, Gallows Pond, 1965, EJC; 17D3, Richmond, ditch nr half-tide lock, 1956, BW, JEL (first record); 17E2, Richmond Park, Bishop's Pond, 1961, WBT, DP; 27A2, Richmond Park, Martin's Pond, 1965, BWu. A native of temperate South America, the spread of this species in Surrey is probably due to aquarists.

Lagarosiphon Harv.
L. major (Ridl.)   Moss.   Alien.
1950 (Lousley, *BSBI Year Book* **1950,** 96.)   Rare.
Ponds and lakes. 94C1, Milford Heath, pond by roadside, 1958, OVP, 1959, JEL, 1961, RWD; 14C5, Bury Hill Lake, 1972, JES, JCG, ECW; 15E5, Island Pond, Ashtead Park, 1964, EJC; 17D1, pond in gravel-pit nr Teddington Lock, 1948, JPMB, JEL, 1950, BW, pond filled in 1950; 17E1, Richmond Park, pond W of Isabella Plantation, 1957, BW, pond cleaned out and plant not seen again; 24D5, Earlswood Common, 1956, HNHS, still abundant in both ponds, 1959, JEL, abundant, 1966, DPY; 25B1, Buckland Village Pond, 1970, JEL; 15C4, Burgh Heath, 1962, DCK; 25D4, Chipstead, Elmore Pond, 1961, DCK; 25E1, Brockham, pond in wood under the chalkpits, 1964, BMCM; 26B5, Cannon Hill Park, 1956, RCW; 26C5, extension of last, 1958, RCW; 26D5, Mitcham Common, cricket green, 1956, RCW; 35C1, Godstone, pond on the green, 1959, EMCI. The rapid extension of this aggressive species is due to deliberate introductions by aquarists to maintain their tanks. In large pools it can soon become so abundant that boating and fishing have to cease. This happened at Earlswood where, in one cleaning operation, the local authority raked 40 tons of (mostly) *Lagarosiphon* from one lake and 15 tons from the other.

JUNCAGINACEAE

Triglochin L.

T. palustris L.   Marsh Arrowgrass.   Native.   1827.   Very rare, 3.
Wet grassland. 95C2, Worplesdon, Littlefield Common, 1964,
WEW; 06D4, Chertsey Mead, 1950, BW, 1960, RAB; 26E5,
Mitcham Common, nr small pond, 1958, RCW, JEL, pond since
filled in and plant destroyed. Earlier records include 05A5, Woking,
abundant in swamp N of canal, 1932, JEL; 26B1, Buckland, field
by Dungate's Farm, 1928, ECW.

[*T. maritima* L. Sea Arrowgrass. Native. 1910. Extinct. Was found on
the bank of the tidal Thames at Putney in 1909 and 1910, but not seen
since, Salmon, 1931.]

POTAMOGETONACEAE

Potamogeton L.

Surrey today is rich in records of pondweeds, but is now far from
being a rich county for living plants. The losses are due mainly to
heavy traffic on the Thames, and the disuse of the Wey and Arun
Canal and the Basingstoke Canal. British material has been
thoroughly revised by J. E. Dandy and Sir George Taylor and we
are greatly indebted to them for many determinations of Surrey
material. Specimens they have named are indicated by an asterisk;
in many cases not so marked, Dandy and Taylor have named
pre-1950 gatherings from the same localities.

P. natans L.   Broad-leaved Pondweed.   Native.
1863.   Frequent, 62.
Ponds, rivers, canals, slow streams, and ditches, in basic or neutral
waters.

P. polygonifolius Pourr.   Bog Pondweed.   Native.
1863.   Frequent, 50.   Map 373.
Ponds, bog-pools and ditches with acid water. Mainly on boggy
heathland towards the W of the county, and there sometimes locally
abundant.

P. lucens L.   Shining Pondweed.   Native.   1847.
Rare, 6.   Map 374.
Ponds and rivers. Now restricted to the SW of the county at 84,
Frensham Little Pond, Stockbridge Pond and R. Wey; 94, Enton,
and 04A3, river between Gosden and Wonersh. It seems to have

gone from the Wey Navigation in which it grew at Send and Ripley in 1931, and from the Basingstoke Canal, in which it was plentiful in 1930 and 1931 from Byfleet to Woking and beyond, and we have not seen it in the R. Mole.

[x *P. natans=P. x fluitans* Roth Recorded by Beeby from the Wey and Arun Canal, about 2 miles below Tickner's Heath in 1888, but Salmon was unable to refind it in 1917, and this stretch of the canal is now drained.]

[x *P. perfoliatus=P. x salicifolius* Wolfg. (*P. decipiens* Nolte ex Koch. Found in the same canal by Beeby in 1884 nr Alford, and above and below Tickner's Heath, but not reported since.]

P. gramineus L. (*P. heterophyllus* Schreb.)
Various-leaved Pondweed.   Native.   1849.   Very rare, 5.
Canal and lake. Thirty years ago this was still plentiful in many stretches of the Basingstoke Canal from Byfleet to Ash Vale; owing to the disuse of the canal it is now very rare. Even the following distribution may now be an overstatement: 85E2, canal nr Ash Vale, 1961, JEL*, 1963, WEW, 1974, JEL; 85E3, Mychett Lake by towpath, 1947, RAB, 1966, BWu; 85E4, canal nr Frimley Green, 1947, RAB; 06B1, Byfleet, in canal towards Woking, 1949, JEL, 1951, RWD; 34C1, Hedgecourt Pond, 1963, RARC*.

x P. natans=P. x sparganifolius Laest. ex Fr. is recorded from 95 by Perring (*Supplement*, 1968, 137).

[x *P. lucens=P. x zizii* Koch ex Roth Only record: 34C1, Hedgecourt Mill Pond, 1883, WHB* (*Rep. BEC*, **1**, 95, 1885 & *J. Bot.* **21**, 348, 1883).]

[x *P. perfoliatus=P. x nitens* Weber was found in the Basingstoke Canal nr Pirbright by Straker (*J. Bot.*, **22**, 300, 1884), and was abundant in many places from Cowshot Bridge to Frimley Lock, Bennett—see Salmon, 1931, 619. Last report: 85E4, Basingstoke Canal nr Frimley, 1931, ECW (*Watson BEC Rep.*, **4**, 142, 1932), as *P. gramineus* var. *fluviatilis* Fries.]

P. alpinus Balb.   Reddish Pondweed.   Native.
1863.   Rare, 7.
Canal and ditches. 85E2, Basingstoke Canal nr Ash Vale, 1931 (as var. *palmeri* Druce), Lousley* (*Rep. BEC*, **9**, 842, 19, 1932) and by many botanists at intervals, until 1966, EJC; 85E3, canal at Mytchett, scarce, 1963, WEW; 85E4, canal nr Frimley Green, 1947, RAB; 95B4, Pirbright, still in canal by camp, 1951, RWD; 05B4, ditch nr Newark Priory, 1932, ECW, JEL, 1954, WEW, 1965, EJC, JM, 1966, DPY; 05D5, Sanway, in ditch, 1932, 1960, JEL.

[*P. praelongus* Wulf. Formerly in the Thames at Walton Bridge but not seen since 1887.]

P. perfoliatus L.   Perfoliate Pondweed.   Native.
1844.   Rare, 12.   Map 375.
Slow running rivers, canals, and ponds. Reported only from
Virginia Water and from the R. Wey and ponds in the SW. It was
never 'common and widely distributed' as stated by Pearsall in
Salmon, 1931, but had a number of scattered localities which are
now lost.

[*P. friesii* Rupr. Always rare and now probably extinct. I have specimens
from 05B4, canal nr Send, 1931, JEL*, and 05, Wey Navigation nr
Ripley, 1930, ECW*.]

P. pusillus L.   Lesser Pondweed.   Native.   1863.   Common.
Ponds. 03D4, Vachery Pond, 1957, JCG*; 04A4, Shalford, village
green pond, 1956, OVP*, 1965, EJC*; 06C1/2, Wey Navigation
between New Haw and Byfleet, 1948, MBG*; 24C5, Reigate
Priory Pond, 1966, DPY; 26D5, Mitcham, Three Kings Pond,
1956, RCW*; 26E3, Beddington Park, 1958, RCW; 26E5, Mitcham
Common, Seven Island Pond, 1969, DPY; 27E1, Tooting Bec,
1958, RCW; 27E3, Clapham Common, 1966, DPY; 34B4, Bletchingley, Langham Lodge pond, 1966, DPY. For a long list of
earlier Surrey records see Dandy & Taylor, *J. Bot.* **78**, 5–7, 1940.

P. obtusifolius Mer. & Koch   Blunt-leaved Pondweed.
Native.   1846.   Frequent, 25.   Map 376.
Basingstoke Canal, lakes, ponds. Locally comon in the NW.

P. berchtoldii Fieb.   Small Pondweed.   Native.
1863.   Rare, 7.
Ponds, lakes and R. Mole. 94C1, Witley, Enton Great Lake, 1955,
OVP; 15D3, R. Mole, S of Leatherhead, 1960, AWJ; 17D4,
ditch by Thames by Mid-Surrey Golf Course, 1966, DPY; 17E2,
Larger Pen Pond, Richmond Park, 1965, EJC; 34D1, nr Wire Mill,
small pond in field at Park Farm, 1963, RARC; 35D4, Woldingham, Slines Pond, 1968, RARC; 44A1, Dormans, Burnpit Wood,
pond, 1958, SFC*. For old records see Dandy & Taylor, *J. Bot.* **78**,
56–7, 1940.

P. trichoides Cham. & Schlecht.   Hair-like Pondweed.
Native.   1894 (Beeby, *J. Bot.*, **32**, 88).   Rare.
Ponds and lakes. 07, A1, Langham Pond, 1957, AWW*; 14D2,
Beare Green, Arnold's Pond, 1965, RARC*; 14E3, pond at Swires

Farm, 1963, ECW; 17E2, Upper Pen Pond, Richmond, 1934, JPMB; 1954, JEL; 27A2, Northern Pen Pond, Richmond, 1961, RWD; 34E3, Lingfield, small pond E of 'Waterside', 1962, RARC*; 34C1, Hedgecourt Mill Pond, 1879, 1886, WHB*, washed up at E edge, BWu, sp. *non vidi*.

[*P. compressus* L. (*P. zosteraefolius* Schumach.) Native. 1869. Extinct. Formerly near the Thames at Chertsey and Walton Bridge, but not reported since 1898; 06E4, Walton on Thames, HWP, Hb. Mus. Brit.]

P. acutifolius Link   Sharp-leaved Pondweed.   Native. 1869.   Very rare.
Only one recent locality: 05B4, ditch nr Newark Mill, 1929, ECW, 1932, JEL*, seen regularly until 1965, EJC, JM, ditch overgrown and failed to find, 1972, JEL. Formerly in several localities nr the Wey and Wey Navigation from Old Woking to Weybridge.

P. crispus L.   Curled Pondweed.   Native.   1724.   Frequent, 48.
Ponds, lakes, canal, rivers, flooded sand and gravel pits.
[x *P. pusillus*=*P. x lintonii* Fryer. Very rare. 04D4, in Tillingbourne at Shere, 1904, 1909, W. Biddiscombe*, 1912, CES; 04C4, in Tillingbourne in Albury, 1932, JEL* (*Rep BEC*, **10**, 452, 1933), as *P. x bennettii*, 1933, ECW* (*Rep BEC*, **10**, 772–3, 1934). It appeared in this locality annually until it was destroyed by an extra-thorough cleaning out of the stream about 1950. 94? Guildford, R. Wey 1881, WHB* 1914, EBB*, Hb. Mus. Brit. Also given for 06 in Perring, 1968, 138. See also W. H. Pearsall, x *Potamogeton bennettii* Fryer, *Rep BEC*, **10**, 118–20, 1933].

P. pectinatus L.   Fennel-leaved Pondweed.   Native. 1838.   Rare, 23.   Map 377.
Ponds, rivers, lakes. A widespread but rather patchy distribution.

Groenlandia Gay
G. *densa* (L.) Fourr. (*Potamogeton densus* L.)
Opposite-leaved Pondweed.   Native.   1824.   Rare, 8.   Map 378.
Ponds, ditches, streams. A decreasing species, but it was never common in Surrey.

ZANNICHELLIACEAE

Zannichellia L.
Z. *palustris* L. (*Z. gibberosa* Reichb.)   Horned Pondweed.
Native.   c1683.   Rather rare, 30.   Map 379.
Ponds and R. Wey.

LILIACEAE

Narthecium Huds.
N. ossifragum (L.) Huds.   Bog Asphodel.   Native.
1640.   Locally frequent, 41.   Map 380.
Bogs and wet heathland, mainly on the large commons in the W,
where it is sometimes plentiful. The three detached localities are:
14B2, sphagnum bog, N of Leith Hill, about 15 spikes, 1962, CPP;
16B1, Oxshott Heath, plentiful, 1957–63, JES, RAB, BW, FR; 27B1,
Wimbledon Common, Farm Bog, 1950, AWJ, 1955, JES.

Convallaria L.
C. majalis L.   Lily-of-the-Valley.   Native (also garden escape).
1785.   Rare, 14.   Map 381.
Dry woods and copses. Certainly native in Dulwich Woods on
London Clay light soils, whence it was first recorded by Curtis in
1785, and on Blackheath Beds on Croham Hurst, Coombe Wood,
Shirley and in Springpark Woods just across the Kent boundary,
from which area records go back for over a century. Lily-of-the-
valley is also thought to be native in various localities in the W
round Witley, Milford and Pirbright. It is still at 95C2, Merrist
Wood, Worplesdon from which it was recorded by E. Capron in
Brewer, 1863. Many records as an obvious garden escape have been
omitted from the map, but we have included eleven where it is well
established or in some cases perhaps native.

Polygonatum Mill.
P. odoratum (Mill.) Druce (*P. officinale* All.).   Garden escape.
35C4, Warlingham, Stuart Road, among bushes on chalky roadside, 1956, DMcC.

P. multiflorum (L.) All.   Common Solomon's Seal.
Native?   1597.   Rare, 11.   Map 382.
Woods and copses. Possibly native in a few stations, of which
those in 84C2, 96E4, 14B4 are the most convincing, but often
bird-sown from gardens. In seven cases careful examination has
shown that the plant is the garden hybrid (see below), these are
marked with an 'x' on the map and no doubt others should be so
referred.
x *P. odoratum*=*P. hybridum* Brügger. This is the usual plant grown
in gardens which is easily established in the wild—see above.

Maianthemum Weber

M. bifolium (L.) Schmidt   May Lily.

94D1, in a wild part of the garden of 'Juniper Hill', Hydon Heath, naturalised, 1966, M. Hills. In a garden and no doubt planted, but the previous owner is sure that it was not planted while she lived there. The house was built by Dawson Turner (1775–1858) for a member of his family and the May Lily is quite a likely plant for this botanist to provide as an interesting plant to grow.

Asparagus L.

A. officinalis L. subsp. *officinalis*.

Garden Asparagus is freely distributed by birds that feed on the fruits and void the seed to grow into plants which appear, and thrive for a time, in hedgerows and other places.

Ruscus L.

R. aculeatus L.   Butcher's-broom.   Native.   1836.
Frequent, 52.   Map 383.

Woods and hedges. In addition to the native localities there are 11 where it was probably planted. Butcher's-broom spreads by creeping rhizomes so that there are often large patches with all the plants of one sex.

Lilium L.

L. martagon L.   Martagon Lily.   Naturalised alien.
1819.   Rare, 4.

Coppiced woods on chalk. (Plate 26). 15, D1, Boxhill, nr Stepping Stones over R. Mole, 1953, PRN, known to WHS for some years but regarded as planted (*Lond. Nat.* **53**, 52, 1954); 15E2, Headley Lane, in oak coppice, 1826, E. Newman, 1840, in abundance, G. Luxford (*Phyt.*, **1**, 62, 1841), extended for 'nearly half a mile', Brewer, 1856, 50, also ¼ mile from copse, 1962, HM-P, recorded regularly, and still plentiful in SNT reserve. John Stuart Mill was another early discoverer of this locality. In a letter to Sir William Hooker dated 26 January 1831 he says he found it about four years earlier and observes, as so many later observers have, that it was abundant after the wood had been coppiced; Nower Wood, found from directions over a century old, c1930, AHC, JEL, about 30 plants, 1948, JESD, 13 plants, 1962, RARC, now in SNT reserve; 25B5, Nore Park, 2 plants in derelict orchard, 1972, BRR; 26E1, 'Copse on grounds of Mr Reid at Woodmanstone ... known to the inhabitants of the village under the name of Turk's-cap

Shaw . . . remembered by the older people of the neighbourhood to have flourished truly wild in that locality for more than half a century', Bromfield in Hooker, *Engl. Bot. Suppl. t.* 2799, 1837; Ruffett, Big Wood, 1947, CNHSS, wood contains cultivated cherries, etc., 1952, CTP, JEL, HB, Smith's Wood, Woodcote (the same), 1968, R. O. Britten comm. HB, see also *Lond. Nat.* **32,** 80, 1953. Martagon, or Turk's-cap Lily, is well known to gardeners as appearing in orchards and the wilder parts of gardens in places where it has not been planted. The seeds are flat papery disks like confetti and are probably wind distributed; but jays will pull the bulbs out of the ground and fly away with them, and some of the occurrences are more likely to be due to bird carriage. It was formerly in other Surrey localities. Widespread in the mountains of Europe it was introduced into English gardens about 1596, and seems to have been an early escape into our three oldest Surrey localities.

L. pyrenaicum Gouan.
Garden escape W of Caterham Barracks, 1959, JPSR.

[*Fritillaria* L.
F. meleagris L. Fritillary. Doubtfully native in Britain. 1762. Grew in Thames meadows between Mortlake and Kew Bridge until 1876, but now extinct. Occasionally established in abandoned gardens.]

[*Tulipa* L.
T. sylvestris L. Formerly established in orchards and parks as a relic of ancient cultivation, but not reported recently.]

[*Gagea* Salisb.
G. lutea (L.) Ker-Gawl. Formerly in meadows nr Godalming but not reported for over a century.]

Ornithogalum L.
O. umbellatum L.   Common Star of Bethlehem.
Denizen.   1548.   Uncommon, 54.   Map 384.
Roadside verges, fields in grass, commons, and copses. On a wide range of soils though at its best on sand and chalk.

[*O. nutans* L. Alien. Formerly well established in at least four places round Reigate, and also elsewhere, but not reported since 1922.]

[*O. pyrenaicum* L. Spiked Star-of-Bethlehem. Alien. 24E2, Horley, Langshott Wood, 1 plant only, 1947, 1948, BMCM; 34A3, Langshott

Wood, in dense woodland, few, 1945, 1947, 1949, BMCM. Not seen since last dates given.]

Scilla L.

S. autumnalis L.   Autumn Squill.   Native. 1666. Very rare. Gravelly pastures and by towpath. 16B5, Thames-side outside Hurst Park Race Course, 1944, BW; 16C5, Hurst Park Race Course, at SE perimeter, 1948, 1949, WEF, 1962, BW; nr old gravel workings on Hurst Park Race Course over area of 45 x 15m, 1962, BW, JES, 1963, JEL. This is the 'Molesey Hurst' site, recorded three centuries ago, and the gravel-pits are specifically mentioned by H. C. Watson as 'near the ferry to Hampton', Brewer, 1863. Much of the area has been destroyed by building and the Autumn Squill may now be extinct. 17D2, on gravelly slope in meadow nr Ham, about 40 plants, 1944, JEL, frequent over small area, 1955, BW, a few plants S of this 1955, BW, above and below Ham Dock, a very few plants, 1958, LJJ. Ham, 1973, NH, another old locality which has greatly decreased during our survey. It is still in Hampton Court Park across the Thames in Middlesex.

Endymion Dumort.

E. non-scriptus (L.) Garcke (*Scilla non-scripta* (L.) Hoffmanns. & Link).   Bluebell.   Native.   Common.   1548.
Woods and coppices, occasionally under bracken on commons and hedgebanks. Still common throughout the county, though decreasing rapidly through trampling in places much used by the public.

E. hispanicus (Mill.) Chouard   Spanish Bluebell.   Alien.
This is the plant supplied as 'Bluebell' for gardens. It is commonly grown and occasionally established on commons and roadsides from garden outcasts.

Muscari Mill.

M. atlanticum Boiss. & Reut. (*M. racemosum* auct.).
Grape Hyacinth.   Alien.   1901.
25A2, Pebblecombe, well-established patch amongst chalk scrub far from habitation, 1965, 1966, CWW; 35D4, Nore Hill gravel workings, 1965, RARC.

[*M. comosum* (L.) Mill was found in a rough grassy field at Thursley by F. Clarke (*Rep. BEC* **8**, 761, 1929), but there are no further records.]

Colchicum L.
C. autumnale L.   Meadow Saffron.
Probably a garden escape.   1838.
35D5, Chelsham, just inside fence of Ledgers Park, 1 plant 'un-
likely to be planted just here', 1967, RARC. From 1838 to 1874
this was plentiful in a meadow on the Wray Park Estate, 'near
London Road', Reigate, 25D1, and in 1923 it was recorded from
Court Wood, Selsdon, 36D2, ABead.

Paris L.
P. quadrifolia L.   Herb Paris.   Native.   1836.
Rare, 13.   Map 385.
Damp woods, mainly on the Gault clay, where it is locally plentiful
and characteristic, but also on chalk, There are old records from
Peper Harow and Albury, but there are no localities at these places
which would seem very suitable, and it is now restricted to the
extreme E of the county.

JUNCACEAE

Juncus L.
J. squarrosus L.   Heath Rush.   Native.   1716.
Frequent, 120.   Map 386.
On acid soils on heath and moorland. Locally common on the
commons in the W.

J. tenuis Willd.   Slender Rush.   Naturalised alien.   1901.
Common, 140.   Map 387.
Footpaths, tow-paths, roadsides, tracks and waste places, especially
in gravelly and sandy places. A North American species first
recorded from Britain in 1883, and from Surrey at Horsell Moor in
1901. The small seeds are spread in mud on footwear and wheels
along tracks, and I have seen it spread during the last 45 years along
the towpath of the Basingstoke Canal from Byfleet to Ash Vale.

J. compressus Jacq.   Round-fruited Rush.   Native.
1838.   Rare, 10.   Map 388.
Banks of the Thames, and Basingstoke Canal, also in gravelly
places nr ponds. Closely associated with the Thames, except for a
few ponds a short distance away, and the canal path of the Basingstoke Canal in 06C2 which communicates with the Thames.

[*J. gerardii* Lois. Saltmarsh Rush. Native. 1916. By tidal Thames. 27B3, above Putney, CEB (*J. Bot.* **54**, 112, 1916), 1931, ECW, Hb. Lousley. Not reported since.]

J. inflexus L. (*J. glaucus* Sibth.). Hard Rush. 1838. Common. Rough grassy places on damp commons and pastures, mainly on stiff basic clays.

J. effusus L. (*J. communis* auct.). Soft Rush. Native. Very common.
Wet commons, poorly drained pastures, marshes, and open woods on acid soils.
x *J. inflexus*=*J.* x *diffusus* Hoppe Believed to be common when the parents grow together, but no special search has been made and we have only the following record: 95E2, Worplesdon, Britten's Pond, 1953, ECW.

J. subuliflorus Drej. (*J. conglomeratus* auct., *J. communis* E. Mey.)
Compact Rush. 1836. Native. Common, 250. Map 389.
Damp places on heathland, rides in acid woodland. We have included only records which have been carefully checked using modern characters and *J. subuliflorus* proved to be more common than we expected, though often only in rather small numbers.

J. subnodulosus Schrank (*J. obtusiflorus* Ehrh. ex Hoffm.)
Blunt-flowered Rush. Native. c1706.
Salmon, 1931, gives eight old records, some of which are probably errors. We have seen it only at 26E5, Mitcham Common, depression by Seven Islands Pond, 1932, ECW, 1935, JEL. The habitat was finally made unsuitable by drainage changes resulting from the construction of a large refuse tip.

J. acutiflorus Ehrh. ex Hoffm. (*J. sylvaticus* auct.)
Sharp-flowered Rush. Native. 1666. Common, 216.
Wet meadows, swampy woods, and wet heathland, on acid soils.
x *J. articulatus*=*J.* x *surrejanus* Druce This hybrid was named for Surrey by G. C. Druce (*Rep. BEC.* **8**, 876, 1929) based on specimens found by W. H. Beeby in three places near Hedgecourt Mill Pond, c34C1 (*Rep. BEC*, **1**, 79, 96, 1884 & 1885). In some other counties it is locally common and it is surprising that it has not been reported for Surrey since.

J. articulatus L.   Jointed Rush.   Native.   1666.   Common, 171.
Boggy places, wet pastures, sides of streams and ponds.

J. bulbosus L. aggr.   Bulbous Rush.   Native.   1839.
Common, 124.   Map 390.
Boggy pools, acid heaths and by acid ponds. Is now sometimes
divided into the two species which follow, the distribution on the
maps being based on determinations by P. M. Benoit and D. P.
Young:
>   *J. bulbosus* L. *sensu stricto.*   Rare, 13.
>   Mainly in the bogs in the NW.
>   *J. kochii* F. W. Schultz   Frequent, 62.   Map 391.
>   More widely distributed over acid soils.

Luzula DC.
L. pilosa (L.) Willd.   Hairy Woodrush.   Native.
1836.   Common.   Woods and hedgebanks.

L. forsteri (Sm.) DC.   Southern Woodrush.   Native.
1809.   Frequent, 103.   Map 392.
Dry woods, and hedgebanks. Absent from much of the N and NE,
but locally more frequent than *L. pilosa* on parts of the chalk,
Weald Clay and Lower Greensand.
x L. pilosa=L. x borreri Bromf. ex Bab. Not rare where the parents
grow together. 93B3, nr Frillinghurst Wood, 1958, OVP; 13D4, nr
Bonet's Farm, 1972, SFC; 15B1, NE corner The Spains, 1958,
BMCM, BW; 15D2, Box Hill, nr Headley Lane, 1961, BW.

L. sylvatica (Huds.) Gaudin (*L. maxima* (Reichard) DC.).
Great Woodrush.   Native.   1746.   Rare, 16.   Map 393.
Woods on acid, peaty soils. Established colonies are sometimes
very difficult to distinguish from native.

L. luzuloides (Lam.) Dandy & Wilmott (*L. albida* (Hoffm.) DC.,
L. nemorosa (Poll.) E. Mey., non Hornem.).
White Woodrush.   Established alien.   1879.
83E2, bank below house on Shottermill to Hindhead road, 1951,
BMCM, seen by many botanists since.

L. campestris (L.) DC.   Field Woodrush.   Native.
1597.   Common.
Dry grassy places on heaths, commons, pastures and lawns.

L. multiflora (Retz.) Lejeune (*L. erecta* Desv.).
Heath Woodrush.   Native.   1763.   Common, 257.   Map 394.
Heathland, acid pastures and peaty places in woods. Apparently avoids the London Clay and does not grow on outcrops of chalk.

[*L. pallescens* Sw. Recorded from Lady Davy's lawn at Pyrford, (*J. Bot.* **48**, 188, 1910, & *Proc. Holmesdale Nat. Hist. Club*, **1910-13**, 90, 1914), but not reported since. Probably introduced in the newly laid turf.]

AMARYLLIDACEAE

Allium L.
A. vineale L.   Crow Garlic.   Native.   1838.
Locally frequent, 74.   Map 395.
Well-drained habitats in pastures, roadsides, nr rivers and on commons. Usually the heads are made up of bulbils only (var. *compactum* (Thuill.) Boreau). Mainly by the Thames and in the N of the county, and in the Tilford-Frensham area.

[*A. oleraceum* L. was persistent in a hedgebank at Coldharbour Lane, Croydon from 1866 to 1872 (Salmon, 1931, 606), but was probably not native. No later records.]

A. roseum L.   Alien.
17D4, towpath nr Isleworth Ferry Gate, 1949, LJJ; 17E4, shrubbery by towpath by gate to Kew Palace, 1958, BW. No doubt an escape from Kew Gardens.

A. paradoxum (Bieb.) G. Don   Few-flowered leek.
Naturalised alien.   Woods and lanesides.
94C4, Compton, well established by bridle road N of Eastbury Manor, 1958, LJJ ,1974, SFC; 96D1, Chobham, small patch on grass verge, 1965, BWu; 05C4, Ripley, woodland E side of Rose Lane, 1966, JFL, 1973, SFC; 35B5, Kenley, plentiful in several places in Hayes Road and Welcomes Road area, 1958, 1959, HB. Naturalised in woods near Edinburgh since 1863 (Balfour, *Trans. Proc. Bot. Soc. Edinb.*, **7**, 458, 1863; Lousley, *Rep. BEC.* **13**, 172, 1947) and recently spreading in the S of England.

A. moly L.
25C1?, Reigate 'much of it grows wild here', 1943, Frazer-Story, Hb. Kew. We have no further information.

A. siculum Ucria
25C1?, Reigate, in a shady hedge and also in an old orchard, 1967,
Mrs J. M. Raven, comm. DMcC, Hb. Lousley.

A. ursinum L.   Ramsons.   Native.   1836.
Frequent, 115.   Map 396.
Damp woods and shady places, at its best nr streams on the heavy
soils of the Weald Clay.

Leucojum L.
L. aestivum L. Summer Snowflake. Native? 1830. Salmon, 1931, treated
it as extinct in the original Woking locality, but said that it still flourished
at one station in the county, where it was known to C. E. Britton. This
was: 26A3, by Hogsmill River nr Tolworth, 1916, W. A. Todd & CEB,
Hb. Mus. Brit., Worcester Park by the Hogsmill River, in copse where
it is crossed by the Tolworth to Ewell road, 1939, F. Ambrose. The road
has been widened and the plant is thought to be extinct. 06E4, Walton
Bridge, water meadows by the Thames, 1969, MBG, Hb. Mus. Brit.

Galanthus L.
G. nivalis L. Snowdrop. Naturalised alien. 1834. Rare, 13.
Damp woods and by streams. The Snowdrop was first recorded for
Surrey from Carshalton by J. Birch before 1834 (Baxter, *Br.
Phaenogamous Botany*, **1**, 34, 1834) and it still grows in 26E3 on
islands in the river, and in the Oaks and Ravensbury Parks, 1957,
RCW. It is of much greater interest by the R. Mole, where it grows
on alluvial soils from 24A5, nr Betchworth in profusion, 1959,
1963, BMCM, down to 15A5, nr Stoke D'Abernon. In 15D2 and
15D3, it is in eight places on both banks of the Mole between
Burford Bridge and the Leatherhead By-pass bridge, 1958, AWJ,
and perhaps the best known and most easily accessible places are in
15D1 between Burford Bridge and the Stepping Stones. The
earliest record I have traced from Box Hill area is 1836 (Cooper,
Fl. Metropolitana, 35), but William Bennett in 1851 wrote an
enthusiastic account of its abundance by the Mole at Brockham,
from which I refound it easily (*Phyt.*, **4**, 107). There is little doubt
that Snowdrop, which was very anciently cultivated in England,
started from a planted colony at Betchworth or higher up the Mole,
and has spread by being carried down in flood water. The remaining
Surrey localities appear to be more recent introductions.

G. elwesii Hook.   Alien.
95B2, Henley Park Lake, planted and semi-naturalised, 1967, JRP,

JM, DT. This locality appears to be the basis of the record 'persists in a bog near Guildford' (DMcC, *Supplement*, 42, undated-? 1957), but it is 5 miles from Guildford, not in a bog, and 'quite obviously planted'. 35D3 & D4, Woldingham, in two places by Church Road, 1972, JEL.

Narcissus L.
N. pseudonarcissus L.   Wild Daffodil.
Native (also alien).   1836.   Rare, 29.   (Plate 27).
Copses and woods, we only know one occurrence in a field where it is likely to be native (13A5, Walliswood). Restricted to Weald Clay.

N. x medioluteus Mill. (*N. x biflorus* Curt). 34D2, meadow E of Wire Mill, 1954, LJJ. Not refound in any of the old localities.

IRIDACEAE

[*Sisyrinchium* L.
S. bermudiana L. (*S. angustifolium* Mill.) In addition to the records in Salmon, 1931, 604, this alien is reported from wet heath, Chobham Common, M. J. Thomas (*Rep. BEC*, **7**, 898, 1926), but we have had no recent records.]

Iris L.
I. foetidissima L. Gladdon.   Stinking Iris.   Native.
1836.   Rather rare, 37.   Map 398.
Hedgerows, woods, copses and scrub, most common on the chalk.

I. pseudacorus L.   Yellow Iris.   Native.
1640.   Common.
Ponds, marshy places, meadows flooded in winter, by rivers and ditches. Plants with primrose yellow flowers (var. *pallidiflava* Sims; var. *bastardii* Boreau) have been known since 1892 from 25A5, Baron's Pond, Epsom, 1925, JEL, 1927, ECW; 1963, DPY, 1972, JEL. They grow with plants with petals of the normal colour.

I. germanica L.
Commonly established on roadsides and commons from garden rubbish, but seldom, if ever, seen in flower.

I. sibirica L.   Garden alien.
94E4, Shalford Common, in roadside ditch, 1963, DPY; 34B2, pondside on waste E of Smallfield Place, 1963, RARC.

Crocus L.

[*C. purpureus* Weston (*C. officinalis* Huds., *C. vernus* auct.) Alien. 1763.
26A5, New Malden, naturalised in an old pasture, 1923, Pugsley (*J. Bot.*,
42, 82–3, 1924); in great abundance, 1937, JEL, destroyed soon after by
building: There were 'feathered' and 'unfeathered' forms but, as Pugsley
pointed out, they were all smaller than the garden crocuses and probably
derived from old cultivated forms.]

C. tomasinianus Herbert
This commonly grown, February-flowering crocus is becoming
increasingly established on commons and roadsides in S England.
45B2, NE of Limpsfield Chart, 1963, RARC.

[*C. kotschyanus* C. Koch (*C. zonatus* J. Gay). An autumn-flowering
garden crocus. 94B3?, Shackleford, edge of wood, not nr houses, 1929,
F. Clarke (*Rep. BEC*, **9**, 37, 1930), still appearing, 1930 (*Rep. BEC*, **9**,
371, 1931). Not reported since.]

DIOSCOREACEAE

Tamus L.
T. communis L.   Black Bryony.   Native.   1725.   Common.
Wood-margins, hedgerows and scrub.

ORCHIDACEAE

Cephalanthera Rich.
C. damasonium (Mill.) Druce (*C. grandiflora* Gray).
White Helleborine.   Native.   1670.
Locally common, 61.   Map 399.
Woods and shady lanesides on calcareous soils and usually under
beech. Restricted to the chalk with the exception of 93B5, a tetrad
with Bargate limestone outcrops. Achlorophyllose plants have been
found at 15D1, Dorking, in little strip of wood, 1970, BMCM;
15D2, Mickleham, by steep path from Headley Lane to White Hill,
1968, DPY.

C. longifolia (L.) Fritsch (*C. ensifolia* (Schmidt) Rich; *Epipactis
ensifolia* Sw.)   Narrow-leaved Helleborine.
Native.   1667.   Very rare.
Beech woods on the chalk. (Plate 28). 04D5 (or E5?), Shere, 1893,
S. T. Dunn, Netley Wood, 1901, CES, HWP, 1960, C. E. Staples,
1961, failed to refind, VSS, NYS, PFH, 1971, not refound, JEL, CPP,

1973, again not refound, SFC; 04E5, Hackhurst Downs, 1920, Hb. Mus. Brit; 05C1, nr Clandon, 1960, C. E. Staples, 1961, 12 plants all sterile, VSS, NYS, PFH, 1973, 2 plants in flower, SFC, 1973, a second small colony, CWW, RARC; 15D2, Norbury Park and Box Hill, George Graves (Hooker, *Curtis's Fl. Lond.*, ed 2, **4**, *t.* 77, 1821), Juniper Hill, Graves, 1834, Hb. Lousley; 15E2?, Sir Lucas Pepys's woods, on Mickleham Downs, Cameron in Luxford, 1838, 77. This was near 'that far famed entomological spot, "The hilly field",' and the site may be just on the SNT Headley Warren Reserve. The plant was at times plentiful and last seen here in 1891 by A. H. Wolley-Dod.

Epipactis Sw.
E. palustris (L.) Crantz (*Serapias longifolia* auct.)
Marsh Helleborine.   Native.   1805.   Now very rare.
Chalk grassland. (Plate 29). 15D2, on Box Hill, T. F. Forster, jun., Turner & Dillwyn, 1805, 591, on or near Box Hill, Smith, Luxford, 1838, 77, failed to refind, Brewer, 1856, 121, Box Hill, 1963, E. A. E. Hills, 1964, DMTE. The rediscovery of this species on Box Hill after such a long interval is a good example of how easily a widely recognised plant may be overlooked on one of the most thoroughly botanised habitats in Britain. It is still there (1974) and, although the species is more often seen on dunes and marshes, there are other habitats on chalk in S England. Salmon, 1931, gives records from eight other localities, but none of them has been confirmed in the present century.

E. helleborine (L.) Crantz (*E. latifolia* All. *E. media* auct.)
Broad-leaved Helleborine.   Native.   1666.
Rather common, 127.   Map 400.
Woods, margins of woods, hedges, scrub and roadsides. Distributed over most of the county on a wide range of soils.

E. purpurata Sm. (*E. sessilifolia* Peterm. *E. violacea* (Dur. Duq.) Bor.).   Violet Helleborine.   Native.   1838.
Frequent, 73.   Map 401.
Woods and hedges on Clay-with-Flints and Weald Clay. Except for two clumps at 84A4, Cooper's Hill, absent from the N of the county.

E. leptochila (Godfrey) Godfrey   Narrow-lipped Helleborine.
Native.   1919.   Rare, 4.
Beech woods and belts on the chalk. (Plate 30). 05E1, W. Horsley,

Sheepleas, 1946, DPY, 1964, CNHSS; 05E2, W. Horsley, *locus classicus*, Godfery, *J. Bot.* **57,** 37–42, 1919, & Godfery, 1933, 72, seen here by many botanists since, and probably source of specimens labelled 'East Horsley' dated 1938 & 1943 in Hb. Mus. Brit., 1967, DPY; 15A2, roadside S of Effingham, 1965, DPY; 35D3, wood E of Marden Park, 1923, onwards, CBT, 1937 onwards, JEL, especially abundant in 1959, 1968, DPY; 36C1, Warlingham, Kingswood Valley, 1924, CBT.

E. phyllanthes G. E. Sm.   Green-flowered Helleborine.
Native.   1952 (Young, *Watsonia* **2,** 253–76).   Very rare, 7.
Road and track sides, usually amongst ivy in deep shade. 83D5, Churt, 1965 onwards, SFC; 84D1, Tilford, 1965, LP; 97E1, Egham, 1965, DPY; 06D2 & D3, Weybridge, 1963, LWL, 1967, DPY, BMCM, 1970, 1974, JEL; 16A1, Fairmile, 1970, AB, JES; 16B3, Mr Spicer's Park nr the railroad, Esher, nd, H. C. Watson, Hb. Kew, *Watsonia,* **5,** 138, 1962; 17E4?, between Kew and Mortlake, 1870, M. Moggridge, Hb. Mus. Brit; 35B5, Kenley, 1968, DPY.

Spiranthes Rich.
S. spiralis (L.) Chevall. (*S. autumnalis* Rich.)
Autumn Lady's-tresses. Native. 1597. Rare, 20. Map 402.
Short grassland on chalk slopes. 96D3, Chobham, side of concrete road to Longcross Halt, 3 plants, 1968, PFleB. Several calcicoles have been found on chalk rubble by this road. This is our only record off the chalk escarpment, but there were several old reports.

Listera R.Br.
L. ovata (L.) R.Br. Twayblade. Native. c1683. Common.
Moist woods and pastures on chalk and base-rich Weald Clay and other soils.

Neottia Ludw.
N. nidus-avis (L.) Rich.   Bird's-nest Orchid.   Native.
1830.   Uncommon, 48.   Map 403.
In deep humus in woods, especially under beech on chalk. In 1947 there were thousands of spikes on the slopes above Headley Lane, 15E2; they have gradually decreased since, were still abundant in 1955, but are now relatively few. Off the chalk, grows under Common Oak and other trees on various soils, but the colonies are usually smaller.

[*Hammarbya* Kuntze

H. paludosa (L.) Kuntze (*Malaxis paludosa* (L.) Sw.) Bog Orchid. Native.
Wet sphagnum bogs. Has been found in only two places: 16B2, Esher,
swamp on N side of Oxshott Hill descending to fir-woods outside Claremont Park, H. C. Watson, Brewer, 1863; 94A4, Puttenham Common,
1847, J. D. Salmon, *Phyt*, **2**, 888, 1904, HWP, 1905, T. A. Dymes, 1925,
ERD. Not reported since.]

Herminium R.Br.

H. monorchis (L.) R.Br.   Musk Orchid.   Native.
1805 (T. F. Forster, jun. in Turner & Dillwyn, 591).
Rare, 8.   Map 404.
Short grassland on shallow soils over chalk. (Plate 31). 04E5,
Hackhurst Downs, 1950, CPP; 14A5, White Down, 1956, GIC;
15D1, D2, E1, E2, many localities in this area; 25A1, nr Betchworth
Chalkpit, 1953, DHB, 1958, BMCM, CLC; 25B2, Pebblecombe,
1964, BMCM. An important experiment is being carried out with
Musk Orchids in Headley Warren reserve by marking plants as they
appear or flower each year.

Coeloglossum Hartm.

C. viride (L.) Hartm. (*Habenaria viridis* (L.) R.Br.)
Frog Orchid.   Native.   1836.   Rare, 5.
Short grassland on chalk. 04B5, Merrow Down, about 8 plants,
1963, JCG, about 80 plants, 1966, RARC, another site, 1966, JFL;
05B1, Merrow Downs, 1966, RARC; 05C1, another site, 1966,
RARC, (also reported from Merrow Downs in 1916, R. M.
Kennedy, and 1956, OVP); 15E2, below Juniper Top, 2 plants,
1958, FR, JHPS (previous record from Box Hill in 1877); 26C1,
Banstead Downs, 1953, DEK (previous records, 1836, 1926, 1929,
1947, 1948).

Gymnadenia R.Br.

G. conopsea (L.) R.Br.   Fragrant Orchid.   Native.
1706.   Locally frequent, 32.   Map 405.
Chalk grassland. Confined to the chalk where it is often plentiful.

[*G. odoratissima* (L.) Rich. One much-discussed record from between
Juniper Hill and Box Hill, 1833, is almost certainly an error as the
habitat is too dry.]

[*Leucorchis* E. Mey.
L. albida (L.) E. Mey. ex Schur. Two old records nearly a century old are
almost certainly errors.]

Platanthera Rich.

P. chlorantha (Custer) Reichb. (*Habenaria chloroleuca* Ridley).
Greater Butterfly Orchid.   Native.   1763.
Rare, 40.   Map 406.
Woods, copses, scrub, and rarely in grassland, on chalk or Weald
Clay. Usually in small numbers and decreasing.

P. bifolia (L.) Rich. (*Habenaria bifolia* (L.) R.Br.)
Lesser Butterfly Orchid.   Native.   1814.   Very rare, 3.
Bogs, boggy meadows and other acid habitats. Old records also
from woods on chalk. 96B1, Lightwater, boggy meadow with
Dactylorhiza maculata, 1961, JSW, West End Common, in sphagnum, 1953, FR; 96D4, Chobham, on edge of patch of *Narthecium*,
1955, 1956, EAB. A decreasing species.

Ophrys L.

O. apifera Huds.   Bee Orchid.   Native.   1666.
Locally common, 55.   Map 407.
Chalk grassland, pastures, chalkpits, embankments, woods. Mainly
on the chalk but also on base-rich clays and gravels. Varies widely
in numbers from year to year, and soon appears on recently disturbed soils on chalk, especially in quarries. An interesting locality
was found in 1974 by Raymond Fry in 84E1, Hankley Common,
where 12 fine plants grew near a broken-down wall in a large area
of otherwise acid soils.

[*O. fuciflora* (Crantz) Moench This rare orchid of SE Kent was recorded
from Coulsdon in 1833 on no less authority than Rev G. E. Smith. If
indeed the identification was correct it has long been extinct.]

[*O. sphegodes* Mill. Early Spider Orchid. Native, 1670. Very rare. Chalk
grassland. Old records, not all of which can be trusted, extend along the
chalk escarpment from Farnham to E of Merstham. The last report
accepted was from near Bletchingley (var. *fucifera* Sm.) confirmed by
C. E. Salmon (*SE Nat.*, **1905**, 68). The report from 15C1, Denbies, above
Dorking, J. G. Lawn (*Rep. BEC*, **11**, 43, 1936) must be regarded as an
error in view of the unreliability of the recorder. The doubts attached to
the following record are therefore particularly disappointing: 45A3,
Titsey, a single plant in chalk scrub above Limpsfield, 1959, BSB &
PRN (*Proc. BSBI*, **4**, 170, 1961). In 1942 Dr F. Rose transplanted
O. sphegodes from Queen Down Warren, Kent, to a down behind Titsey
Church, and Mr Brookes' discovery is thought to be one of the progeny.
Thus Kent has lost the root of a rarer orchid, Surrey has gained a doubtful
record, and science confused by the unknown history of an abandoned
root. See also *Orchis purpurea*.]

O. insectifera L. (*O. muscifera* Huds.)   Fly Orchid.
Native. 1778. Locally frequent, 34. Map 408.
Woods and wood borders, lanes, scrub, and abandoned quarries, always on chalk. An inconspicuous plant which varies greatly in quantity from year to year, but is exceptionally plentiful in Surrey.

[*Himantoglossum* Spreng.
H. hircinum Spreng. Lizard Orchid. Very erratic in appearance, and subject to peaks and troughs in Britain generally. The first record was from Box Hill in 1821 and it was reported from there again in 1927 (G. H. Spare, Hb. Kew)—a gap of 106 years. Most of the remaining Surrey records belong to the period 1921–7, which was a peak for Britain as a whole (Good, *New Phyt.* **35**, 142–70, 1936), and fall into three groups: (1) the Hog's Back, nr Guildford, probably the vicinity of Monk's Hatch (2) Merrow Downs, 1924, and (3) Denbies, Dorking (=Ranmore Common, CBT) 1924–6, and probably the colony which Tahourdin said produced 13 plants in 1925. The Surrey occurrences could be due to seed carried by high-level air currents from France.]

Orchis L.
O. purpurea Huds.   Lady Orchid. Native. 1832. Very rare.
Woods on the chalk. 25B2, Buckland Hills, last record 1838; 35A4, Coulsdon, in natural grass by a drive, 1945–7, Lady Goodenough, AJW (*Rep. BEC*, **13**, 309, 1948), seen several years, HB; 35E3, Titsey, a single plant, 1958, Miss D. Smith, 1959, T. A. Smith, DPY, JEL, 1968, no longer to be found, T. A. Smith; 35D4, Warlingham, 1911, 1 plant 1916, flowered 1931 (*J. Bot.*, **70**, 336, 1932), gone 1920, A.Bead; 36C1?, Selsdon, 1948, 1949, 1950, 1954, HB. *O. purpurea* has its headquarters in Kent and appears to be making attempts to spread westwards; these take the form of small numbers of plants appearing on the E side of Surrey and usually soon dying out. It is therefore most unfortunate that in 1942 Dr F. Rose sowed seed from Kent near the main road up Titsey Hill and failed to keep his experiment under close observation. The site is so near to that of the plant found by Miss Smith in 1959 that it is impossible to say whether this is a natural appearance or not—see *Ophrys sphegodes* above.

[*O. militaris* L. and *O. simia* Lam. There are old and dubious records for the Military and Monkey Orchids but, if they occurred, it is over a century since either has been reported.]

O. ustulata L.   Burnt Orchid.   Native.   1821.
Chalk grassland.   Very rare.
14A5, White Downs, L. G. Payne, c1930, Salmon, 1931, known

there to several botanists, and seen recently; 25B2, above Buckland, small colony, 1946, JEL; 25B4, Walton Downs, 1927, ECW, JEL, seen regularly, in some years plentiful, until 1939, 1 plant, 1966, M. Bell, the site is now overgrown with scrub; 25D2, Wingate Hill, 1965, DH. This was never as frequent as Salmon, 1931, suggests; it has always been erratic in appearance and is decreasing.

O. morio L.　　　Green-winged Orchid.　　　Native.　　　1827.
Rare, 15.　　　Map 409.
Usually in small numbers on chalk grassland, but most abundant in pastures which have never been ploughed, on wet clay soils. This species has suffered a spectacular decrease in the last 30 years due to the ploughing of old grassland. This has destroyed some of our best colonies on the chalk, such as 35B2, Quarry Hangers, but even more of the larger ones on clay.

O. mascula (L.) L.　　　Early-purple Orchid.　　　Native.
1799.　　　Frequent, 93.　　　Map 410.
In woods with bluebells, hazel coppices, lanesides, rough pastures. Still fairly frequent in woods on the Gault in the S of the county, decreasing on the chalk, and absent from large areas of the county. A rapidly decreasing species.

Dactylorhiza Nevski
　D. fuchsii (Druce) Soó subsp. *fuchsii* (*Orchis maculata* auct.)
　Common Spotted-orchid.　　　Native.　　　1792.
　Common, 214.　　　Map 411.
　Woods and grassy slopes on the chalk, and other base-rich soils, marshes, meadows, sandpits and lanes.
　x *D. maculata* (*Orchis transiens* Druce). 83C5, Churt, marshy meadow in valley NE of Crosswater, with both parents, 1964, SFC.
　x *D. praetermissa* = *D. x grandis* (Druce) P. F. Hunt 94E4, Shalford Common, 1944, JEL; 95, Worplesdon, boggy field off Gooseye Road, 1968, DHB, det. PFH; 34E2, S of Lingfield, 1958, DPY; 35A4, below Farthing Downs, 1964, DCK, JEL.

　D. maculata (L.) Soó subsp. *ericetorum* (E. F. Linton) Hunt & Summerhayes (*O. elodes* auct.)　　　Heath Spotted Orchid.
　Native.　　　1899.　　　Locally common, 49.　　　Map 412.
　On acid peaty ground, mainly on heaths and commons. Frequent or common on the large commons in the W and in a few scattered acid areas elsewhere.

x *D. praetermissa* (*Orchis hallii* Druce). 95A2, Ash, moist hollow on No 4 range, 1968, DPY.

D. incarnata (L.) Soó (*Orchis incarnata* L.; *O. latifolia* auct.; *O. strictifolia* Opiz)   Early Marsh Orchid.   Native.
1856 (see below).   Bogs and water meadows.
Very rare, 5.   Map 413.
subsp. *incarnata*: 06D4, Chertsey Mead, 1957, BW, 1960, RAB. The first evidence for the county is a specimen from 25B1?, nr Reigate Heath, 1856, Hb. Holmesdale NHS, det. DPY—it has not been seen there for many years. The first printed record is 27A1?, Wimbledon Common, Trimen (*J. Bot*, **2**, 90, 1862), and the plant there was in full flower as early as 27 May, 1883, de Crespigny (*Rep. Bot. Record Club*, **1883**, 47, 1884). There were many records until 1934 but H. W. Pugsley altered the labels of the specimens in his herbarium to *D. praetermissa* and C. E. Britton suggested that the latter was the only marsh orchid on the Common (*J. Bot*, **70**, 336, 1932). Subsp. *pulchella* (Druce) Soó: 94A1, Thursley Common, JEL (*Rep. BEC*, **9**, 447, 1931), 1960, DPY, 1967, PFH, 1974, JEL; 95B2, Henley Park Pond, DCK; 95B5, Colony Bog, one colony entirely white-flowered, 1968, JES; 96B1, Hagthorne Bog, several white-flowered plants on higher ground, 1972, SFC. Subsp. *incarnata* is the plant of water meadows, and subsp. *pulchella* of bogs.

D. praetermissa (Druce) Soó (*Orchis praetermissa* Druce).
Southern Marsh Orchid.   Native.   1913.   Rare, 36.
Map 414.   1915 (*Rep. BEC*, **4**, 75).
Marshes, wet meadows and commons; also on chalk downs. Wet habitats may be basic or neutral but are often markedly acid on commons in the W, as, for example, by Peatmoor Pond, 95B2. On the chalk escarpment it grows in apparently dry habitats in small scattered colonies which are sometimes persistent and accompanied by hybrids with *D. fuchsii*. A good colony appeared on 27B2, Wimbledon Common, nr the Windmill, 1969, RHJ, and a very large one on 16E1, Epsom Common, on ground ploughed during the war, 1968, D. Parr, 1969, DPY, 1971, ECW, JEL.

Dactylorhiza x Gymnadenia
D. fuchsii x G. conopsea (*Orchigymnadenia heinzeliana* (Reichardt) Camus. 35A1, nr Farthing Downs, 1954, HB, 1964, DCK. Probably under-recorded.

D. praetermissa x G. conopsea (*Orchigymnadenia wintoni* (Druce)
Tahourdin). 25A2, Pebblecombe, 1961, Mrs Dawe, confirmed
VSS, comm. BMCM.

Aceras R.Br.
A. anthropophorum (L.) Ait.f.   Man Orchid.   Native.
1798.   Locally common, 32.   Map 415.
Chalk grassland, scrub, wood borders and chalk pits. Extends the
full length of the chalk and often in large colonies.

Anacamptis Rich.
A. pyramidalis (L.) Rich.   Pyramidal Orchid.   Native.
1821.   Frequent, 32.   Map 416.
Chalk grassland, tracksides and chalkpits. Has been found off the
chalk at 94B1, Witley Common, 1964, comm. OVP (Witley Common is on Sandgate Beds, but calcareous material has been brought
in to make up tracks for the army camps); 94C3, Godalming,
Bargate escarpment SW of Charterhouse, 1950, OVP; 94D2,
Tuesley, Bargate Stone Quarry, 1950, OVP.

ARACEAE

Acorus L.
A. calamus L.   Sweet Flag.   Denizen.   1805.
Rather rare, 40.   Map 417.
River banks, ponds, lakes. Sweet Flag, a native of southern and
eastern Asia, is known to have been sent from Asia Minor to
Prague in 1557 and from Istanbul to Europe in 1574. It was first
recorded in Britain from Gerarde's garden in Holborn in 1597 and
Merrett's record 'found . . . by Mr Brown of Oxford near Hedly in
Surry' of 1666 (*Pinax*, 2) has been generally accepted as the first
record for our county. The 'Mr Brown' was William Browne of
Magdalen College, an excellent botanist who died in 1678, but
unfortunately the 'Hedly 'referred to is not the well-known Headley
in 25A3, which has no very likely habitats for Sweet Flag, but
another Headley now in Hampshire and about 2 miles from our
boundary at Churt (see Raven, C. E., 1947, *English Naturalists
from Neckam to Ray*, 318). From Tudor times *Acorus* was in
demand for medicines, and for strewing on the floors of the dwell-
ings of the wealthy and it was probably established in the N of the
county from an early date. The earliest printed record I have traced
is 'By the Thames at Walton, and Hampton. Mr Borrer' (Turner &

Dillwyn, 1805, 582). It now grows along the whole length of the Thames and Surrey Canal, along the Wey and Wey Navigation and Basingstoke Canal, and by many ornamental lakes and ponds. It is rather a shy flowerer and fruit is unknown in Britain.

Calla L.
C. palustris L.   Bog Arum.
Alien, persistent and spreading where planted.
First recorded as 'seemingly wild in N. Surrey' where it was planted in 1861 (*J. Bot.*, **11**, 339–40, 1873). This locality was apparently 16B2, Black Pond, Esher, where it was found again by C. A. Wright in 1884, and by H. P. Guppy (*Sci. Gossip*, **1895**, 109), but is now extinct. The locality well known to present-day botanists at 05D5, Boldermere (=Wisley Pond=The Hut Pond=swamp among fir-woods near Cobham) was probably the one described by H. C. Watson in *Topographical Botany*, 411, 1874, and is certainly the one known to W. H. Beeby who died in 1910, to C. E. Britton (*J. Bot.*, **53**, 177, 1915), and many later observers. It fluctuates in quantity from year to year and is still there. Also reported from 93C5, Buss's Common, well established by boggy stream, 1966, JES, and 35E1, Oxted, E of Coltsford Millpond, 1966, RARC.

Lysichiton Schott
L. americanus Hulten & St John.   Skunk Cabbage.
Alien, spreading down streams.   1951 (*Watsonia*, **2**, 109).
83E3, Nutcombe, by stream, 1964, DWF; 04A4, Shalford Common, by stream nr Bradstonebrook, 1967, SFC, Wonersh Common, small pond on N side, 1967, SFC; 05D5, Wisley Common, by streamside in wood near Haslemere, 1947, and Chobham, Woking, 1948, in Hb. Mus. Brit.

Arum L.
A. maculatum L.   Lords-and-Ladies.   Native.
1836.   Common.
Woods, copses and hedgebanks throughout the county, but most common on base-rich soils.

A. italicum Mill.   Italian Lords-and-ladies.   Garden alien.
1956 (Kent & Lousley, 280).   Rare.
93D3, roadside ditch NW of Hazel Bridge, 1962, JCG; 94E5, Guildford, R. Wey towpath, 1969, JFL; 14B5, Westcott, shady laneside, 1950, CTP & DPY; 15D2, bank of R. Mole, 1962, AB;

17D4, Thames towpath between Kew & Richmond, 1950, BW, RAB & DHK—first record; 25D2, Reigate Hill in enclosed woods, 1967, CWW; 25C1, Reigate Heath, in bracken nr cottages, 1970, EMCI; 34C5, Godstone, in vegetable garden of 'Norbright', 1965, RARC.

Dracunculus Mill.
 D. vulgaris Schott   Dragon Arum.
24E5, Redhill, a plant survived for several years in the hedge of Philanthropic Road, 1958, FGC & DPY.

LEMNACEAE

Lemna L.
 L. polyrhiza L. (*Spirodela polyrhiza* (L.) Schleiden)
Greater Duckweed.   Native.   c1706.   Rare, 35.   Map 418.
Stagnant or very slow-moving water of ponds and ditches. Frequent in the Basingstoke Canal, but otherwise scattered localities over the county, excluding the NE.

 L. trisulca L.   Ivy-leaved Duckweed.   Native.   1597.
Frequent, 81.   Map 419.
Still water of ponds and ditches in low-lying areas and especially on Weald Clay.

 L. minor L.   Common Duckweed.   Native.   1836.
Very common throughout the county.
Still water of ponds, ditches, water-tanks. Often appears very soon after the construction of new or temporary ponds.

 L. gibba L.   Fat Duckweed.   Native.   1836.   Rare, 25.   Map 420.
Still water of ponds, and ditches. Scattered over the county but perhaps most plentiful in the Basingstoke Canal.

Wolffia Hork. ex Schleid.
[*W. arrhiza* (L.) Hork, ex Wimm.   Rootless Duckweed.
Native.   c1816.   Extinct.
Salmon, 1931, gave six records from the county and one from Middlesex included in error. Our only reports are from 25B4, a small pond on Burgh Heath, 1937, 1943, 1944, 1954, JEL, and 1957, AWJ. Here it was introduced from N. Stoke, Sussex, by A. E. Ellis; it persisted until the con-

dition of the pond deteriorated and is now lost. This smallest of British plants is easily detected at a distance from the characteristic pea-green colour of ponds where it is plentiful. It may reappear.]

SPARGANIACEAE

Sparganium L.

The following is based on the revision by C. D. K. Cook ('Sparganium in Britain', *Watsonia*, **5,** 1–10, 1916) and determinations of Surrey material by D. P. Young.

S. erectum L.   Branched Bur-reed.   Native.   1800.

Common in shallow water or on exposed mud, by ponds, ditches, rivers, reed- and watercress-beds. The following subspecies occur:

subsp. *microcarpum* (Neuman) Domin   Rare, 8.

Probably under-recorded but seems to be most frequent on the clays of the SE.

subsp. *neglectum* (Beeby) Schinz & Thell. (*S. neglectum* Beeby). Frequent, 43.   Map 421.

The most common subspecies in the county. *S. neglectum* was first noticed by Beeby in October 1883 at 'Albury Ponds, near Guildford' (04C4) and in the basins of the rivers Blackwater, Wey, Mole and Arun, and he described it as a new species with an illustration drawn from a Reigate specimen (*J. Bot.*, **23,** 26, 193–4, 1885). It still occurs at Albury.

subsp. *oocarpum* (Celak) C. D. K. Cook   Rare, 10.

Specimens from 34C1, Hedgecourt Pond, 1962, RARC & DPY were determined by Dr Cook, the remainder by Dr Young.

S. emersum Rehm. (*S. simplex* Huds. pro parte)

Unbranched Bur-reed.   Native.   1597.   Frequent, 60.   Map 422.

Shallow water in ponds, ditches, streams and rivers. Most frequent in the valleys of the Mole and Wey, and in the extreme SE of the county.

[*S. angustifolium* Michx. and *S. minimum* Wallr. have been recorded for Surrey (Salmon, 1931, 614), but have not been seen for many years and are regarded as extinct.]

TYPHACEAE

Typha L.

T. latifolia L.   Reed Mace.   Native.   1805.   Common.

In reed-swamps in lakes, ponds, canals and backwaters of rivers.

Appears very quickly in suitable new ponds, and especially those in gravel-pits, being dispersed by large numbers of small windborne seeds.

T. angustifolia L.   Lesser Reed Mace.   Native.   1821.
Frequent, 34.   Map 423.
In similar habitats to the last species but much less common.

CYPERACEAE

Eriophorum L.
E. angustifolium Honck.   Common Cottongrass.   Native.
1782.   Locally common, 56.   Map 424.
Bogs and acid heathland, mainly on the heathy commons in the W, where it is locally abundant. Perhaps now extinct on the commons round London.

E. gracile Roth.   Slender Cottongrass.   Native.
1844 (W. Borrer in Sowerby, *English Botany Suppl. t.* 2886).
Very rare.
Spongy sphagnum bogs. (Plate 3). 85E2, sphagnum bog nr Aldershot, 1885, WHB (*J. Bot.*, **23**, 311, 1885). There were numerous records from this area, variously described as Ash Vale, North Camp, etc., probably involving slight changes in locality as the habitat changed, until 1932 (Lousley in *Rep. BEC*, **10**, 453, 1933), when it was still plentiful. I saw it there at intervals until about 1950. 95B2, Pirbright Common, 1961, R. Lancaster, still plentiful, 1973, JEL; 95E2, Whitemoor (=Whitmore) Pond nr Worplesdon, 1842, W. Borrer, Hb. Mus. Brit.—see first record. The pond was being drained when the plant was discovered and it was last seen in 1848 by J. A. Brewer.
E. gracile is rare because it can exist only under a narrow range of ecological conditions which arise for a short time in the development of some bogs. It requires very liquid shallow sphagnum bogs without even a crust of more solid material and, unless there are exceptional conditions, such places rapidly develop into firmer bogs where other plants, often including *E. angustifolium*, take its place. It follows that most reported stations for *E. gracile* are ephemeral and that records for longer periods are due to new habitats opening up in the same area. With careful management it is hoped to maintain the present excellent Surrey locality.

[*E. latifolium* Hoppe   Broad-leaved Cottongrass. Recorded from 25A2, near Reigate Heath, in Brewer, 1863, 262, and 44A3, swamp draining into the Eden Brook, 1937, ABead (*J. Bot.*, **76**, 23, 1938). Both habitats are unlikely for this plant of basic swamps. Salmon rejected the first, and the second record is almost certainly an error as I found *E. angustifolium* at a place answering to the description.]

E. vaginatum L.   Harestail Cottongrass.   Native.
1724.   Rare, 14.   Map 425.
Peaty places on heathland, sometimes in rather dry habitats. In small quantity in the NW about Bisley and Chobham, in the SW about Frensham and Thursley, and 14B2, Leith Hill, Spring Bog, 1952, FR, 1957, 1963, CPP.

Scirpus L.
S. cespitosus L. (*Trichophorum cespitosum* (L.) Hartm.).
Deergrass.   Native.   1666.   Locally common in the W, 45.
Map 426.
Peaty places on bogs and heaths. Only subsp. *germanicus* (Palla) Brodesson is known for Surrey.

S. maritimus L.   Sea Clubrush.
Native near tidal Thames, probably introduced elsewhere.   1836.
Rare, 2 native+3, probably alien.
Muddy banks of the Thames and ditches, but also by inland ponds and a pool in a gravel working. There are many old records of Sea Clubrush from Battersea to Richmond but, owing to the embankment of the river now restricting possible habitats, we have seen it by the Thames only from 17D4, Kew, 1966, DPY, and further S opposite S boundary of Kew Gardens, 1950, BW, 1961, DPY, and 27B4, Barnes, river-wall S of Harrods warehouse, 1960, BW. Introduced at 96E5, Egham, pond nr Ulvercroft, 1958, WEW; 04A4, pond at Rice's Corner, behind farm, 1958, JCG— this is nr the 'Shalford Common' locality known to ECW and JEL in 1938, now gone; 35D4, Nore Hill gravel workings, 1968, RARC.

S. sylvaticus L.   Wood Clubrush.   Native.   1778.
Frequent, 123.   Map 427.
Marshes, wet places in woods, and streamsides, usually in shade.

[*S. triquetrus* L. Triangular Clubrush. Native. 1807. Extinct. Tidal mudbanks by Thames. This rare species of estuaries grew formerly by the Thames at intervals from Lambeth to about 2km above Kew Bridge.

In 1930–1 I knew three patches: 17E4, just above Kew Bridge; 17D3, opposite Old Deer Park, and 27A4, Mortlake. The one near Kew Bridge survived until about 1946 when a major reconstruction of the river-wall destroyed the last patch of estuarine mud on which the plant grew.]

S. lacustris L.   Common Clubrush, Bulrush.   Native.
1836.   Rare, 30.   Map 428.
Shallow water in rivers, lakes and ponds. A decreasing species, still to be seen in perfection in the R. Mole at Cobham, Stoke d'Abernon, Sidlow Bridge and elsewhere.
[x *S. triquetrus=S. carinatus* Sm. Doody's Clubrush. Native. 1716. Extinct. Described as new to science from Battersea, this interesting hybrid grew on estuarine mud with *S. triquetrus* nearly to Richmond, and its habitats have been similarly destroyed. It persisted at A4, Mortlake until 1937; 17D3, Old Deer Park until about 1936; and 17E4, above Kew Bridge, until about 1946. See Lousley, 1931.]

S. tabernaemontani C. C. Gmel.   Grey Clubrush.   Native.
1836.   Rare.
Formerly native on the banks of the Thames but now only in ponds and lakes where it was probably introduced. 05A1, Guildford, old sewage bed, 1970, JFL; 06A1, Horsell Common, sandpit, 1972, WEW; 13D5, pond in wood, 1962, JCG; 24D2, Horley, extensive colonies in flooded brickworks on Weald Clay, 1964, RARC; 25D2, Gatton Park, the Lake, Brewer, 1856, 1961, with *S. lacustris*, DCK; 35C1, Godstone Bay Pond, 1965, RARC.

S. setaceus L.   Bristle Clubrush.   Native.   1836.
Rare, 36.   Map 429.
Grows in a wide range of habitats, the most characteristic being bare sand which has been wet in the winter and spring; also tracks and open spaces in woods, damp roadsides, tips, streamsides and arable fields, dried up ponds, pits, drainage ditches and arable fields wet in winter. There is some evidence that in Surrey the seeds are conveyed on the wheels of vehicles.

[*S. cernuus* Vahl. The only record is from 27A1, Wimbledon Common, 1922, JEL, Hb. Lousley conf. HWP (*J. Bot.*, **74**, 200, 1936). The plant was gathered near the Beverley Brook, and as the importance of the record was fairly soon realised, I searched for it again repeatedly but failed to refind. The occurrence of this usually maritime plant is difficult to explain, though at Lane End, Bucks, the species has been found even farther from the sea.]

S. fluitans L.　　Floating Clubrush.　　Native.　　1724.
Rare, 28.　　Map 430.
Ponds and ditches, usually peaty, and on heathland. Still locally plentiful, but absent from the W and most of the S of the county.

Eleocharis R.Br.
E. acicularis (L.) Roem. & Schult.　　Needle Spikerush.
Native.　　1838.　　Rare, 10.　　Map 431.
Canals and ponds, submerged and sterile for most of the time, but flowers when the water-level is low and it is exposed on the sandy or muddy margins. Still plentiful in parts of the Basingstoke Canal, though not seen recently in others which are overgrown or permanently dry.

E. quinqueflora (F. X. Hartmann) Schwarz *Scirpus pauciflorus* Lightf.)　　Few-flowered Spikerush.　　Native.
1884.　　Very rare, 2.
Wet peaty ground in small ponds or on margin of a large one. 84C1, Frensham Little Pond, 1947, JEL, Hb. Lousley, 1948, OVP, Hb. Young, 1949, WEW—did not appear after pond was refilled with water; 96B1, nr Donkey Town, in pool on bog, 1971, JES & ECW, 1972, SFC.

E. multicaulis (Sm.) Sm.　　Many-stalked Clubrush.
Native.　　1801.　　Rare, 21.　　Map 432.
Wet peaty places on the heathy commons, of the NW and SW of the county, where it is sometimes locally common.

E. palustris (L.) Roem. & Schult.　　Common Spikerush.
Native.　　1824.　　Common.
Round ponds, marshes and ditches. Subsp. *vulgaris* Walters is the common plant in Surrey; subsp. *palustris* (subsp. *microcarpa* Walters) has not been reported recently, but it was collected at Woking, 05?, in 1882, and at Kew, 17E5, in 1928 (Hb. Kew).

Blysmus Panz.
[*B. compressus* (L.) Panz. ex Link　　Native.　　Extinct.
The only recent record is from 05A5, Woking, bogs nr Basingstoke Canal, 'Beadell', (*Rep. BEC*, **7**, 218, 1924), W. Biddiscombe (correctly), c1923, Hb. Salmon in Hb. Mus. Brit. Salmon, 1931, 625 gives two very old records in addition and also six records without authority which cannot be accepted.)

Plate 30 *Narrow-lipped Helleborine,* Epipactis leptochila, *Horsley (1973), from near the place where it was described as new to science*

Plate 31   *Musk Orchid,* Herminium monorchis, *Box Hill (1965)*

Plate 32   *Bulbous Meadowgrass,* Poa bulbosa, *a viviparous grass, Frensham (1973)*

Cyperus L.
C. longus L.   Sweet Galingale.   Established alien.
1887 (in Salmon, 1931, 623).   Rare.
Streams, ditches, wet meadows and pond margins, sometimes in very wild-looking situations. 84B2, Frensham Heights, 1960, JEL; 84C5, Farnham Park, by stream, 1965, AJS; 93C5, Witley, 1887, ESM in Salmon, 1931, 623; 04B4, Blackheath Common, nr house, 1966, JCG; 14E3, Holmwood Common, by pond, 1935, H. N. Ridley (*J. Bot.*, **73**, 362); 16C2, Arbrook, nr car park, 1958, JES— still there in 1972; 26A2, Ewell Court grounds, by stream, 1966, EJC; 26D5, Morden Hall Park, in wet meadow, 1957, RCW.

C. fuscus L.   Native.   1846.   Very rare.
This has been recorded from only one station in Surrey: 94E4, by a pond on Shalford Common. Since its discovery by J. D. Salmon in 1846 the plant has fluctuated very widely from year to year, from great abundance in years like 1949, when the pond was almost dry, to complete absence when the water is high or the pond over-grown with tall vegetation. Its requirements are a large area of black mud enriched by the droppings of geese and ducks, and freedom from competition. The pond was probably never 'peaty' in the strict sense as described in some records. It has not been seen for over 10 years, but there is little doubt that it would reappear if the pond were cleaned out and the *Typha* and rubbish removed. *C. fuscus* was discovered in the Staines area of Middlesex by T. W. J. D. Dupree in 1957 and has been plentiful there in recent years. This is now within the administrative area of Surrey though outside the Watsonian vice-county.

C. eragrostis Lamk. (*C. vegetus* Willd.)
Alien, which persists elsewhere in Britain.
26E4, Mitcham, Beddington Lane pits, casual, 1956, DPY; 35C5, Warlingham, clay-chalk bank below garden, 1969, RARC, Hb. Lousley; 36C3, Shirley, allotments, believed to have been introduced with onion-seed, 1947, CTP, Hb. Lousley.

Schoenus L.
S. nigricans L.   Black Bogrush.   Native.   1793.   Very rare, 3.
Wet peaty ground. Restricted to a small area of Bagshot Heath and West End Common, 95B5, West End Common, 1953, RAB & FR; 96A2, Bagshot Heath, Lightwater Bog, 1956, FR; 96B1, Light-

water, Folly Bog, 1957, WEW & JEL; Hagthorne Bog, 1971, JES & ECW, 1972, SFC.

Rhynchospora Vahl
R. alba (L.) Vahl   White Beaksedge.   Native.   1666.
Local, 27.   Map 433.
Wet peaty places on acid heathland, on the commons in the W, sometimes occurring in drifts.

R. fusca (L.) Ait.f.   Brown Beaksedge.   Native.   1882.
Very rare, 1.
Peaty places in an area of bog. Known only from 94A1, Thursley Common, where it is sometimes plentiful but varies in quantity and locality over the years. The habitats which suit it best are those of very moist bare peat free of other vegetation (Lousley, *Rep. BEC*, **9**, 529, 1931), but in recent years it has appeared less plentifully under conditions with greater competition. It probably appears every year somewhere in the bog, but during the 45 years I have known it at Thursley there have been several years when I was unable to find it.

Carex L.
C. laevigata Sm. (*C. helodes* Link).   Native.   1836.
Locally frequent in the S of the county, mainly on Weald Clay or Lower Greensand, a few scattered records elsewhere, 42. Map 434.
In marshes and swampy places in woods. Still on 27B1, Wimbledon Common, in wet swampy wood on golf course, 1951, RAB.

C. hostiana DC. (*C. hornschuchiana* Hoppe, *C. fulva* auct.)
Native.   1884. (Beeby, *J. Bot.*, **22**, 300, 1884, *Rep. BEC*, **1**, 118, 1885).   Very rare and decreasing.
In base-rich swamps. 96B1, Bagshot, Folly Bog, 1955, RAB & FR; bog nr Lightwater, 1956, RAB.

C. binervis Sm.   Native.   1716.
Locally frequent, 71.   Map 435.
Heaths and rough pastures on acid soils. Most common in the W of the county.

C. demissa Hornem. (*C. tumidicarpa* Anderss., (*C. oederi* var. *oedocarpa* and most of the *C. flava* of Salmon, 1931).

Common Yellow-sedge.   Native.   Frequent, 65.   Map 436.
By ponds, marshy fields woodland rides.
x *C. hostiana* (as *C. hornschuchiana x oederi* var. *oedocarpa*), 95B3
(?), Pirbright Common, E. S. Marshall (*J. Bot.*, **37**, 251, 1899).

C. serotina Mérat (*C. oederi* auct.)   Native.   Very rare, 3.
Sandy margins of ponds exposed through fall in water level. 84C1,
Frensham Little Pond, Biddiscombe in Salmon, 1931, 1935, JEL,
1952, E. W. Davies (*Watsonia*, **3**, 67); 85E3, Mytchett Lake, where
sandy edge meets muddy margin, 1959, WEW, 1973, JEL & JES;
06B1, sandy boggy heathland nr Sheerwater, 1949, DPY, conf.
EN—site now built over.

C. sylvatica Huds.   Wood-sedge.   Native.
1836.   Common.
Woodland rides throughout the county but especially on clay soils.

C. depauperata Curt. ex With.   Native.   1794.
Very rare, perhaps extinct, 1.
Dry woods on Bargate Stone. 94D2, Godalming district. During the
last 180 years this has been found in a number of places near
Godalming and was sometimes fairly plentiful after the woods had
been felled or coppiced. I have seen it in two places—one below
Charterhouse in 1938 but it disappeared during the war, and another
where it has steadily decreased. Here there were 5 large plants in
1949, JEL, 5 in 1961, RWD, 1 in 1967, SFC; this has been watched
annually getting smaller and ceasing to flower, until in 1973 we
failed to refind it, JEL & JES. Efforts have been made to refind it in
15, chalk-pit near Effingham, whence it was recorded in 1874 by
W. W. Reeves (*J. Bot.*, **12**, 1874), but without success. It may be
found yet in a wood on chalk, probably on a steep slope.

C. pseudocyperus L.   Native.   1836.
Frequent, 76.   Map 437.
By canals, ditches, ponds, and slow-flowing streams, often in the
shade of trees. This handsome sedge is especially plentiful by the
Basingstoke Canal and R. Wey and its branches.

C. rostrata Stokes (*C. inflata* auct.)   Bottle Sedge.
Native.   1794.   Very local, 26.   Map 438.
In open water in acid pools, ponds by canal, bogs, water meadows.

x *C. vesicaria=C. x involuta* (Bab.) Syme   16, Esher, 1866,
Hb. Watson, *Rep. BEC.*, **5**, 405, 1919.

H. C. Watson recorded *C. vesicaria* from several places round
Esher, but I have not traced a record for *C. rostrata*. If the two
parents were present the hybrid is not unlikely as it is easily found
by the lochs in Scotland where the parents are abundant.

C. vesicaria L.   Bladder-sedge.   Native.
1836 (Cooper, *Fl. Met.*, 15)   Locally frequent, 43.   Map 439.
By ponds, swamps in woods, marshes. Most frequent on the Weald
Clay.

C. riparia Curt.   Native.   1782.   Frequent, 58.   Map 440.
By slow streams and in marshy meadows, also by ponds and lakes.

C. acutiformis Ehrh. (*C. paludosa* Gooden.)   Native.   1782.
Rather common, 90.   Map 441.
By slow-moving streams, marshy meadows, swamps in woods,
roadside verges, rarely by ponds.

C. pendula Huds.   Native.   1732.   Common, 185.   Map 442.
Damp woods on clay, of which it is a characteristic species, also
planted as an ornamental plant and rather frequent in places where
its status is in doubt. Mainly on Weald Clay, Gault, Clay-with-
Flints and London Clay, but there are a few records in areas where
clay is not known to be exposed.

C. strigosa Huds.   Native.   1838.
Locally common, 49.   Map 443.
Rides and by streams in damp woods on clay. Characteristic of
woods on Gault and Weald Clay, but also on Clay-with-Flints.
The first record gives it from Walton, Weybridge Common and
'boggy woods between Peslick and Ewhurst,' and has been accepted
by authors of earlier *Floras*, but it is not known now from any of
these places.

C. pallescens L.   Native.   1838.
Locally frequent, 54.   Map 444.
Damp woods, mainly on Weald Clay in the S of the county. The
record for 26E4, Mitcham Common, is not supported by specimens seen by Dr Young or myself.

C. filiformis L. (*C. tomentosa* auct.)   Native.
1905 (*Rep. BEC.*, **2**, 6, 1904).   Very rare, 1.
06D4, Chertsey Mead. Seen here regularly since its discovery in 1904, but great changes in the last 25 years have reduced the area of hay and drained it so that the sedge is much scarcer than before. It was 'in some quantity' in 1967, DPY, and may have benefited slightly from the spraying of the meadow with selective weed-killer in 1970. *C. filiformis* is a calcicole which, in addition to limestone pastures elsewhere, grows in hay-meadows flooded in the winter with water containing calcium carbonate. Prevention of flooding at Chertsey has changed the character of the habitat. Another, and probably similar, locality was reported from 06, nr Thorpe, by E. F. Shepherd in 1906, but this has probably been destroyed by the extensive gravel-winning activities in that area.

C. panicea L.   Carnation Sedge.   Native.   1836.
Local, 69.   Map 445.
Wet places ranging from acid heathland to basic swamps, but most common on the acid soils of the commons in the W.

C. flacca Schreb. (*C. glauca* Scop., *C. diversicolor* auct.)
Glaucous Sedge.   Native.   1799.
Common throughout the county.
Chalk grassland, clayey woods, marshes, wet roadsides and bogs.

C. hirta L.   Hairy Sedge.   Native.   1836.
Very common throughout the county.
Grassy places, such as damp meadows, waste ground, and roadsides, also woods and heaths.

[*C. lasiocarpa* Ehrh. Reported from 17D1, by Thames nr Kingston Bridge, with *Acorus*, 1966, MN & KMM, det. ACJ, but seems to be *C. hirta* var. *sublaevis* Hornem., a subglabrous form of very wet habitats.]

C. pilulifera L.   Pill Sedge.   Native.   1762.
Common, 183.   Map 446.
Peaty or sandy places on heathy commons, roadsides and occasionally in acid woodland rides.

C. caryophyllea Latourr. (*C. praecox* auct.)   Native.   1802.
Frequent, 60.   Map 447.
Chalk grassland, dry sandy grassland, and a wide range of habitats in meadows, roadsides and heathland which are difficult to classify.

[*C. elata* All. (*C. hudsonii* A. Benn.) None of the records in Salmon, 1931, 630, can be accepted. The species grows in Sussex, Hampshire and Berkshire, and may yet be found in Surrey.]

C. acuta L. (*C. gracilis* Curt.)   Native.   c1783.
Rare, 17.   Map 448.
By rivers, ponds, old canals, and in marshy meadows. Probably under-recorded but certainly rare.

C. nigra (L.) Reichard (*C. goodenowii* Gay).   Native.   1836.
Common, 127.   Map 449.
Grows in a very wide range of acid and basic habitats—by canals and ponds, marshes, wet pastures and heaths. The species is exceedingly variable in most of its characters and the name may cover several taxa. A caespitose plant at 34C1, Hedgecourt Mill Pond has attracted the interest of several botanists (Lousley, *Rep. Watson BEC*, **4**, 236–7, 1935), and other notable variants have occurred in Surrey bogs.

C. paniculata L.   Greater Tussock Sedge.   Native.   1666.
Common, 98.   Map 450.
In humus-rich waterlogged soils, especially in alder holts.
x *C. remota*=*C. x boenninghausiana* Weihe 03B3, Lakers Green, 1957, 1961, JCG det. EN; 25B1, Reigate Heath, alder swamp to E, 1933, JEL (*Rep. BEC.*, **10**, 777–8, 1934), and seen again to 1944 or later, JEL.

[*C. diandra* Schrank 1837. Several records given by Salmon, 1931, 628, but they require confirmation which is not forthcoming. The one from 05A5, Maybury, 1923, Lady Davy, is represented by a specimen in Hb. Kew which is *C. disticha*.]

C. otrubae Podp. (*C. vulpina* auct.)   False Fox-sedge.
Native.   c1800–30.   Common.
On heavy soils in damp meadows, marshes, ditches and by canals and ponds.
x *C. paniculata* 93?, Witley, 1889, E. S. Marshall (*J. Bot.*, **35**, 491–2, 1897).
x *C. remota*=*C. x pseudoaxillaris* K. Richt. (*C. axillaris* Gooden. non L.).   Locally frequent, 14.   Map 451.
Roadside ditches on clay where parents are plentiful. Mainly on Weald Clay in SE, also 15C5, Leatherhead Common, 1965, JES; 17E1, Richmond Park, 1959, 1961, BW.

C. vulpina L.   True Fox-sedge.   Native.
1947 (*Watsonia*, **1**, 57).   Very rare, 2.
Wet pasture and roadside ditch. 34E3, Lingfield, roadside ditch E
of Waterside, 1959, RARC; 44A3, Lingfield, field W of Cernes,
1947, FR (see first record), 1963, 1970, RARC, 1973, JEL. This
species is decreasing owing to the habitat becoming drier.

C. vulpinoidea Michx.   Alien, recently established.
1881 (*Rep. BEC.*, **1880,** 40.)
17E4, Kew, nr the Thames, 1880, G. Nicholson (as above); 25B3,
Banstead Heath, round Workhouse Pond, 1950, CNHSS, CTP,
JEL (*BSBI Year Book*, **1951**, 125, *Lond. Nat.* **30**, 5, 1951). When
discovered there were about 20 large clumps and the population
was maintained, or even increased, to 1961 when the pond had
become a grassy hollow, but it was still plentiful in 1971 although
much shaded by bushes. In 1973 I failed to find it, but the sedge had
persisted for at least 21 years.

C. disticha Huds.   Native.   1835.   Rare, 21.   Map 452.
Marshes and moist meadows.

C. arenaria L.   Sand Sedge.   Native.   1846.
Very local, 7.   Map 453.
On loose sandy ground. *Carex arenaria* is plentiful over a strip for
about 5 miles E from Frensham Great Ponds, and this is one of the
major inland areas in Britain for this usually maritime sedge. We
have not had recent records from 84D3, Waverley, or from 24D5,
Redhill Common, from which it has been reported.

C. divisa Huds.   Native.   1802.   Now very rare, 1.
A species of coastal marshes, found formerly in Surrey in rather dry
grassland in the Thames Valley and now restricted to one place on the
river embankment. 27B4, between Putney and Barnes on the river
embankment, 1909, CEB, 1932, 1943, JEL, 1956, 1957, 1960, BW.
Earlier records include: 26E5, Mitcham Common, 1932, ECW
(*Rep. Watson BEC*, **4,** 190, 1933), 1935, JEL, destroyed by the
construction of a refuse tip about 1950; 17E2?, Richmond Park, on
grassy slope, 1921, JF, as *C. chaetophylla* Steudel (*Rep. BEC*, **7,**
219–20, 1924; 601, 1925; **8**, 281, 1927); 1935, JEL—a slender
form of *C. divisa*.

C. divulsa Stokes   Native.   1809.   Common, 150.   Map 454.
Hedgebanks, roadsides and wood borders. Widespread but especially common on chalk and Weald Clay.

C. muricata L. (*C. pairaei* F. W. Schultz)   Native.   1914.
Common, 88.   Map 455.
Dry grassy places on sand or gravel, mainly off the chalk.
subsp. *leersii* Aschers. & Graebn. (*C. polyphylla* auct.)   Native.
1898.   Rare, 27.   Map 456.
Dry grassy places on basic soils, mainly on the chalk.

C. spicata Huds. (*C. contigua* Hoppe).   Native.   1843.
Common, 99.   Map 457.
Hedgebanks, rough grassy places, gravel pits, pathsides, on sand, gravel, chalk and other soils.

C. elongata L.   Native.   1844.   Very rare, 1.
By ditch in marshy meadow. 05B4, Newark Mill, known to many botanists since 1920, 1934, 1943, JEL, 1951, RWD, 1967, CNHSS, 1973, ECW, 1974, JEL. This is the only locality known to us since 1950, but a little earlier: 96E1, Horsell, one tuft in Carthouse Lane, 1932–43, not seen since; 05C5, Wisley, by stream, 1943, WEW, not refound.

C. echinata Murr. (*C. stellulata* Gooden.)   Star Sedge.
Native.   1836.   Locally frequent in the W, 49.   Map 458.
Damp meadows, woods and bogs on acid humus-rich soils.

C. remota L.   Native.   1827.   Common.
Damp woods, hedgerows, and sides of ditches in partial shade.

C. curta Gooden. (*C. canescens* auct.)   White Sedge.   Native.
1804.   Rare, and mainly in the W, 28.   Map 459.
Calcifuge, on peaty soil by ponds, in woods, meadows and bogs.
A decreasing species, now lost in the extreme E of the county.

C. ovalis Gooden. (*C. leporina* auct.)   Oval Sedge.   Native.
1824.   Very common, 237.   Map 460.
Rough grassy places on heaths and pastures, especially on clay.
Throughout the county, with the exception of the chalk and other

basic areas. An interesting dwarf plant with congested panicle and
bracts 7–12cm long, much overtopping the spikes (var. *malvernensis*
(Gibson) Lousley) has been found at 94, Milford, IAW (*Rep. BEC.*,
8, 590, 1928, where its constancy is discussed), and 94E4, Peasemarsh, 1938, ECW (*Rep. BEC.*, **12**, 203, 1942, where the above
name is suggested). The character of Peasemarsh has changed with
the withdrawal of grazing and the plant has not been refound.

C. pulicaris L.   Flea Sedge.   Native.   1762.
Very rare, 9.   Map 461.
In base-rich wet marshy ground in pastures and flushes. A de-
creasing species owing to the drainage of suitable habitats.

C. dioica L.   Native.   1883.   Very rare, 1.
Spongy bogs. This was first found in Surrey on Bisley Common in
95 or 96, and it still existed in the same area, 1933, ECW. There
are other records from 95B4, nr Furzehill Pond, Brookwood, and
near here, but last reported in 1899.

GRAMINEAE

Leersia Sw.
L. oryzoides (L.) Sw.   Cut-grass.   Native.   1851.   Very rare, 3.
Banks of disused Basingstoke Canal. We have found this only by
the canal in 95C4, W of St John's, locally common, 1963, WEW;
95E5, W of Goldsworth Bridge, 1962, WEW, 1965, RARC & JRP.
Canal now getting dry and overgrown; unable to find *Leersia* in
these two tetrads, 1970, WEW. 06B1, Sheerwater, long known here
in derelict barge, 1932, JEL. The barge has gradually disintegrated
and sunk and *Leersia* has moved down the bank as the water level
fell, still there, 1973, JEL. It has not reappeared at the old localities
at Shalford Common, Brockham by the R. Mole, Molesey, or by
the Wey & Arun Canal nr Tickner's Heath.

Phragmites Adans.
P. australis (Cav.) Trin. ex Steud. (*P. communis* Trin.).
Common Reed.   Native.   1797.   Uncommon, 40.
Shallow water by rivers, lakes, ponds, reservoirs and swamps,
often in habitats such as ornamental lakes where it may have been
planted. (Plate 11). Mainly in the W of the county and by the
Thames, but scattered records elsewhere.

Molinia Schrank
  M. caerulea (L.) Moench   Purple Moor-grass.   Native.
  1666.   Locally abundant, 187.   Map 462.
  Damp heathland, bogs, marshes and peaty open woods. Abundant
  wherever such habitats occur.

Sieglingia Bernh.
  S. decumbens (L.) Bernh.   Heath-grass.   Native.   1716.
  Common, 190.   Map 463.
  Heaths and commons, sometimes on sand, rarely on leached chalk
  grassland. Often on heavily trodden ground and by tracks.

Glyceria R.Br.
  G. fluitans (L.) R.Br.   Floating Sweet-grass.
  Native.   Very common.
  Ponds, lakes, ditches, canals and rivers—in shallow water and on
  the banks.
  x *G. plicata*=*G. pedicellata* Townsend.
  Not rare, 35.   Map 464.
  The records reflect the enthusiasm of our recorders, but it is probab-
  ly more frequent than the map suggests. In some localities it was
  found in the absence of one or other of the parents.

  G. plicata Fr.   Plicate Sweet-grass.   Native.   1852.
  Frequent, 78.   Map 465.
  In similar situations to *G. fluitans*.

  G. declinata Bréb.   Small Sweet-grass.   Native.   1864.
  Frequent, 121.   Map 466.
  On muddy margins of ponds, in ditches and by streams, many
  stations are where water stands in the winter and mud is exposed in
  the summer. Widespread, but the density of records in the SE, round
  Richmond and Claygate, and round Leith Hill reflect the work of
  recorders who made a special search for the grass.

  G. maxima (Hartm.) Holmberg (*G. aquatica* (L.) Wahlb.)
  Frequent, 107.   Map 467.
  Bank of canals, rivers, large ponds and lakes, and in meadows
  subject to winter flooding. Distribution almost continuous along
  the Basingstoke Canal, frequent along the Thames, Wey and Tillingbourne valleys, and also the Wandle. A hortal form, with attractive
  leaves striped with green and very pale yellow, is commonly

grown and very invasive. It persists on commons and waste places
and has been recorded, for example, from 03A4 and 14B5.

Festuca L.
F. pratensis Huds.   Meadow Fescue.   Native.   1790.
Very common, 249.   Map 468.
Meadows and pastures, especially when wet, roadsides and river-
banks. Absent from areas of acid heathland. Sown in leys and
pastures.

F. arundinacea Schreb.   Tall Fescue.   Native.
Common, 234.   Map 469.
Most common in meadows and roadsides on clay, and by the
Thames. Characteristic of these wet habitats and yet grows also
on well-drained chalk and sand.

F. gigantea (L.) Vill. (*Bromus giganteus* L.)   Giant Fescue.
Native.   c1791.   Very common, 364.   Map 470.
Damp woods and shady hedgebanks, especially on heavy soils.

F. heterophylla Lam.   Naturalised alien.   1889.   Rare, 2.
15C1, Polesden Lacey, margin of Freehold Wood, 1966, BSBI Exc;
17D4, naturalised in Queen's Cottage grounds, Kew Gardens,
1936, CEH, Hb. Lousley, 1961, BW.

F. rubra L.   Red Fescue.   Native and also introduced.
1762.   Common in grassland, meadows, heaths and roadsides.
Subsp. *rubra* is the common plant. Subsp. *commutata* Gaudin,
Chewing's Fescue, is thought to be not uncommon on well-drained
soils, especially on chalk, sand and gravel as a native. It is also
sown extensively from seed imported from Europe and the United
States, and appears as an alien on new roadsides and elsewhere.

F. ovina L.   Sheep's Fescue. Native.   1762.   Very common.
On well-drained shallow soils on chalk grassland, heathy commons
and pastures.

F. tenuifolia Sibth. (*F. capillata* auct.).
Fine-leaved Sheep's Fescue.   Native.   1887.
Frequent, 32 (under-recorded).   Map 471.
Mainly on commons on sand or gravel, but also on other wet or
dry grassland and in open places in woods.

F. longifolia Thuill. (*F. trachyphylla* (Hack.) Krajina).
Hard Fescue.   Naturalised alien.   No previous record.
Uncommon, 9.
On sand or other light well-drained soils on roadsides, waste
ground and gardens. 93B5, Witley Park, site of old mansion, 1959,
JEL; 06D1, Old Byfleet, in sandy field, 1932, JEL; 16B1, Oxshott,
sandy roadside, 1962, JES; 24E5, Redstone Hill in sandpit, 1959,
DPY; 25B1, Reigate Heath, on roadside bank, 1970, EJC; 25B5,
Epsom Downs, Great Burgh, as garden weed, 1955, DPY (believed
to have been sown as lawn grass about 1928); 26E5, Streatham, old
railway platform, 1957, RCW; 36A1, Purley, roadside verge, 1959,
DPY; 38A1, by Royal Festival Hall on waste ground, 1966, JEL.

F. glauca Lam. (*F. caesia* Sm.) A doubtful Surrey plant; although it was
accepted in Salmon, 1931, 646, it is now regarded as restricted in Britain
to E. Anglia and Lincolnshire. Smith's original description was based on
a plant from Bury St Edmunds and he adds 'Mr Dickson informed
Mr Crowe that he had long known this plant on dry ground about
Croydon, but could not find any description of it'. (*English Botany*, *t*
1917, 1808). It is likely that the plant found by Dickson was a very
glaucous form of *F. ovina* which still forms large patches conspicuous
from a distance on Mitcham Common Golf Course, 26E4, named for
me recently by Dr C. E. Hubbard.

Festuca x *Lolium*=*Festulolium* Aschers. & Graebn.
F. pratensis x *L. perenne* = x *F. loliaceum* (Huds.) P. Fourn.
(*F. loliacea* Huds.)   Native.   179.   Rather rare, 20.   Map 472.
Wet meadows and near streams and lakes. A characteristic species
of the Thames water meadows.
F. arundinacea x *L. perenne* = x *F. holmbergii* (Dorfl.) P. Fourn.
17D2(?), Ham Pits, 1947, NYS and CEH—the only record.
F. arundinacea x *L. multiflorum*. 06A5, Thorpe, edge of sown field of
L. multiflorum, *F. arundinacea* close by, 1965, EJC, conf. CEH—
the only record.

Lolium L.
L. perenne L.   Perennial Rye-grass.   Native.   1838.
Abundant throughout the county.
Pastures, meadows, waste ground, often by tracks where it with-
stands trampling. Commonly sown in grass and clover mixtures,
and very variable.

L. multiflorum Lam. (*L. italicum* A. Braun)
Italian Rye-grass.   Alien, possibly naturalised.
1840.   Common.
Sown in meadows, and persisting round the sides of fields, on roadsides and waste places. We have no evidence of persistence for a
long period.

L. temulentum L.   Darnel.   Alien.   1836.
Formerly this persisted as a colonist in arable ground, but now
restricted to appearances as a casual on refuse tips and waste
ground, where it is not rare.

Vulpia C. C. Gmel.
V. membranacea (L.) Dumort. (*Festuca uniglumis* Ait.)
Dune Fescue.   Native.   1974 (this *Flora*).   Very rare, 1.
Loose sand with other usually maritime species. 84C1, Frensham
Great Pond, locally plentiful on sandy margin, 1960, JEL & VML,
Hb. Lousley; N of Frensham Village Green at base of wall, 1960,
VML, det. FR. *V. membranacea* is a characteristic species of
coastal dunes, and is found on dunes in Sussex some 25–30 miles to
the S. It has also been found inland as a wool and cotton alien. At
Frensham it grew with, or very near to, *Poa bulbosa, Carex arenaria,
Vulpia ambigua* and other species more often seen near the sea,
and I am inclined to treat it as a native. It has not been reported
recently but the species is easily overlooked.

V. bromoides (L.) Gray (*Festuca sciuroides* Roth)
Squirrel-tail Fescue.   Native.   1856 (Brewer, p 12).
Common, 221.   Map 473.
Dry places, commons, heaths, roadsides and walls, plentiful on
open sandy ground on our commons.

V. myuros (L.) C. C. Gmel. (*Festuca myuros* L.)
Rat's-tail Fescue.   Established alien.
1856 (Brewer, p 12).   Frequent, 74.   Map 474.
Railway tracks, goods yards, sandy and gravelly fields and roadsides. Most of our records are from railways, some from gravel and
sand pits, and a very few from places where it might be native, as
for example, 84C1, sandy ground near Frensham Great Pond,
1961, VML, 1973, JEL, and 24E5, Redhill, 1856, Brewer. It is
possible that this is a rare native which has used the railway network
to spread widely.

V. megalura (Nutt.) Rydb.
26E4, Beddington Corner, old gravel pit, 1957, RCW, Hb. D. P.
Young, conf. AM. Alien, established temporarily. This differs on
such trifling characters that it hardly seems worth keeping up as a
species.

V. ambigua (Le Gall) More (*Festuca ambigua* Le Gall)
Bearded Fescue. Native. 1959 (*S.E. Nat.*, **1958,** 13).
Very rare, 1.
Loose sandy ground. 84C1, near Frensham Great Pond, 1958, JEL
& WEW, Hb. Lousley, 1967, PLT. An earlier record by H. F.
Parsons from Mitcham Common, 1911, was treated by Salmon
as an error.

Puccinellia Parl.
P. distans (Jacq.) Parl.
Is found occasionally in Surrey as a casual, and usually associated
with horse-manure. Only one native locality: 38C1, on gravelly soil
by Lavender Dock, Surrey Commercial Docks, Rotherhithe, 1972,
JEL.

Catapodium Link
C. rigidum (L.) C. E. Hubbard (*Desmazeria rigida* (L.) Tutin;
Festuca rigida (L.) Rasp.) Fern Grass. Native. 1804.
Frequent, 67. Map 475.
On chalk where free from competition, such as chalk-pits, wood
borders and track-sides, and on and below mortared walls, also
rarely on sand. A decreasing species.

Nardurus (Bluff, Nees & Schau.) Reichb.
N. maritimus (L.) Murb. Native. 1961 (*Lond. Nat.*, **40,** 18).
Very rare, 1.
On shallow chalk soil almost free from other vegetation. 25D4,
Chipstead Valley, discovered here by J. F. & P. C. Hall, 7 June
1960, and seen by many botanists since (*Proc. BSBI*, **4,** 253, 1961).
It fluctuates in quantity widely from year to year, being plentiful
when first found, decreased until very scarce in 1970, 1972, and
more plentiful in 1973. The Surrey locality is a not unexpected
westward extension of the sites on the chalk in Kent discovered by
the Halls, but it is surprising that more Surrey localities have not
been found. With us it grows in the company of *Teucrium botrys*,

a very rare species with which it is also associated in Kent and in a locality in Gloucestershire (Holland, *J. N. Glouc. Nats.*, **23**, 115–16, 1972).

Poa L.

P. annua L.　　Annual Meadow-grass.　　Native.　　1802.
Abundant throughout the county.
Gardens and arable land, lawns, meadows, paths and roadsides, by ponds and ditches and in marshes, wherever there is bare soil on which the seedlings are not crowded out by other vegetation.

P. bulbosa L.　　Bulbous Meadow-grass.　　Native.
1762.　　Very rare, 1.
Open loose sandy ground. (Plate 32). 84C1, Frensham Great Pond, 1957, JH. Varies in quantity from year to year, and 1973 was a very good one. It is sometimes recorded as 'viviparous' and sometimes as 'non-viviparous', but this seems to depend on the date and the season. In my experience early flowers never appear to be viviparous, and wet weather encourages vivipary, but this needs to be tested by experiments. The grass has not spread from the area where it was first found by John Hodgson. Hudson's record of 1762 was from near 'Clapham in Com. Surriensi', which at that time offered habitats which might be suitable.

P. nemoralis L.　　Wood Meadow-grass.　　Native.　　1830.
Very common.
Woods, shady lanesides and banks, and under trees, on sand, chalk and clay.

P. compressa L.　　Flattened Meadow-grass.　　Native.
1849.　　Frequent, 63.　　Map 476.
Dry well-drained places on and below walls, pavements, paths, railway yards, and in gardens. The frequency with which this is recorded from roadside, railway and garden habitats is probably due to the presence of suitable habitats rather than to an alien origin.

P. pratensis L.　　Smooth Meadow-grass.　　Native.
1838.　　Common.
Dry places, pastures, roadsides, wall-tops, waste places, most frequent on light soils.

P. angustifolia L. (*P. pratensis* var. *angustifolia* (L.).)   Native.
1863.   Frequent, 105.   Map 477.
Mainly on chalk grassland, but also dry places on sand and gravel,
and wall-tops.

P. subcaerulea Sm. (*P. pratensis* var. *subcaerulea* (Sm.))
Spreading Meadow-grass. Native. 1910. Apparently uncommon.
Pond margins, roadsides, wall-tops, sandy meadows, shady places
on chalk, railways, heaths. This species is not understood by Surrey
workers, and the habitats reported follow no recognisable pattern.
We have records identified by CEH, DPY and AM from the following tetrads: 84C1, 94B1, 94C1, 06B3, 06C3, 06D1, 14D5, 15B2,
16E3, 25A2, 26C1, 26D3, 26D5, 26E4.

P. trivialis L.   Rough Meadow-grass.   Native.   1838.
Very common.
Meadows, pastures, woods, waste-ground and roadsides. More
common than *P. pratensis*.

P. palustris L.   Swamp Meadow-grass.
Native, but also introduced. 1880 (Nicholson, *J. Bot.*, **18**, 381).
Rare.
By the Thames, along a canal, pondside and damp places. 95C4,
Brookwood, at frequent intervals along canal between A322 bridge
and road from Pirbright to Bisley Camp, 1955, RWD; 15C4
Fetcham Millpond, dried-up bed, 1944, JEL, Hb. Lousley, 1963,
DPY, probably exterminated here when pond was reconstructed
shortly afterwards; 15D4, Leatherhead, in two places, by a stream
and on verge of footpath, E of millpond; 17E4, Kew, banks of
Thames, 1879, Nicholson, (*Rep. BEC*, **1879**, 24); 27A4, Mortlake,
Nicholson, *loc. cit;* 27B4, bank of Thames below Hammersmith
Bridge, 1917, CEB, Hb. Mus. Brit., river wall between Putney and
Barnes, 1910, CEB. In these three areas—by the Thames from
Hammersmith to Kew, by the Basingstoke Canal at Brookwood,
and about Fetcham Millpond—*P. palustris* is, by analogy with
occurrences elsewhere, almost certainly native. It may also have
been so in 05 and 06, where Lady Davy found it at Pyrford (Salmon,
1931, 644) and at Byfleet in 1931 (*Rep. BEC*, **9**, 677, 1932), on
building sand and nr allotments but in an area of streams and
canals. Probably under-recorded.

P. chaixii Vill.   Broad-leaved Meadow-grass.
Naturalised alien.   1853.   Rare.
In woodland on large estates, where sown as an ornamental grass.
94D5, Dean Bottom, in woodland, 1966, CPP; 14B5, Westcott,
wood adjoining the Rookery, 1959, 1960, BMCM; 17D4, Kew
Gardens, Queen's Cottage Grounds, well known as established here
and collected in 1933, CEH, Hb. Lousley. Recorded from Kew by
'Dr Hooker' in 1879, (Watson, *Comp. Cyb. Brit.*, **3,** 594).

Catabrosa Beauv.
C. aquatica (L.) Beauv.   Whorl grass.   Native.   1847.
Rare, 14.   Map 478.
Muddy margins of ponds and ditches and by small streams. A
decreasing species in Britain generally, but always scarce in Surrey.

Dactylis L.
D. glomerata L.   Cocksfoot.   Native.   1838.
Common throughout the county.
Meadows, pastures, roadsides, waste places and occasionally in
woods.

Cynosurus L.
C. cristatus L.   Crested Dog's-tail.   Native.
1826.   Abundant.
Meadows, chalk grassland, roadsides, pastures, usually on dry soils.

C. echinatus L. Rough Dog's-tail. Alien. 1866. Found occasionally,
but not persisting in Surrey.

Briza L.
B. media L.   Common Quaking-grass.   Native.   1795.
Locally abundant, 126.   Map 479.
Abundant in calcareous grassland and basic alluvial meadows by
the Thames. Elsewhere scattered colonies on railway embankments
and roadsides made up with chalk foundations, and in old meadows
and marshes which are basic or neutral.

Melica L.
M. uniflora Retz. Wood Melick. Native. 1836. Very common.
Woods and shady hedgebanks, especially abundant in some of our
beechwoods on the chalk.

Bromus L. (including *Ceratochloa* Beauv.; *Serrafalcus* Parl.; *Anisantha* C. Koch; and *Zerna* Panz.)

B. erectus Huds.   Upright Brome.   Native.
Before 1807 (James Crowe MS in Huds, *Fl. Angl.* Ed 2 at Linn. Soc.).   Locally abundant, 128.   Map 480.
A calcicole which is all too plentiful on chalk, also on railway embankments and roadsides where chalk has been used for foundations, in calcareous hay-meadows by Thames, and on Bargate Stone. Rarely on sandy or loamy soils.

B. ramosus Huds. (*B. asper* Murr.)   Hairy-brome.   Native.
1802.   Frequent.
Woodland, hedgebanks, and roadsides in partial shade. Probably most common on moist soils, but nevertheless frequent on chalk.

B. benekenii (Lange) Trimen   Lesser Hairy-brome.   Native.
1966 (see below).   Rare, 4.
Wooded slopes on the chalk, usually under beeches. 15A2, Effingham at White Hill, 1968, CEH; 15B1 & 15B2, Polesden Lacey, 1966, CEH; 94C5, Compton, near Monks Hatch, 1968, EJC, SFC, 1970, JEL. Salmon, 1931, cites a specimen from near the 'Plough', Camberwell, 1802 in Hb. Sowerby as the first record for Surrey (see Trimen, *J. Bot.*, **8**, 376, 1870), but Dr Melderis has examined this specimen in Hb. Mus. Brit. and identified it as *B. arvensis*. It seems that the first reliable record was made on a BSBI Field Meeting to Polesden Lacey in 1966 but has not been published.

B. sterilis L.   Barren Brome.   Native.   1836.
Abundant in built-up as well as country parts of the county.
Roadsides, hedgerows, field borders and dry waste ground.

[*B. inermis* Leyss. Hungarian Brome. Alien. Was well established at 27A1, Coombe, 1910, CEB (*Rep. BEC*, **2**, 607), and said to be naturalised, but has not been reported recently.]

[*B. madritensis* L. was treated in Salmon, 1931, as a native and seems to have been established near Battersea Old Church, 27D4, whence Curtis reported it in 1794. There are later records from elsewhere (not all reliable) as a casual, and it has not been seen recently.]

B. diandrus Roth (*B. gussonii* Parl.; *B. maximus* auct.)
Great Brome. Colonist. 1906. Rare, but increasing, 5.
On loose sandy ground by tracks and roads. 84C1, Frensham Great
Pond, 1957, VML, 1958, TBR, JEL, onwards; 95D5, Knaphill
Nurseries, small patch on sandy soil, 1966, WEW; 96B2, Windles-
ham, by track in nursery, 1965, WEW; 06E3, Hersham, Burwood
Park, established for 90m along roadside verge by A317, 1965,
EJC, Walton-on-Thames, by sandy road, 1955, NYS & RAB
(probably the same locality); 25C1, Reigate, railway siding Rush-
worth Road, 1928 for many years, gone before 1963. Of the earlier
records given by Salmon, 1931, 648, two were casuals, but Miss
Saunders' 1906 record from Maybury, Woking, may perhaps have
persisted for a time.

B. rigidus Roth   Established alien.
Reported 04B4, border of cultivated field adjoining trackway below
Chilworth Manor, 1957, JCG, Hb. Gardiner det. JKO'B, 1963,
still there, JCG.

B. hordeaceus L. (*B. mollis* L.)   Soft-brome.   Native.
1836.   Abundant throughout the county.
Meadows, hay-fields, waysides, and waste ground.

B. thominii Hardouin   Lesser Soft-brome.   Alien.
1942 (Kent & Lousley, *Handlist*, 320, but no doubt some of Salmon's
records of *B. hordeaceus* var. *glabratus* (Doell.) were really this).
Common, 98.   Map 481.
Roadsides, meadows and cultivated land, but also on sandy commons. Most of our '*B. thominii*' is probably the 'inland plant'
introduced with grass-seed mixtures, which Philip Smith has
described as *B. x pseudothominii*, being a hybrid derived from
B. hordeaceus L. and *B. lepidus* Holmberg. We may also have his
B. hordeaceus L. subsp. *thominii* (Hardouin) Hylander on the sandy
commons round Frensham and elsewhere, which he accepts as a
native on the coast. (*Watsonia* 6, 327–44, 1968). Our recorders have
not found his differences sufficiently clear-cut to apply in their
field work.

B. lepidus Holmberg (*B. britannicus* I. A. Williams).
Slender Brome.   1929 (as *B. britannicus* Williams, *J. Bot.*, 67, 68–
9).   Established alien.   Frequent, 84.   Map 482.
Meadows sown with grass mixtures, cornfields, roadsides and waste

places. Williams cites Surrey records going back to 1867 and had found it himself at Horsell Common and Thursley. It seems that it was introduced with imported seed of Italian and Perennial Rye-grass grown for fodder and has persisted on roadsides, etc.

B. racemosus L.   Smooth Brome.   Native.   1802.
Rare, but under-recorded.
Water meadows, hayfields, roadsides and waste land.

B. commutatus Schrad.   Meadow Brome.   Native.
1844.   Rare, 12.
Water-meadows, hayfields, rough grassland on commons and roadside verges. Under-recorded; still quite plentiful in meadows by the Thames at Chertsey Mead and perhaps still at Runnymede, but not seen recently in meadows by the R. Mole at Esher, and at some of the localities shown on the map only 1 or 2 plants were seen. With softly hairy spikelets (var. *pubens* Wats.) at 95C1, Worplesdon, nr Cleygate Barn on rough heavy soil, 1966, WEW, Hb. D. P. Young det. CEH.

[*B. interruptus* (Hack.) Druce. This easily recognised species was de-scribed by G. C. Druce in 1895 and recorded from Surrey in 1904 (*J. Bot.*, **42**, 66) without locality. Fraser found it abundant in fields of sainfoin nr Selsdon, 36C1? (Salmon, 1931, 649). In 15E3, it was found nr Headley in a sainfoin field in 1931 by I. A. Williams (*Rep. BEC*, **9**, 846, 1932), and the following year by Lousley in, it is believed, the same field, which adjoined the present Surrey Trust Headley Warren Reserve, and it continued here until at least 1936 when it was collected by N. D. Simpson (*Nature in Cambs.*, **5**, 29, 1962). Apart from England the species is known only in the Netherlands, where it is introduced, but from the close association with sainfoin crops it is thought to come from some un-known overseas source with the seed. It has not been reported from Surrey for nearly 40 years.]

[*B. arvensis* L. First found in Surrey in 1802 (see *B. benekenii*), this grass has been reported several times as a casual and persisted as a colonist in fields round Box Hill from 1809 until at least 1849 (Salmon, 1931, 649). Not seen recently.]

B. secalinus L.   Rye Brome.   Colonist.   c1800–30.
Rare, 9.   Weed in cornfields, usually oats or barley.
Formerly more common.
From 25A4, Ashtead and 25A5, Woodcote Park, D. P. Young reported var. *hirtus* (Schultz) Asch. & Graebn. in 1958 and 1949; this is the usual plant in Britain.

B. carinatus Hook. & Arn.   Californian Brome.
Naturalised alien.   1938 (Hubbard, *Rep. BEC*, **11**, 672).
Locally abundant, 14.   Map 483.
Riverwall and towpath, waste ground and roadsides. 17 and 27, since Hubbard reported this grass as naturalised on Thames bank at Kew in 1938, it has spread rapidly along the towpath below the place where it was first noticed to below Barnes, and above to Ham; 25A5, Epsom Downs, edge of chalkpit nr waterworks, 1938, AEE, Hb. Lousley; 25B5, Epsom Downs, 1974, BRR, conf. JEL; 35C5, 1972, RARC; 35D5, Chelsham, on roadside dump nr Slines Oak, 1962, FGC, 1973, RARC & JEL.

B. willdenowii Kunth (*B. unioloides* Kunth).
Rescue Grass.   Colonist.
94E4, Littleton, field border, 1935, JEL, Hb. Lousley. Here it was known to G. M. Ash 'for many years' but has not been seen recently. There are also old records as a casual from Ham, Wimbledon Common and Ewell.

Other species of *Bromus* given by Salmon, 1931, occurred only as casuals.

Brachypodium Beauv.
B. sylvaticum (Huds.) Beauv.   Wood False-brome.   Native.
1838.   Very common throughout the county.
Woods and copses and in shady places in hedgerows, roadsides and by tracks on commons on a wide range of soils.

B. pinnatum (L.) Beauv.   Tor Grass.   Native.
Before 1807 (J. Crowe).   Locally abundant, 47.   Map 484.
Dominates large areas of chalk grassland from the boundary with Kent westward, with some gaps, to Guildford, but not W of this. Off the chalk it occurs: 84C1, by Frensham Pond, a fine patch on road bank, 1961, WEW, 1962, SFC, and 17C2, Petersham Common, roadside at southern tip, 1962, BW. These may be due to the use of calcareous material for foundations of the roads.

Agropyron Gaertn.
A. caninum (L.) Beauv.   Bearded Couch.   Native.   1724.
Frequent, 106.   Map 485.
Wood borders, hedges, shady lanes. Most frequent on the chalk,

but we are unable to suggest a convincing explanation of the irregular distribution of the remaining records. Plants with glaucous leaves, stems and spikes occur at 15C1, nr Tanner's Hatch, and elsewhere.

A. repens (L.) Beauv.   Common Couch.   Native.   1838.
Far too abundant throughout the county, a pest in rural and urban areas. Arable land, gardens, and allotments, roadsides, and waste land. A variable species.

A. pungens (Pers.) Roem. & Schult.   Sea Couch.   Native.
1915.   Very rare, 1, perhaps extinct.
Riverwall by Thames. 27B4, above Putney, 1915, CEB (*J. Bot.* **54,** 112, 1916); 1932, JEL, Hb. Lousley, 1956, BW; Barnes, 1945, BW, 1956, BW. Formerly also in 17E4, Kew, 1915, CEB (*Rep. BEC*, **4,** 288, 1916), but not reported since. A maritime species, still plentiful on some of the N. Kent coast, which has suffered in Surrey from repeated alterations and repairs to the riverwall.

Elymus L.
E. arenarius L.   Lyme Grass.
A maritime grass sometimes found inland on builder's sand; the following reports probably have some such origin: 85E2, Ash Common, Ash Vale, sandy ground on Rifle Range, a nice patch, 1961, VML; 96D3, Chobham Common, on disturbed (? dumped) ground SW of Longcross Halt, a large patch not flowering, 1965, WEW, 1965, JEL & WEW, Hb. Lousley.

Hordeum L.
H. secalinum Schreb. (*H. pratense* Huds., *H. nodosum* auct.).
Meadow Barley.   Native.   1716.
Locally abundant, 155.   Map 486.
Alluvial meadows by the Thames, and pastures on clay, usually plentiful where it occurs. Most common on Thames-side alluvium, London Clay and Weald Clay, less so on Clay-with-Flints.

H. murinum L.   Wall Barley.   Native.   1836.
Very common throughout the county, in rural and urban areas alike. Disturbed ground, especially roadsides and the base of walls, waste ground and field borders.

Hordelymus (Jessen) Harz

H. europaeus (L.) Harz (*Hordeum europaeum* (L.) All.; *Hordeum sylvaticum* Huds.)   Native.   1838.   Rare, 6.

In woods and shady lanes, usually on chalk, and often associated with beech. 14B5, Ranmore Common, ride opposite 'Ranmore Lodge', 1963, JCG; 15A1, S of Effingham, 1957, FR, nr Dunley Hill, 1962, SFC; 15B1, Pigdon Valley and other lanes in this tetrad, many records; 25D4, Chipstead, Long Plantation, 1916, CES, 1941, JEL, 1951, DPY, 1973, JEL; 25C2, Margery Wood, 1960, CPP, 1972, BWu; 27B2, Wimbledon Common, in wood, 1951, LMPS.

Koeleria Pers.

K. arenaria Dumort. var. *anglica* Ujheli (*K. cristata* auct.; *K. gracilis* auct.).   Native.   1688.

Locally common, 66.   Map 487.

Calcareous grassland, calcareous alluvial soils by Thames, and sandy and gravelly soils elsewhere. The strong preference for calcicolous habitats is clear from the density of reports from the chalk escarpment, and the occurrences in meadows by the Thames. The following are examples of places with no obvious calcium carbonate connection: 84C1, nr Frensham Great Pond, 1961, VML; 27B1, Wimbledon Common, Ceasar's Camp, 1961, DWF; 36B2, Croham Hurst, 1911, HFP, 1955, AWJ.

[*Gaudinia* Beauv.

G. fragilis (L.) Beauv. The first evidence of this in Britain was from 16, new reservoir works towards Molesey, A. H. Wolley Dod, Hb. Mus. Brit. (*Watsonia*, **9**, 144, 1972). Our first printed record is from 05, Pyrford, Lady Davy (*Rep. BEC*, **6**, 158, 1921). Both records seem to have been casuals.]

Trisetum Pers.

T. flavescens (L.) Beauv.   Yellow Oat-grass.   Native.
1801.   Very common.

Dry pastures, roadsides, especially on chalk.

Avena L.

A. fatua L.   Common Wild-oat.   Colonist.   1776.

Still occurs in arable fields but we have very few records of this decreasing species.

[*A. ludoviciana* Durieu, established in midland counties, has not been recorded from Surrey. By far our commonest Oat is *A. sativa* L., but this is only a casual.]

Helictotrichon Bess.
H. pratense (L.) Pilg. (*Avena pratensis* L.)
Meadow Oat-grass.   Native.   1793.
Common on the chalk, 41.   Map 488.
Restricted to chalk grassland.

H. pubescens (Huds.) Pilg. (*Avena pubescens* Huds.)
Downy Oat-grass.   Native.   1763.
More common and widespread than the previous species, 99, especially on chalk and in meadows by the Thames. Also on roadsides, towpath, and a clay common in 95 and 96, and on railway banks made up with chalk in 44.   Map 489.

Arrhenatherum Beauv.
A. elatius (L.) Beauv. ex J & C Presl   False Oat.
Native.   1838.   Abundant.
Mainly in disturbed ground, such as land which has been cultivated and abandoned or dug up for roadworks and left, in waste places, roadsides, rough grasslands and hedges. In var. *bulbosum* (Willd.) Spenner (*A. tuberosum* (Gilib.) F. W. Schultz) the lower nodes are bulbous. This occurs occasionally, and mainly on light soils, but is less common than it is near the coast.

Holcus L.
H. lanatus L.   Yorkshire Fog.   Native.   1836.   Abundant.
Meadows and pastures, rough disturbed grassland, waysides and woods, on virtually all soils, wet or dry.

H. mollis L.   Creeping Soft-grass.   Native.
c1785.   Very common.
Mainly in open woodland, heathland, poor neglected farming land, and roadsides, characteristic of poor soils, especially where sandy or on a slope.

Deschampsia Beauv.
D. cespitosa (L.) Beauv.   Tufted Hair-grass.   Native.
1836.   Very common.
Badly drained soils in marshes and rough grassland, heaths, commons, roadsides and woods.

D. flexuosa (L.) Trin.   Wavy Hair-grass.   Native.
1762.   Common, 254.   Map 490.
Heaths, commons, open woodland, roadside banks, on sandy or
peaty soils, usually in dry situations. A lovely grass with its shivering
silvery panicles, to be seen to perfection on our western commons.

D. setacea (Huds.) Hack. (*Aira setacea* Huds.)   Bog Hair-grass.
Native.   1883 (J. L. Warren in H. C. Watson, *Topographical
Botany*, ed. 2, 480).   Very rare, 4.
Very wet peaty places by pools, or on the sites of pools, on acid
commons. 94A1, Thursley Common, 1933, JEL, Hb. Lousley, seen
here at intervals by many botanists, and still present in 1973;
94B2, Royal Common, ESM, but not seen for many years; 95E5,
Horsell Birch, on common, 1938, WEW & ECW, 1961, WEW,
seen here in most years since, 1973, JEL & JES; 96D3, Chobham
Common near Birch Hill, 1940, WEW, 1960, WEW, destroyed by
construction of M3 motorway, WEW; 05D5, Wisley Common,
1960, WEW & AJS, 1961, BW, 1964, JEL, etc. *D. setacea* is an
elusive species, difficult to see in flower on account of the way it
harmonises with the background, almost impossible to find in
years when it fails to flower. We owe much of our present knowledge
of this rare plant in Surrey to Mr W. E. Warren.

Aira L.
A. praecox L.   Early Hair-grass.   Native.   1780.
Common, 202.   Map 491.
Sandy acid soils on heaths, commons, fields, and hedgebanks.

A. caryophyllea L.   Silver Hair-grass.   Native.   1838.
Uncommon, 117.   Map 492.
Dry heaths, fields on sand or gravel, well-drained banks and wall-
tops.

Calamagrostis Adans.
C. epigejos (L.) Roth   Bush Grass.   Native.   1666.
Rare, 15.   Map 493.
Rough bushy places, wet or dry, in woods, gravel pits, on chalk on
escarpment, and in a peaty bog. The characteristic habitat of Bush
Grass elsewhere is in open woods or scrub on clay, but in Surrey it
grows under a wide range of conditions. Three localities: 17D2,
Ham Pits, 27A2, Richmond Park, and 37C5, Surrey Commercial
Docks, have been taken as introductions, but it may have arrived
at these places by natural means.

C. canescens (Weber) Roth (*C. lanceloata* Roth)
Purple Small-Reed.   Native.   1805 (see below).   Very rare, 1.
Under fen-like conditions in wet wood. 84E2, 'Near Elstead, Godalming', GMA, in *Rep. BEC*, **10**, 115, 1933 & 549, 1934, nr Hankley Farm, 1934, JEL, nr Westbrook Farm, 1951, GMA, and seen there by many observers since. The first record was in 1805, by Martyn, as *Arundo calamagrostis*, 'In a lane at Camberwell, between the Grove and Dulwich' (Turner & Dillwyn, *The Botanist's Guide*, **2**, 579) and it was not reported again until the present locality was found 129 years later.

Agrostis L.
A. setacea Curt.   Bristle-leaved Bent.   Native.   c1795.
Locally abundant, 25.   Map 494.
Dry poor sandy and peaty soils on the commons of NW Surrey, and on roadsides in areas now built over. Locally this occurs in great abundance, and especially so round Chobham Ridges and about Bagshot, but the records for many tetrads depend on a few plants surviving from 'development'. Outside the main area it is still at 94A1, Thursley Common, where there are two small patches. We have not refound it at 05B5, Pyrford Common, or 05D5, Wisley Common, from which there are records within the present century.

A. canina L.   Brown Bent.   Native.   1780.
Very common, 236.   Map 495.
Wet and dry habitats—see below.
subsp. *canina*, a creeping grass of damp or wet places, in the meadows, marshy ground, by ponds and in swamps, occurring throughout the county.
subsp. *montana* (Hartm.) Hartm. (*A. coarctata* Ehrh. ex Hoffm. subsp. *coarctata*), a tufted grass found on drier soils on heaths, commons and roadsides. Less common than subsp. *canina*.

A. tenuis Sibth.   Common Bent.   Native.
1762.   Very common.
Pastures, heaths and dry roadsides, often an indicator of poor acid soils.

A. gigantea Roth (*A. nigra* With.)   Native.   1931.
Too abundant, 256.   Map 496.
A serious weed in arable land and gardens, roadsides, waste ground,

and also wood borders and hedgerows. The long, deep and tough rhizomes are difficult to eradicate and make it a hard task to reclaim land where cultivation has been neglected.

A. stolonifera L.   Creeping Bent.   Native.   1838.
Chalk grassland, pastures, sandy commons, and field borders. The var. *palustris* (Huds.) Farw. (*A. palustris* Huds.) occurs occasionally in wet places.

A. scabra Willd.   Alien.
27E4, Nine Elms railway yard, abundant on sidings, 1966, det. CEH, AB, RARC, EJC, DPY. Apparently established but the heavy charge imposed on visitors by British Transport prevented us from keeping this under observation.

Apera Adans.
A. spica-venti (L.) Beauv.   Loose Silky-bent.   Colonist.
1793.   Locally frequent, 113.   Map 497.
Cornfields, persisting for a time on roadsides, field borders and waste places. Sometimes appears in great quantity in cereal crops, as in 1958 when, in a cornfield on chalk between West Humble and Bagden Farm, it was almost more abundant than the crop, JEL, and it again appeared in abundance in another field nearby in 1962, BMCM. After such outbursts it may be completely absent from the field the following year.

Polypogon Desf.
P. monspeliensis (L.) Desf.   Annual Beard-grass.   Alien.
25D1, Redhill, building land on site of old tannery, 1973, BMCM & EMCI. In considerable quantity, some of it round small pools. This is a common hide and wool alien, but also grown as an ornamental grass. There are very old Surrey records as an alien and it is possible that this maritime species, and its hybrid with *Agrostis stolonifera*, x *Agropogon littoralis*, may have extended in brackish marshes by the Thames into the county.

Mibora Adans.
M. minima (L.) Desv. (*M. verna* Beauv.).   Early Sand-grass.
Alien.   1838.
96D1 (?), Chobham, Hilling's Nurseries, probably introduced with shrubs from another nursery 'during war', 1939–45, comm. BW; Hilling's Home Nursery, among frames in rock garden section,

1968, MRM. The 1838 record was from a sandpit 'behind the Fox', Wimbledon Common.

[*Gastridium* Beauv.
G. ventricosum (Gouan) Schinz & Thell. (*G. lendigerum* (L.) Desv.) Nit-grass. Colonist. 1852. Very rare. Weed in cornfields and by rides in woods on stiff soils. Found by ECW well established about Alford, but not reported during our survey. Last record: 03B3, by Wey & Arun Canal, Alford Crossways, 1936, JEL, Hb. Lousley.]

Phleum L.
P. bertolonii DC. (*P. nodosum* auct.)   Smaller Cat's-tail.
Native.   1931.   Very common, 365.   Map 498.
Pastures and other grasslands, mainly on chalk or sand, but also on gravel and clay.

P. pratense L.   Timothy Grass.
Mainly introduced, but perhaps native in a few old meadows.
1724.   Very common, 342.   Map 499.
Pastures and meadows, roadsides and waste places. This valuable agricultural grass is widely sown and persists, but may be native in hay-meadows by the Thames and by streams where it grows with *Hordeum secalinum* and *Festuca pratensis*.

Alopecurus L.
A. myosuroides Huds. (*A. agrestis* L.)   Slender Foxtail.
Native.   1838.   Uncommon, 122.   Map 500.
Weed in cornfields and other arable land, roadsides and waste places, always on disturbed ground. The map shows the tetrads in which our recorders have found it since 1950, but the frequency of the species in Surrey has decreased dramatically over recent years.

A. pratensis L.   Meadow Foxtail.   Native.
1838.   Very common.
Meadows and pastures, especially by rivers and streams on rich alluvial soils.

A. geniculatus L.   Marsh Foxtail.   Native.   1836.
Common, 214.   Map 501.
Wet places, by ponds, ditches, streams and in marshy meadows. Absent from the chalk and other high ground where there are no suitable areas of shallow water, but otherwise throughout the county.

x A. pratensis=A. x hybridus Wimm.
Was reported by C. E. Salmon from 04B4, E of Lockner Farm, Chilworth, with parents, in 1918 (Salmon, 1931, 637) and is likely to be still in the county though not reported since.

A. aequalis Sobol. (*A. fulvus* Sm.)   Orange Foxtail.   Native. 1838.   Rare, 8.   Map 502.
Ponds, ditches, dried-up canal. At Whitmoor Pond it was collected by J. E. Bicheno (c1810–30), survived the threat of enclosures feared by H. C. Watson (Brewer, 1863, 275) and is now being destroyed by the embankments and trampling of fishermen.

Milium L.
M. effusum L.   Wood Millet.   Native.   1836.
Locally frequent, 98.   Map 503.
Beech woods on the chalk on deep humus in shade, and in oak and other woods on clay. Occasionally planted in woods for ornamental reasons.

Anthoxanthum L.
A. odoratum L.   Sweet Vernal-grass.   Native.   1840.
Abundant throughout the county.
Meadows, pastures, open woodland and heaths.

A. puelli Lecoq & Lamotte. (*A. aristatum* auct.)
Annual Vernal-grass.   Colonist.   1880.   Rare, 8.
Weed, sometimes persistent, in sandy arable fields, and roadsides. 84D4, Waverley Woods, on old nursery site, 1964, AJS; 95C3, nr Pirbright, Burner's Heath, 1954, BW & VML, WEW, JEL, 1955, VML, lost when field got overgrown; 95E4, nr Smart's Heath, 1961, WEW, JEL, persisted here; 96C3, Chobham Common, by sandy path, 1964, WEW; 96D2, Chobham, on newly widened road, 1964, WEW; 96E4, Wentworth Estate, on roadside, 1965, JES & WEW; 96E5, Callowhill, in sandpit in which rubbish had been dumped, 1965, EJC; 14B4, Wotton, in stubble field opposite Wotton Hatch Hotel, 1961, BMCM & BW.

Phalaris L.
P. arundinacea L.   Reed Canary-grass.   Native.   1797.   Common.
Wet places round lakes and ponds, by rivers and streams, usually growing in water or in places which are under water during the winter. An attractive variation with the leaves striped with white on

green (var. *picta* L.) is an aggressive garden plant grown as 'Gardener's Garters', and becomes easily established on commons and roadsides from roots rejected by disillusioned growers—for example, at 15C1, Polesden Lacey, on margin of Freehold Wood, 1966, BSBI Exc.

P. canariensis L., Canary Grass, is a very common casual introduced as bird-seed. *P. minor* Retz and *P. paradoxa* L., from the same source, occur occasionally on refuse tips.

Nardus L.
N. stricta L.   Mat-grass.   Native.   1666.
Locally abundant, 124.   Map 504.
On poor, often acid, soils on heaths and commons. Especially abundant on some of the large commons in the W, but scattered over much of the county.

Cynodon Rich.
C. dactylon (L.) Pers.   Bermuda Grass.   Alien.
1844 (*Phytol.* **1**, 870).   Rare, 4.
Roadsides and commons in short grass, usually on light soils. 04A4, Shalford, E side of A281, 1970, RMB, LNHS Exc.; 25C1, Reigate Heath, roadside verge, 1954, BMCM, 1956, BMCM; 25C2, Lower Kingswood, roadside verge, 1949, EMCI, 1970, EMCI; 27B4, Barnes Common, 1948, MWh, JEL, BW, 1962, BW. The 1844 record was from 17E4, Kew Green, where it persisted for a time; it still occurs on several lawns in the Royal Botanic Gardens.

Echinochloa Beauv.
E. crus-galli (L.) Beauv. (*Panicum crus-galli* L.)   Alien.
1778.   Frequent as a casual on refuse tips, where it generally originates from bird seed, but occasionally more persistent, as at: 04D4, carrot field at Pond's Farm, 1957 & 1958, JCG; 17D3, Richmond, Old Deer Park in wartime allotments, 1943 to 1950, when area grassed down, BW; 17E3, Sheen Common, on edge of path, 1950, 1952, BW.

Digitaria P. C. Fabr.
D. ischaemum (Schreb.) Muhl. (*Digitaria humifusa* Pers., *Panicum glabrum* Gaudin)   Smooth Ginger-grass.
Colonist.   1829.
Sandy cultivated fields and paths. 95E4, nr Smart's Heath, in sandy

nursery fields, 1965, WEW, JEL, EJC, etc., 1970, JEL, known to WEW much earlier; 05C5, Pyrford, seen by many botanists in the sandy area NE of Pyrford Church from 1918 onwards, *Rep. BEC*, **9,** 242, 1930, 1948, JEL, 1950, BW, 1964, BWu, 1970, TBR, in recent years restricted to part of one large field, and the adjoining path and not appearing every year; 05D5, Wisley, weed in RHS Gardens (*Proc. BSBI*, **4,** 489, 1962).

Setaria Beauv.
S. viridis (L.) Beauv., *S. verticillata* (L.) Beauv. and *S. lutescens* (Weigel) Hubbard occur rather frequently on refuse tips, allotments and on roadsides, being introduced mainly with bird-seed. Occasionally they appear in a second year at the same place but are not known to persist.

BAMBUSEAE

Several species of Bamboo are recorded from Surrey but all are on abandoned gardens, just outside gardens from which they have spread vegetatively, or on refuse tips, or likely to have been planted. The most common is *Arundinaria japonica* Sieb. & Zucc., but *Sasa palmata* E. G. Camus is sometimes rampant and often in flower. Others reported include *Arundinaria anceps* Mitf. and *A. fastuosa* Mak.

EXPLANATION OF DISTRIBUTION MAPS

These are based on records collected by the Surrey Flora Committee from 1957 to 1974. The standard symbol to indicate the presence of a species in a tetrad is a solid dot (•). Where it is believed that the species has become extinct in the tetrad since the record was made, then a smaller solid dot has been used (·).

Recorders were asked to indicate cases where plants appeared to have been deliberately planted, or were garden escapees, casuals or on refuse tips, and these records are shown with open circles (○).

In a few cases a special sign (⚹) has been used to indicate records of hybrids (eg map 382), and when it has been necessary to introduce pre-1950 records for special reasons (eg *Hieracium*) these are shown by a cross (·±).

Where two 'splits' are shown on one map, and there are instances where they both occur in the same tetrad, they are shown by the initial letters of their names (eg *Monotropa*, map 232).

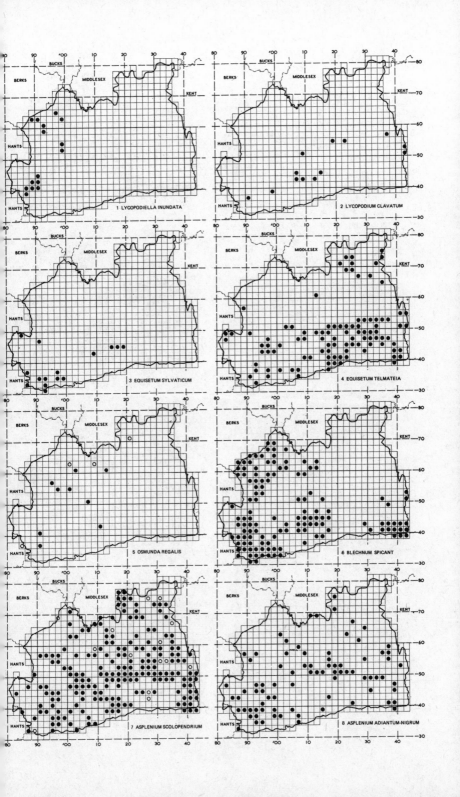

1 LYCOPODIELLA INUNDATA

2 LYCOPODIUM CLAVATUM

3 EQUISETUM SYLVATICUM

4 EQUISETUM TELMATEIA

5 OSMUNDA REGALIS

6 BLECHNUM SPICANT

7 ASPLENIUM SCOLOPENDRIUM

8 ASPLENIUM ADIANTUM-NIGRUM

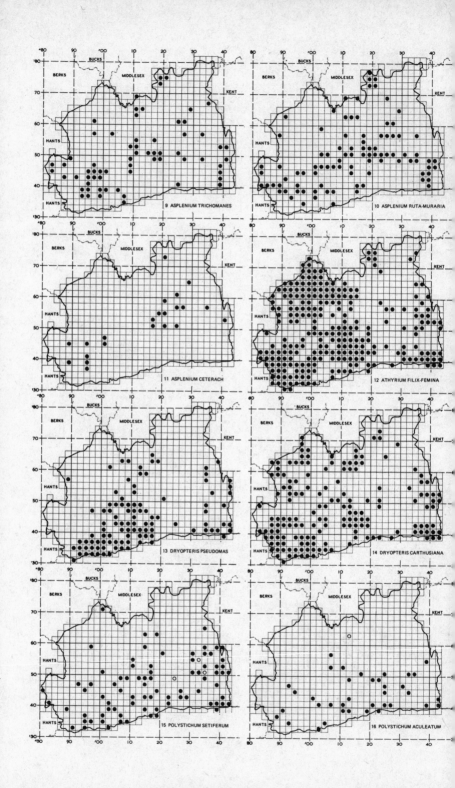

9 ASPLENIUM TRICHOMANES

10 ASPLENIUM RUTA-MURARIA

11 ASPLENIUM CETERACH

12 ATHYRIUM FILIX-FEMINA

13 DRYOPTERIS PSEUDOMAS

14 DRYOPTERIS CARTHUSIANA

15 POLYSTICHUM SETIFERUM

16 POLYSTICHUM ACULEATUM

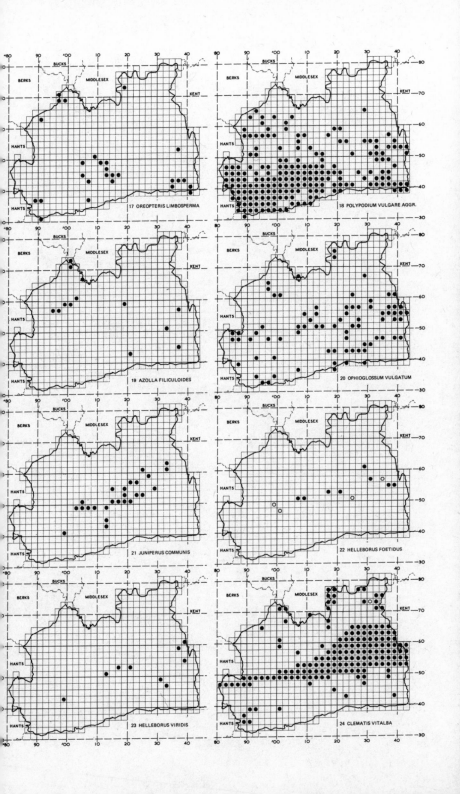

17 OREOPTERIS LIMBOSPERMA

18 POLYPODIUM VULGARE AGGR.

19 AZOLLA FILICULOIDES

20 OPHIOGLOSSUM VULGATUM

21 JUNIPERUS COMMUNIS

22 HELLEBORUS FOETIDUS

23 HELLEBORUS VIRIDIS

24 CLEMATIS VITALBA

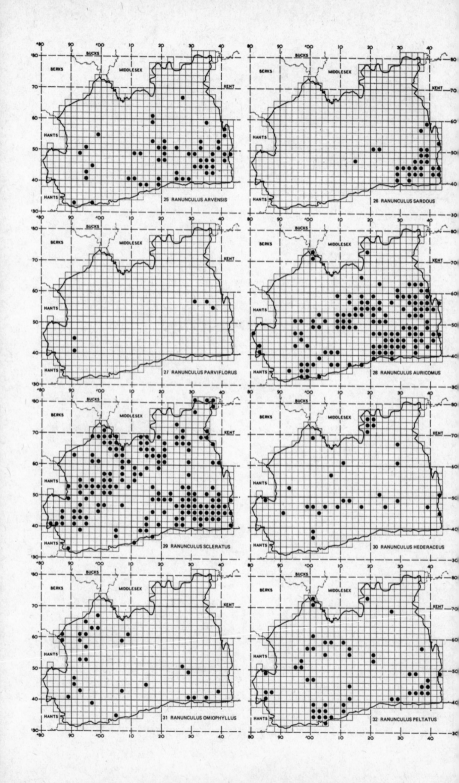

25 RANUNCULUS ARVENSIS

26 RANUNCULUS SARDOUS

27 RANUNCULUS PARVIFLORUS

28 RANUNCULUS AURICOMUS

29 RANUNCULUS SCLERATUS

30 RANUNCULUS HEDERACEUS

31 RANUNCULUS OMIOPHYLLUS

32 RANUNCULUS PELTATUS

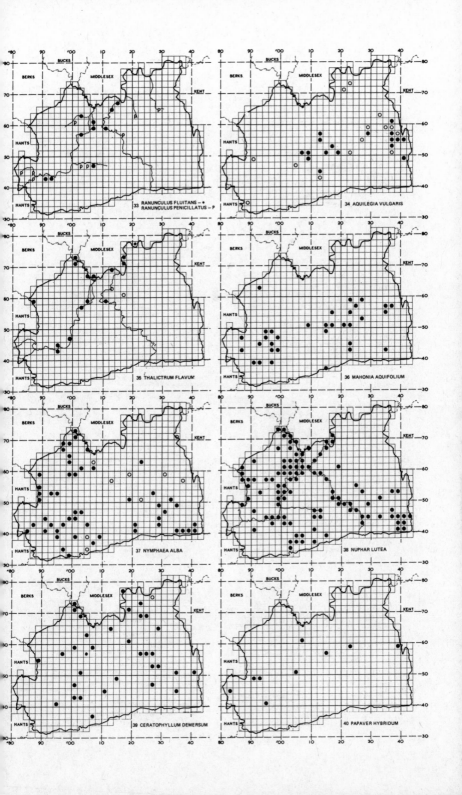

33 RANUNCULUS FLUITANS — ●
RANUNCULUS PENICILLATUS — P

34 AQUILEGIA VULGARIS

35 THALICTRUM FLAVUM

36 MAHONIA AQUIFOLIUM

37 NYMPHAEA ALBA

38 NUPHAR LUTEA

39 CERATOPHYLLUM DEMERSUM

40 PAPAVER HYBRIDUM

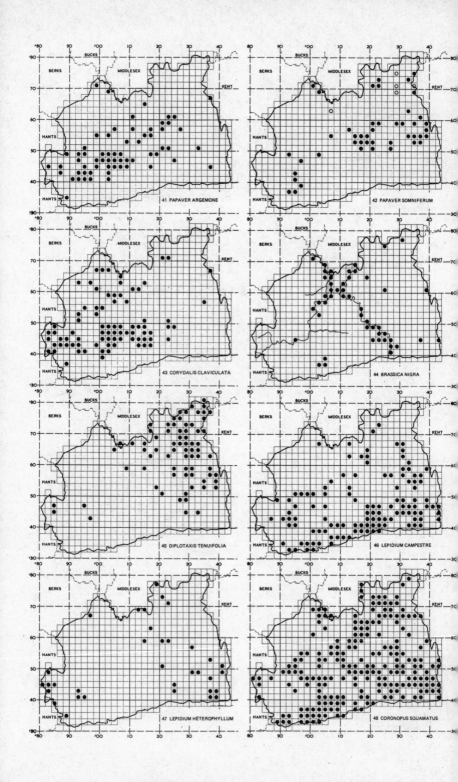

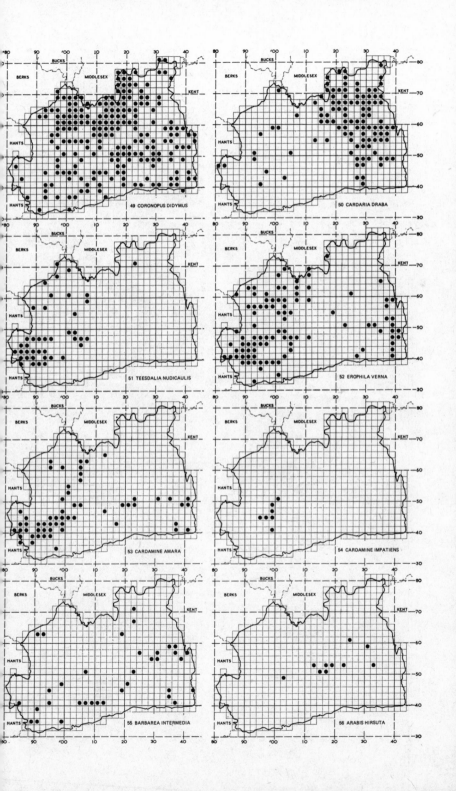

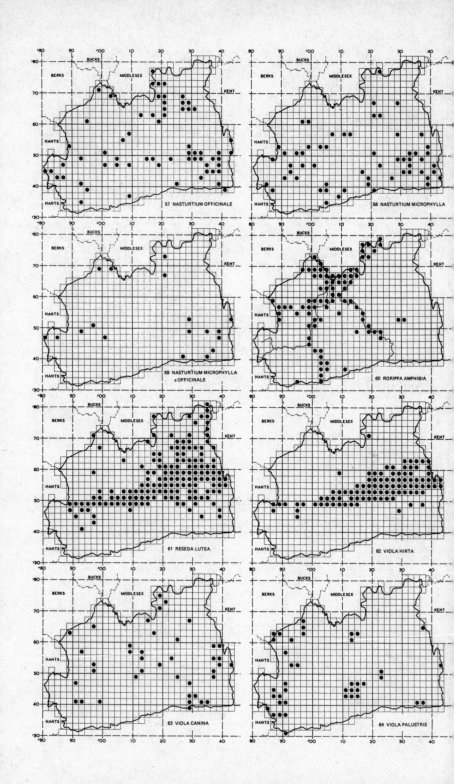

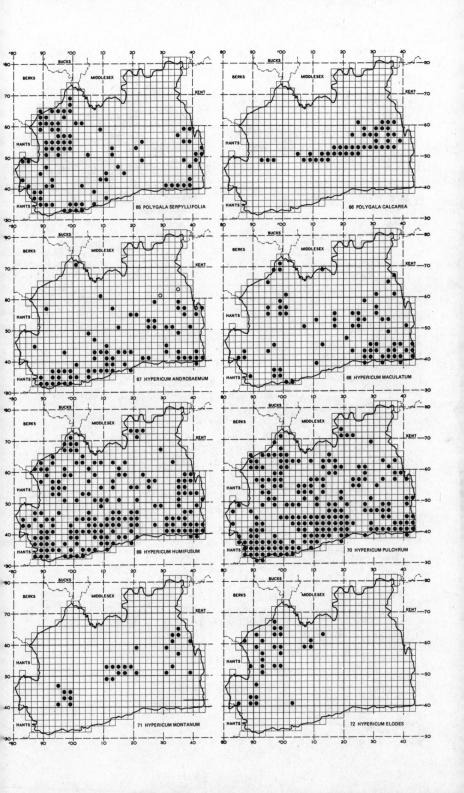

65 POLYGALA SERPYLLIFOLIA

66 POLYGALA CALCAREA

67 HYPERICUM ANDROSAEMUM

68 HYPERICUM MACULATUM

69 HYPERICUM HUMIFUSUM

70 HYPERICUM PULCHRUM

71 HYPERICUM MONTANUM

72 HYPERICUM ELODES

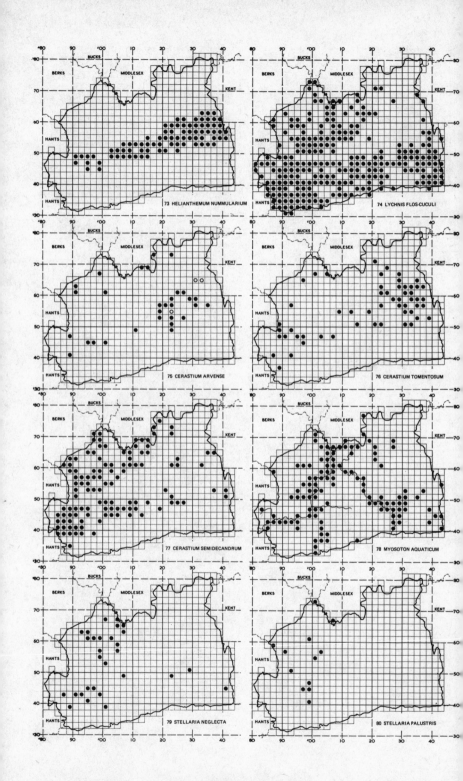

73 HELIANTHEMUM NUMMULARIUM

74 LYCHNIS FLOS-CUCULI

75 CERASTIUM ARVENSE

76 CERASTIUM TOMENTOSUM

77 CERASTIUM SEMIDECANDRUM

78 MYOSOTON AQUATICUM

79 STELLARIA NEGLECTA

80 STELLARIA PALUSTRIS

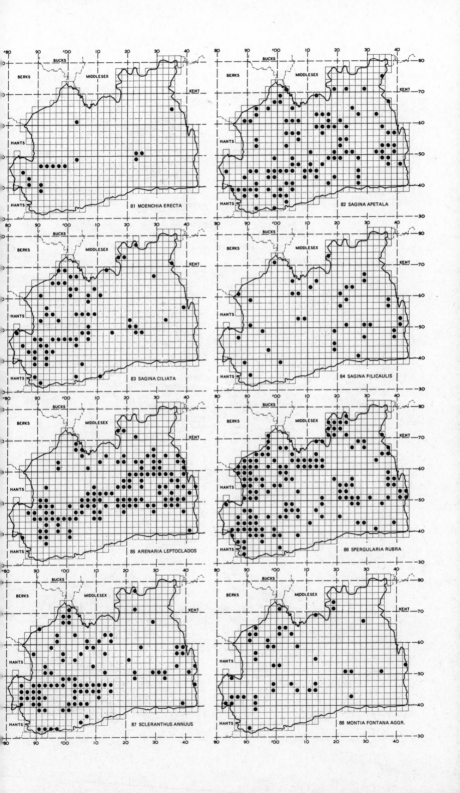

81 MOENCHIA ERECTA

82 SAGINA APETALA

83 SAGINA CILIATA

84 SAGINA FILICAULIS

85 ARENARIA LEPTOCLADOS

86 SPERGULARIA RUBRA

87 SCLERANTHUS ANNUUS

88 MONTIA FONTANA AGGR.

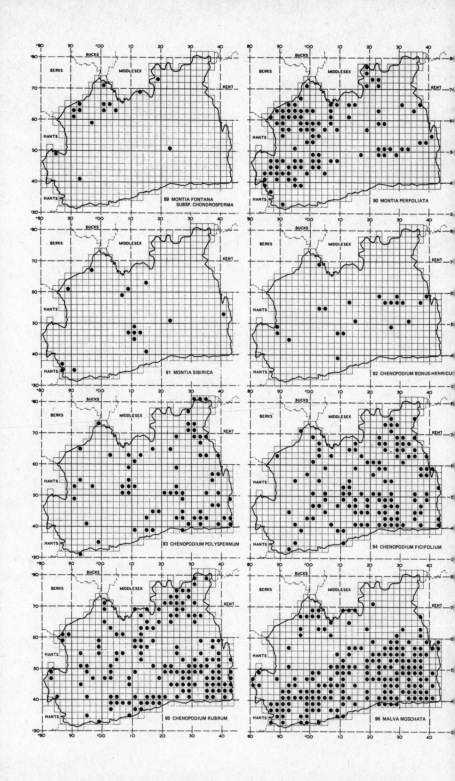

89 MONTIA FONTANA SUBSP. CHONDROSPERMA

90 MONTIA PERFOLIATA

91 MONTIA SIBIRICA

92 CHENOPODIUM BONUS-HENRICUS

93 CHENOPODIUM POLYSPERMUM

94 CHENOPODIUM FICIFOLIUM

95 CHENOPODIUM RUBRUM

96 MALVA MOSCHATA

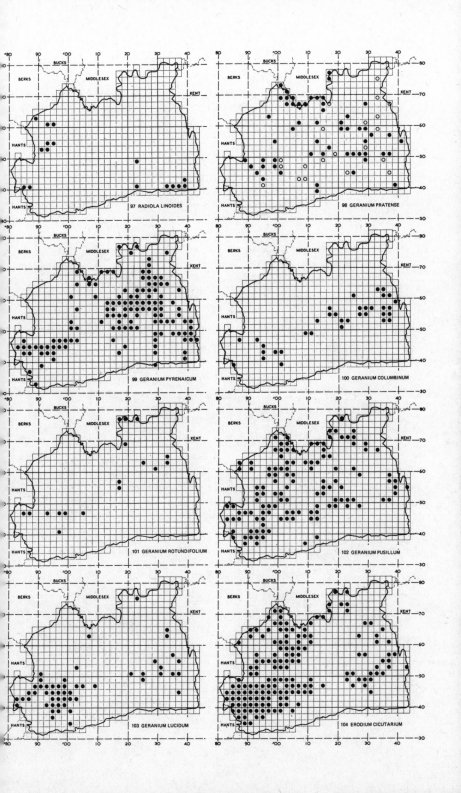

97 RADIOLA LINOIDES

98 GERANIUM PRATENSE

99 GERANIUM PYRENAICUM

100 GERANIUM COLUMBINUM

101 GERANIUM ROTUNDIFOLIUM

102 GERANIUM PUSILLUM

103 GERANIUM LUCIDUM

104 ERODIUM CICUTARIUM

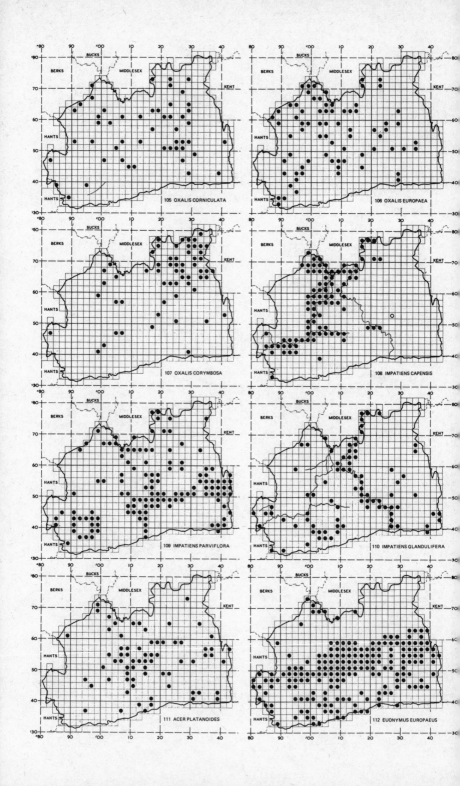

105 OXALIS CORNICULATA

106 OXALIS EUROPAEA

107 OXALIS CORYMBOSA

108 IMPATIENS CAPENSIS

109 IMPATIENS PARVIFLORA

110 IMPATIENS GLANDULIFERA

111 ACER PLATANOIDES

112 EUONYMUS EUROPAEUS

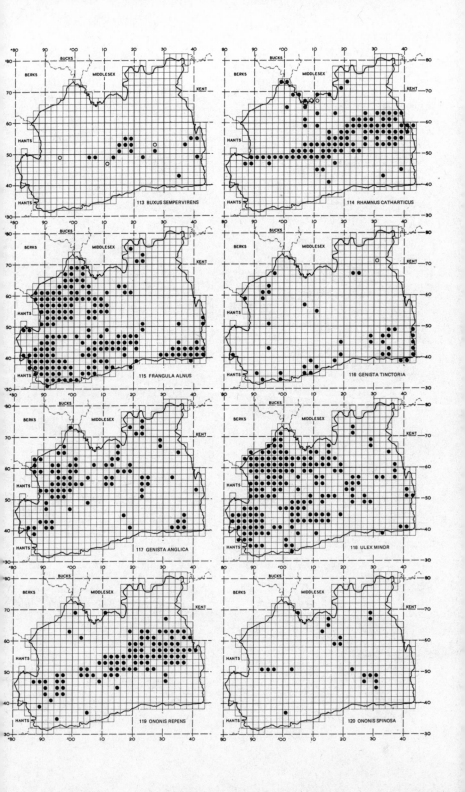

113 BUXUS SEMPERVIRENS

114 RHAMNUS CATHARTICUS

115 FRANGULA ALNUS

116 GENISTA TINCTORIA

117 GENISTA ANGLICA

118 ULEX MINOR

119 ONONIS REPENS

120 ONONIS SPINOSA

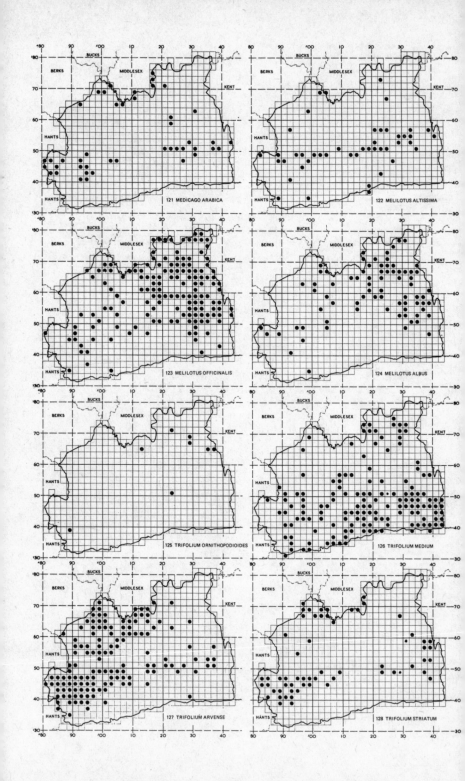

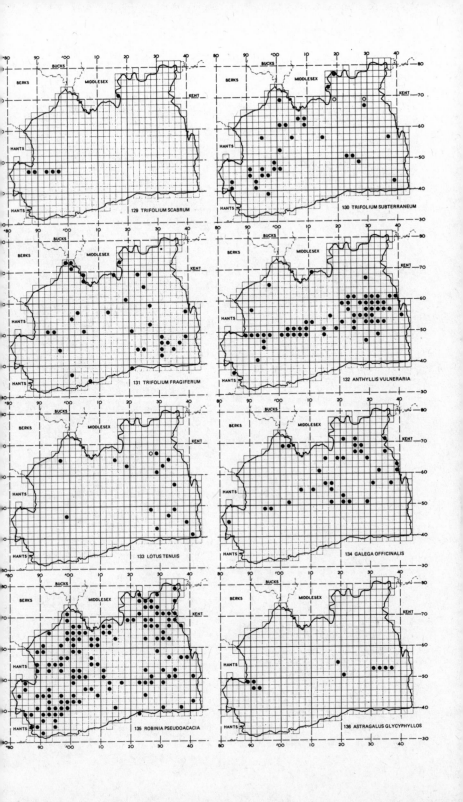

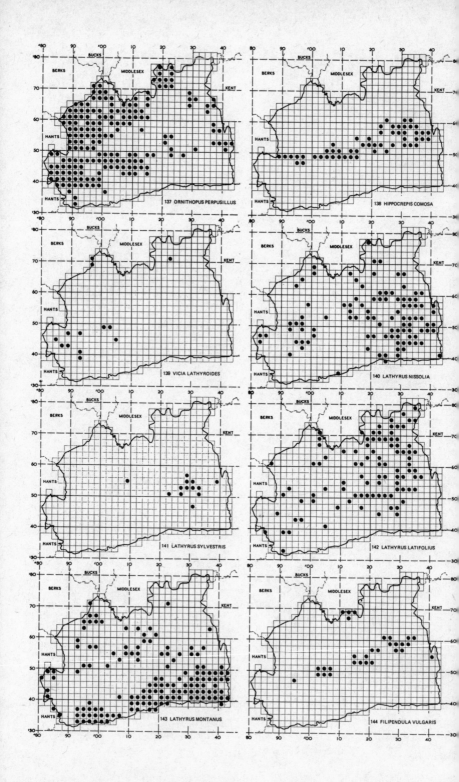

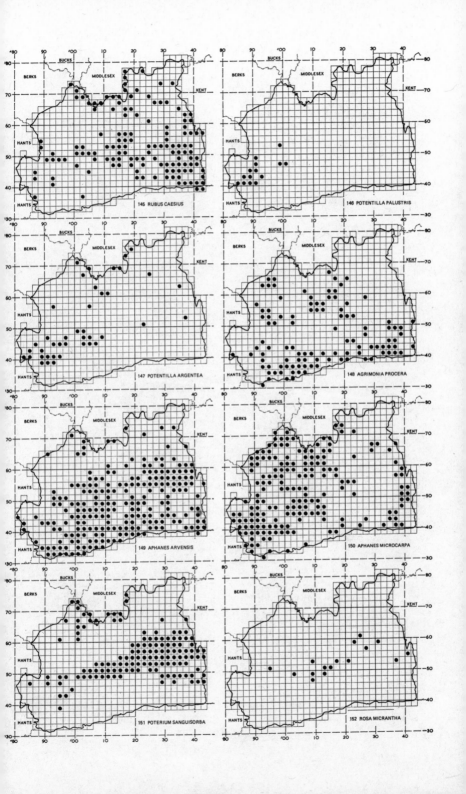

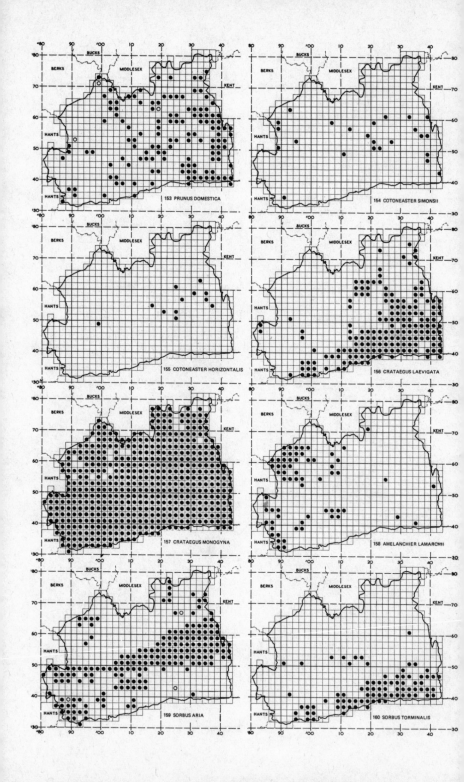

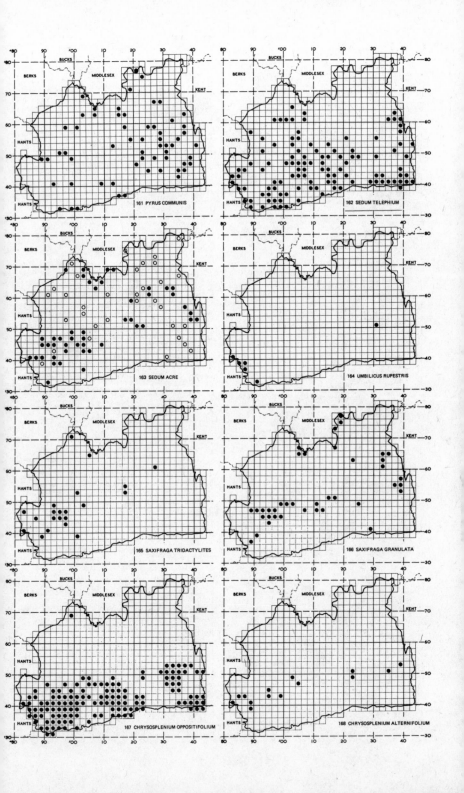

161 PYRUS COMMUNIS

162 SEDUM TELEPHIUM

163 SEDUM ACRE

164 UMBILICUS RUPESTRIS

165 SAXIFRAGA TRIDACTYLITES

166 SAXIFRAGA GRANULATA

167 CHRYSOSPLENIUM OPPOSITIFOLIUM

168 CHRYSOSPLENIUM ALTERNIFOLIUM

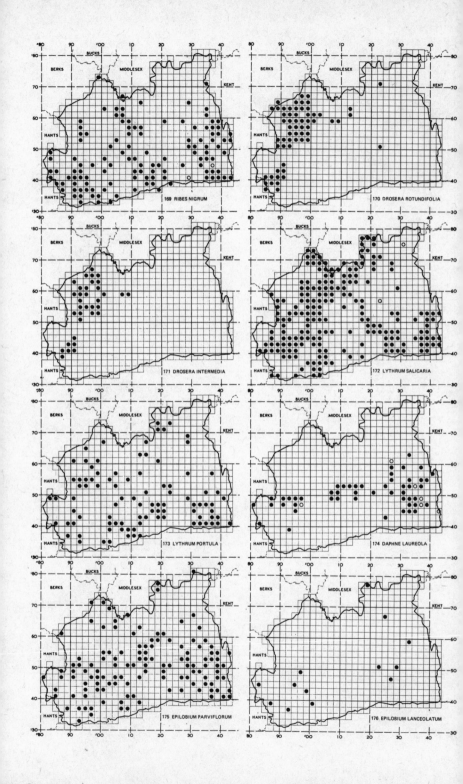

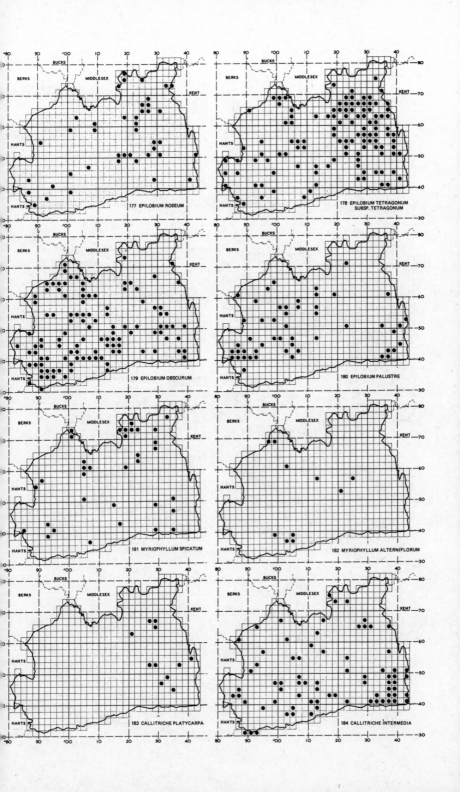

177 EPILOBIUM ROSEUM

178 EPILOBIUM TETRAGONUM
SUBSP. TETRAGONUM

179 EPILOBIUM OBSCURUM

180 EPILOBIUM PALUSTRE

181 MYRIOPHYLLUM SPICATUM

182 MYRIOPHYLLUM ALTERNIFLORUM

183 CALLITRICHE PLATYCARPA

184 CALLITRICHE INTERMEDIA

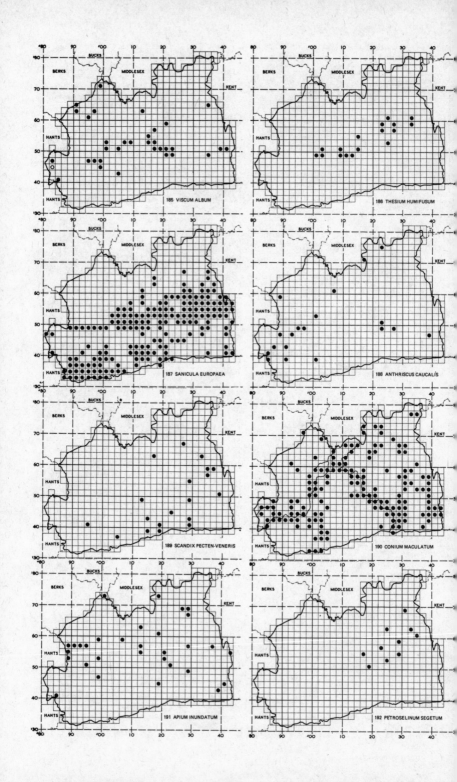

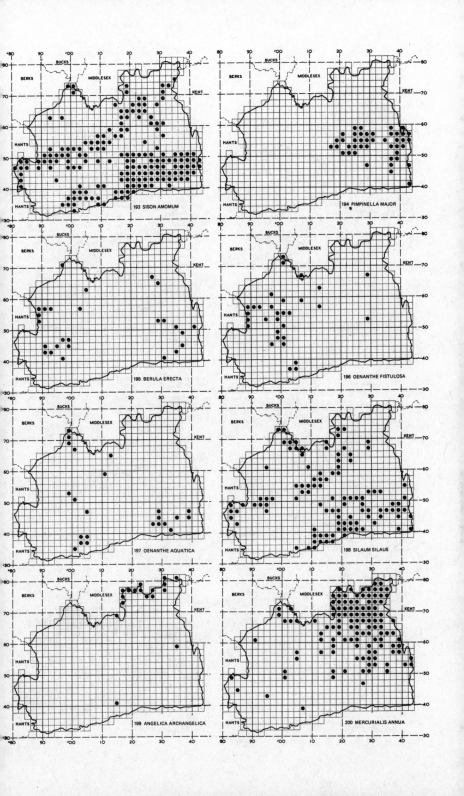

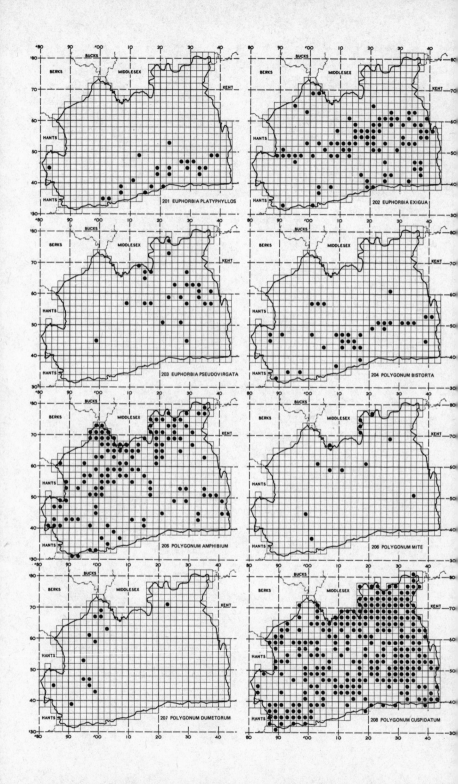

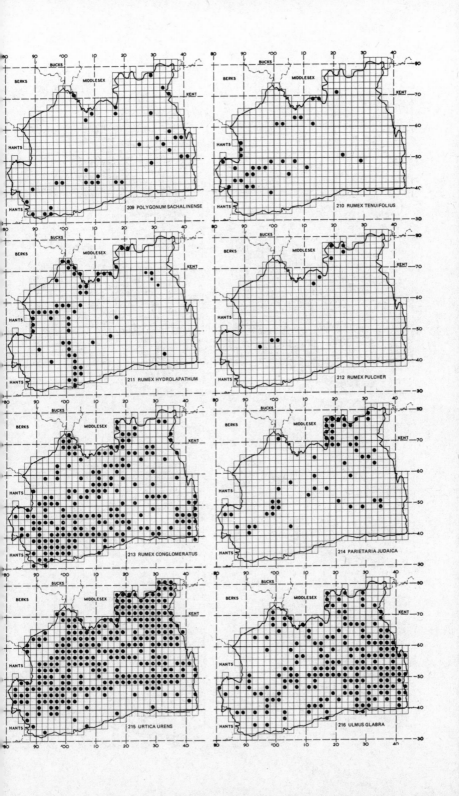

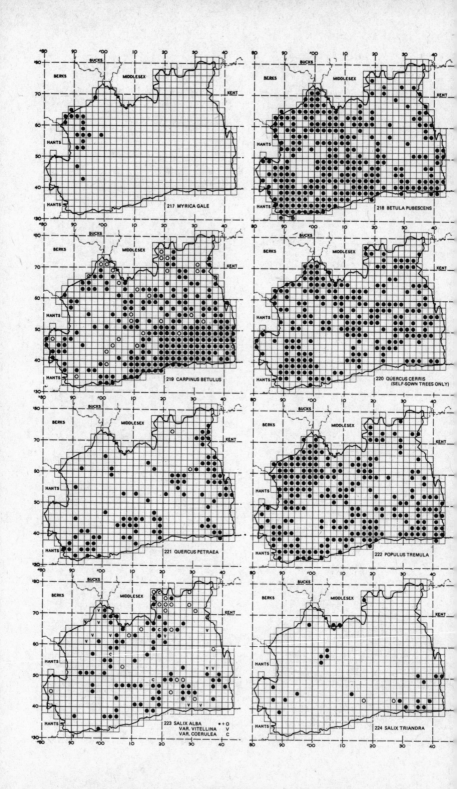

217 MYRICA GALE

218 BETULA PUBESCENS

219 CARPINUS BETULUS

220 QUERCUS CERRIS
(SELF-SOWN TREES ONLY)

221 QUERCUS PETRAEA

222 POPULUS TREMULA

223 SALIX ALBA
VAR. VITELLINA • + O
VAR. COERULEA V
C

224 SALIX TRIANDRA

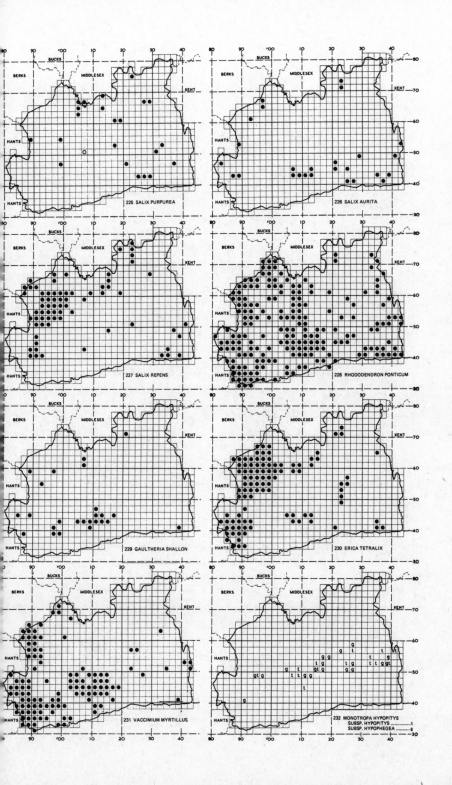

225 SALIX PURPUREA

226 SALIX AURITA

227 SALIX REPENS

228 RHODODENDRON PONTICUM

229 GAULTHERIA SHALLON

230 ERICA TETRALIX

231 VACCIMIUM MYRTILLUS

232 MONOTROPA HYPOPITYS
SUBSP. HYPOPITYS ....t
SUBSP. HYPOPHEGEA ....g

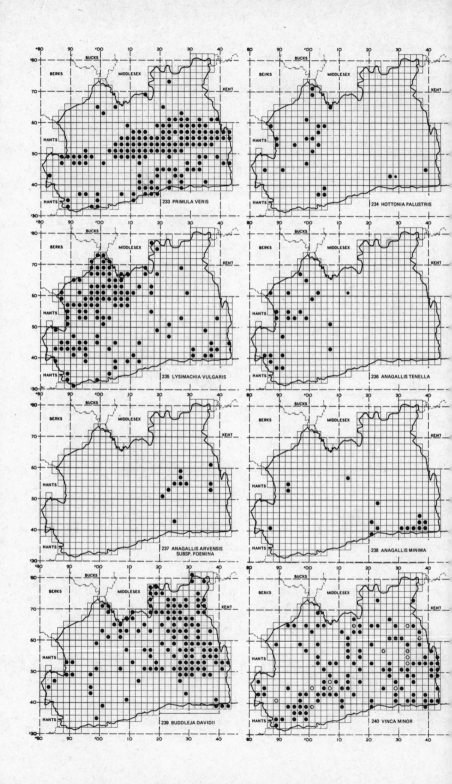

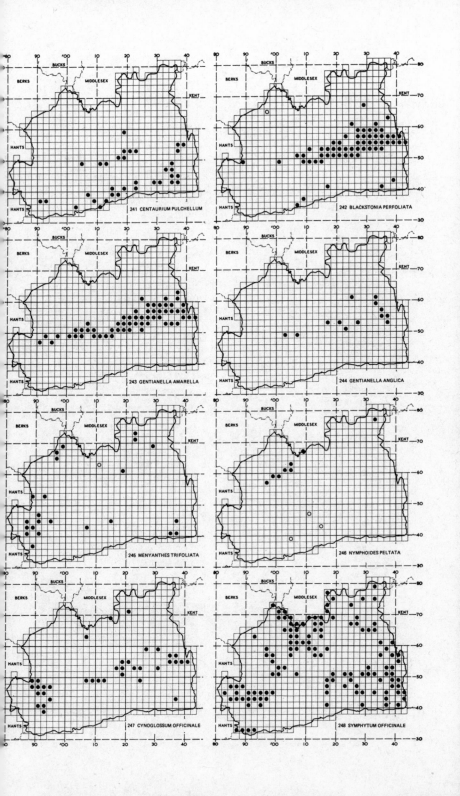

241 CENTAURIUM PULCHELLUM

242 BLACKSTONIA PERFOLIATA

243 GENTIANELLA AMARELLA

244 GENTIANELLA ANGLICA

245 MENYANTHES TRIFOLIATA

246 NYMPHOIDES PELTATA

247 CYNOGLOSSUM OFFICINALE

248 SYMPHYTUM OFFICINALE

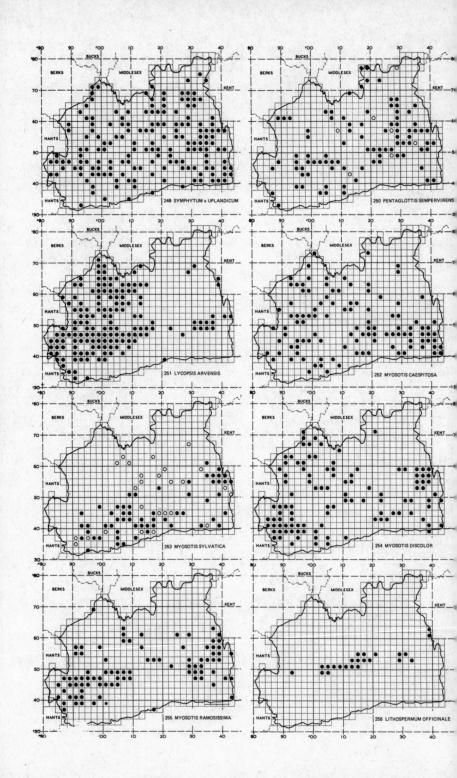

249 SYMPHYTUM x UPLANDICUM

250 PENTAGLOTTIS SEMPERVIRENS

251 LYCOPSIS ARVENSIS

252 MYOSOTIS CAESPITOSA

253 MYOSOTIS SYLVATICA

254 MYOSOTIS DISCOLOR

255 MYOSOTIS RAMOSISSIMA

256 LITHOSPERMUM OFFICINALE

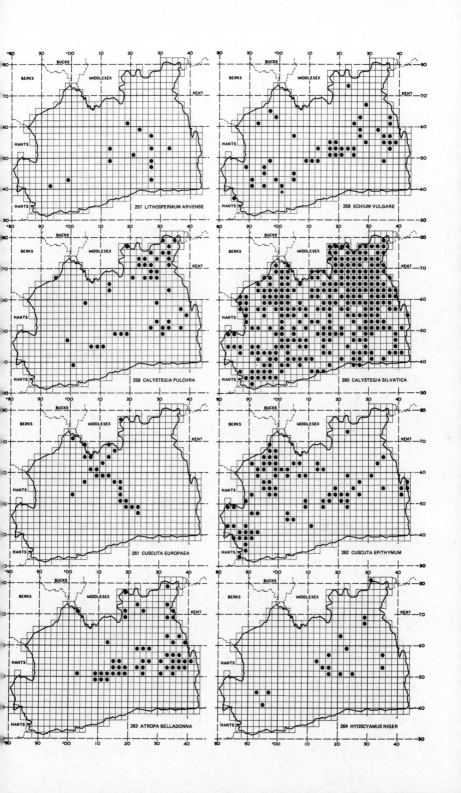

257 LITHOSPERMUM ARVENSE

258 ECHIUM VULGARE

259 CALYSTEGIA PULCHRA

260 CALYSTEGIA SILVATICA

261 CUSCUTA EUROPAEA

262 CUSCUTA EPITHYMUM

263 ATROPA BELLADONNA

264 HYOSCYAMUS NIGER

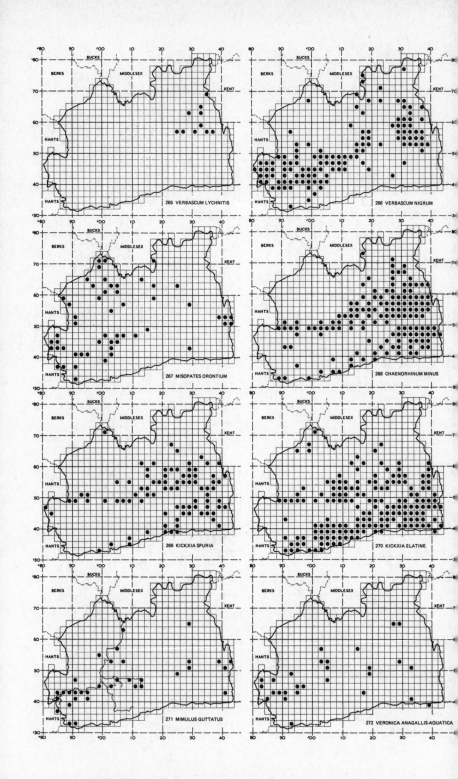

265 VERBASCUM LYCHNITIS

266 VERBASCUM NIGRUM

267 MISOPATES ORONTIUM

268 CHAENORHINUM MINUS

269 KICKXIA SPURIA

270 KICKXIA ELATINE

271 MIMULUS GUTTATUS

272 VERONICA ANAGALLIS-AQUATICA

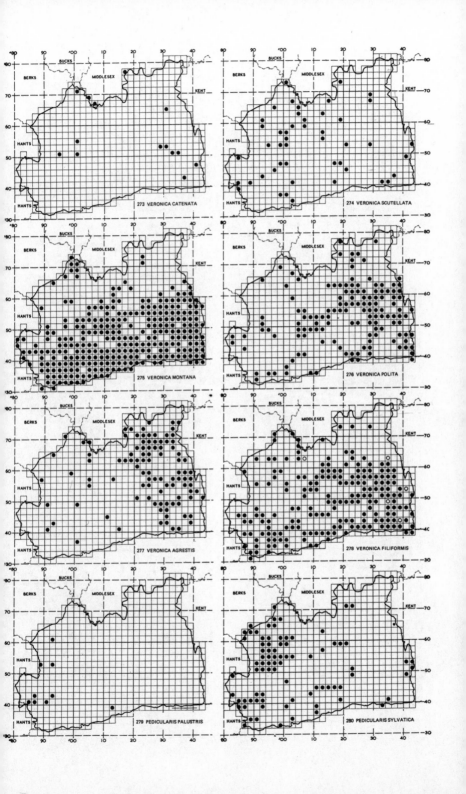

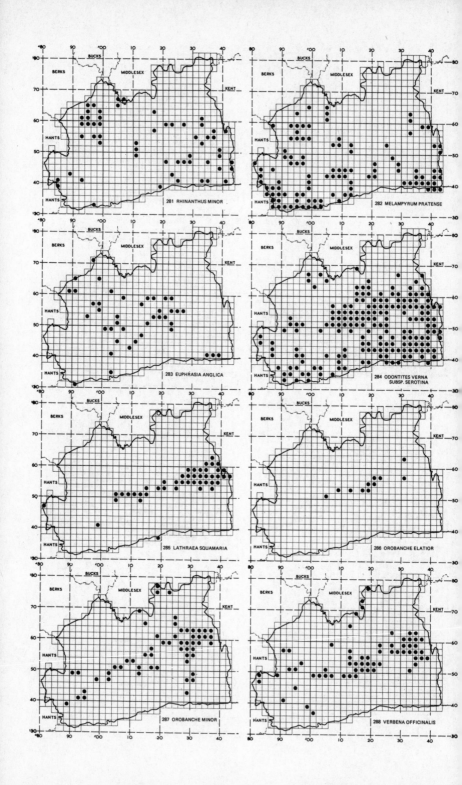

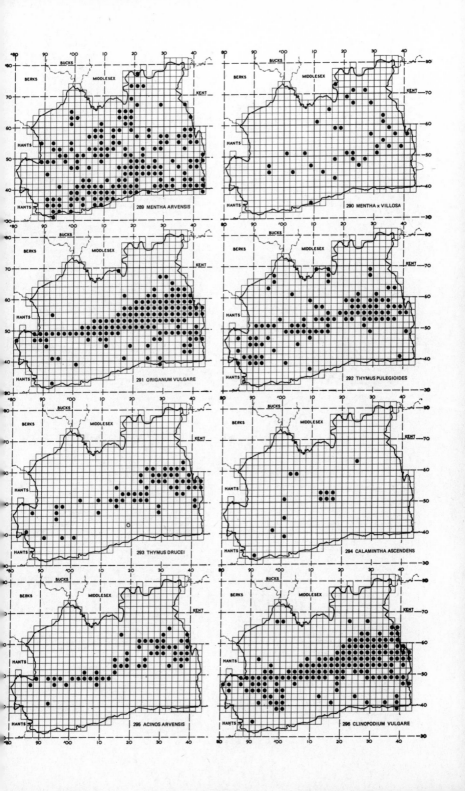

289 MENTHA ARVENSIS

290 MENTHA x VILLOSA

291 ORIGANUM VULGARE

292 THYMUS PULEGIOIDES

293 THYMUS DRUCEI

294 CALAMINTHA ASCENDENS

295 ACINOS ARVENSIS

296 CLINOPODIUM VULGARE

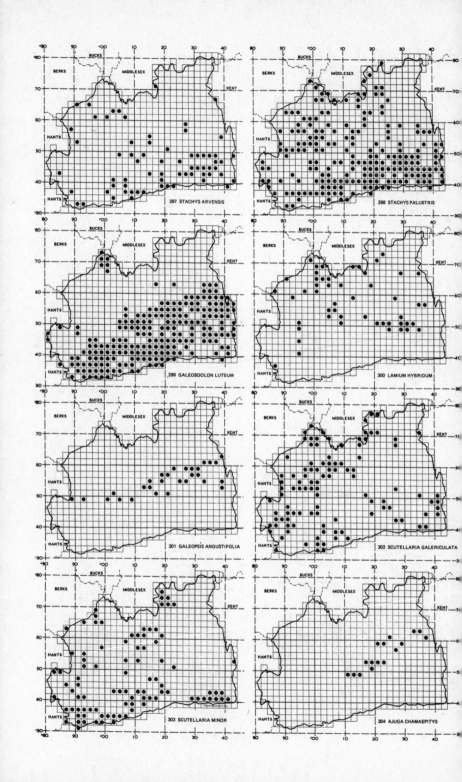

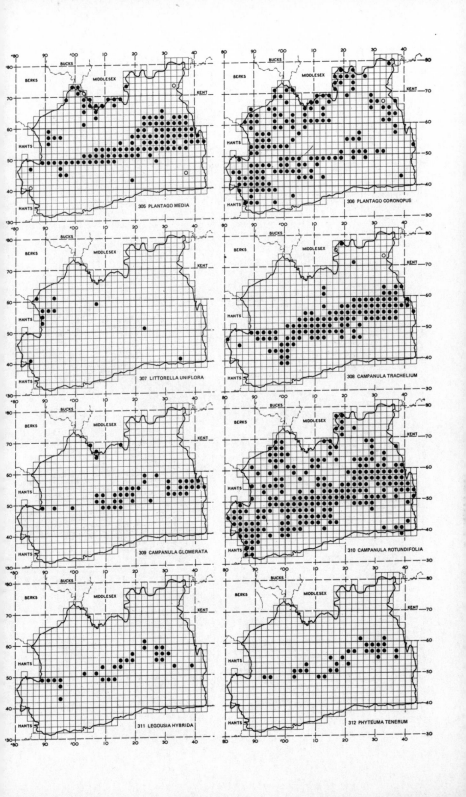

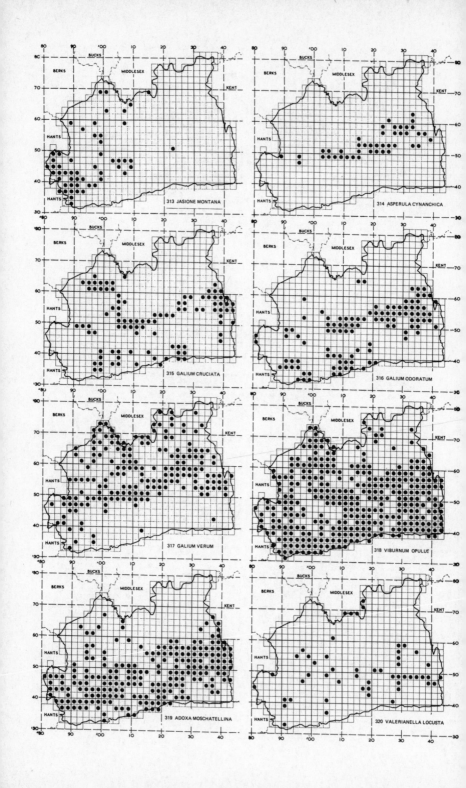

313 JASIONE MONTANA

314 ASPERULA CYNANCHICA

315 GALIUM CRUCIATA

316 GALIUM ODORATUM

317 GALIUM VERUM

318 VIBURNUM OPULUS

319 ADOXA MOSCHATELLINA

320 VALERIANELLA LOCUSTA

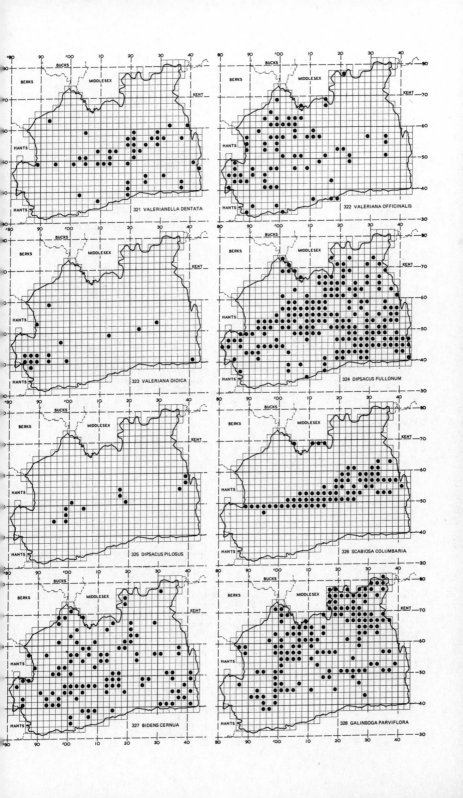

321 VALERIANELLA DENTATA

322 VALERIANA OFFICINALIS

323 VALERIANA DIOICA

324 DIPSACUS FULLONUM

325 DIPSACUS PILOSUS

326 SCABIOSA COLUMBARIA

327 BIDENS CERNUA

328 GALINSOGA PARVIFLORA

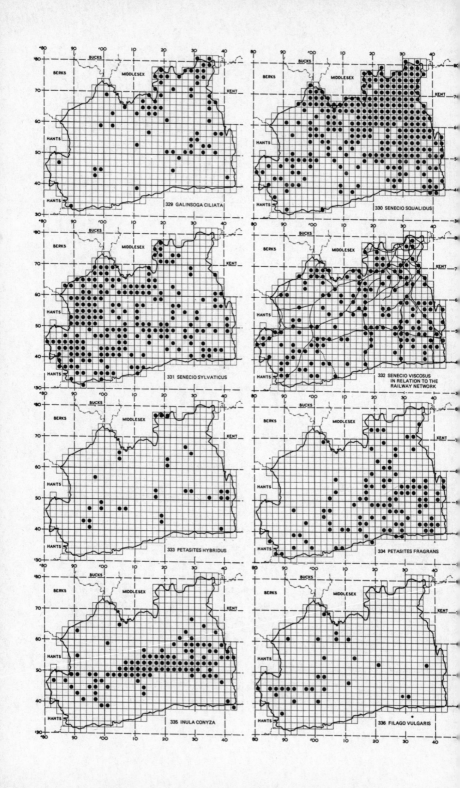

329 GALINSOGA CILIATA

330 SENECIO SQUALIDUS

331 SENECIO SYLVATICUS

332 SENECIO VISCOSUS
IN RELATION TO THE
RAILWAY NETWORK

333 PETASITES HYBRIDUS

334 PETASITES FRAGRANS

335 INULA CONYZA

336 FILAGO VULGARIS

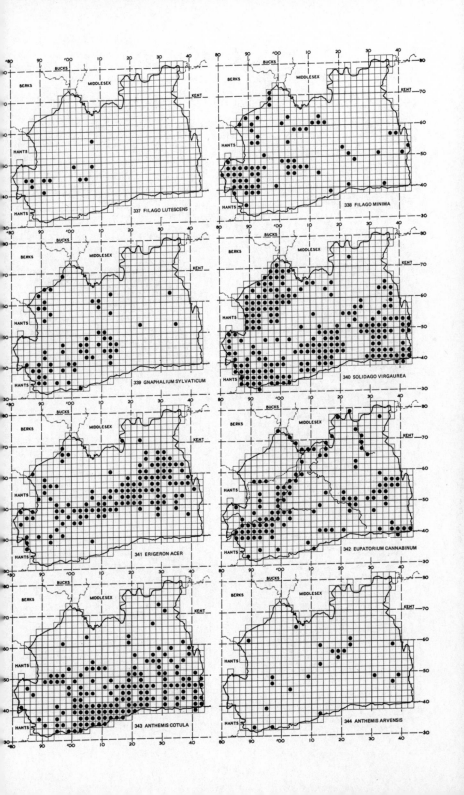

337 FILAGO LUTESCENS

338 FILAGO MINIMA

339 GNAPHALIUM SYLVATICUM

340 SOLIDAGO VIRGAUREA

341 ERIGERON ACER

342 EUPATORIUM CANNABINUM

343 ANTHEMIS COTULA

344 ANTHEMIS ARVENSIS

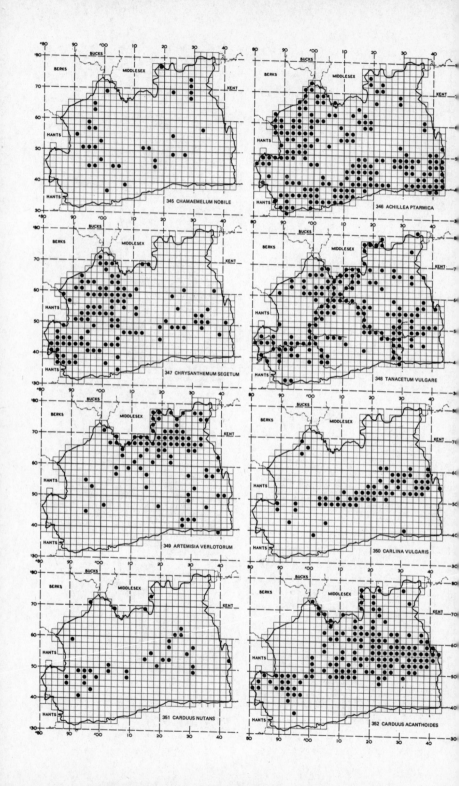

345 CHAMAEMELUM NOBILE

346 ACHILLEA PTARMICA

347 CHRYSANTHEMUM SEGETUM

348 TANACETUM VULGARE

349 ARTEMISIA VERLOTORUM

350 CARLINA VULGARIS

351 CARDUUS NUTANS

352 CARDUUS ACANTHOIDES

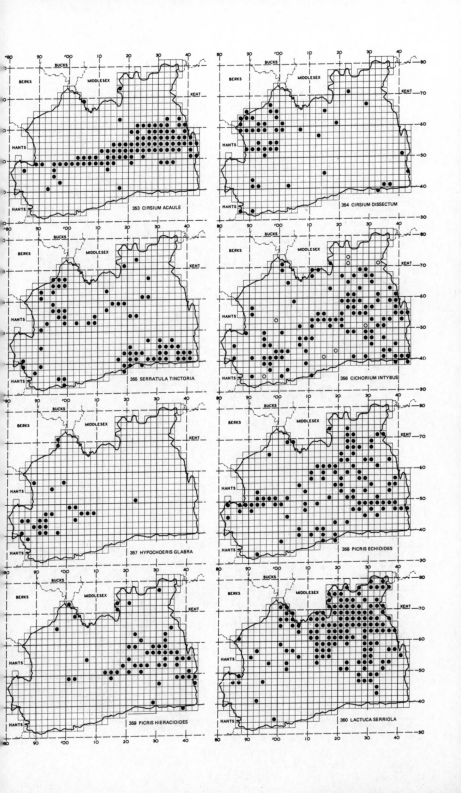

353 CIRSIUM ACAULE

354 CIRSIUM DISSECTUM

355 SERRATULA TINCTORIA

356 CICHORIUM INTYBUS

357 HYPOCHOERIS GLABRA

358 PICRIS ECHIOIDES

359 PICRIS HIERACIOIDES

360 LACTUCA SERRIOLA

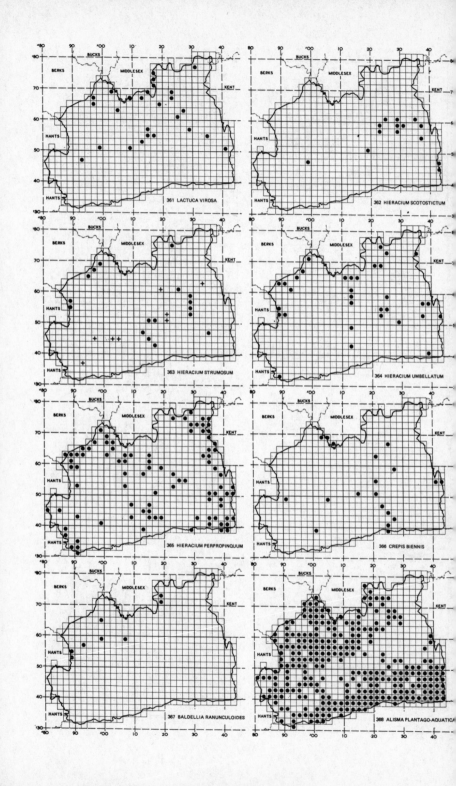

361 LACTUCA VIROSA

362 HIERACIUM SCOTOSTICTUM

363 HIERACIUM STRUMOSUM

364 HIERACIUM UMBELLATUM

365 HIERACIUM PERPROPINQUUM

366 CREPIS BIENNIS

367 BALDELLIA RANUNCULOIDES

368 ALISMA PLANTAGO-AQUATICA

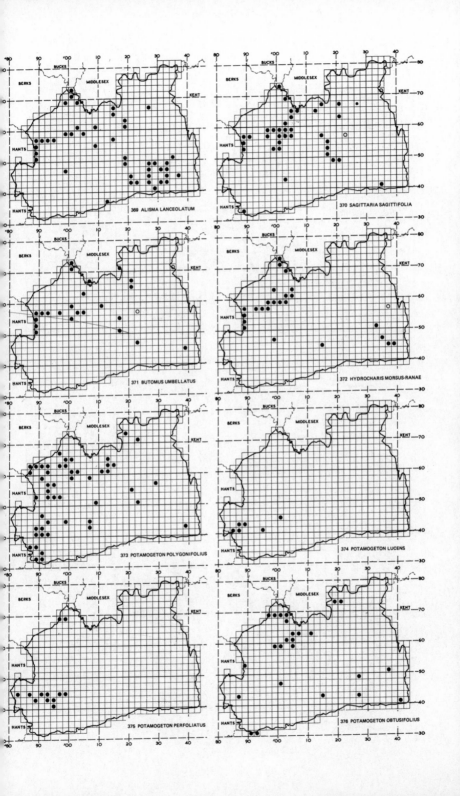

369 ALISMA LANCEOLATUM

370 SAGITTARIA SAGITTIFOLIA

371 BUTOMUS UMBELLATUS

372 HYDROCHARIS MORSUS-RANAE

373 POTAMOGETON POLYGONIFOLIUS

374 POTAMOGETON LUCENS

375 POTAMOGETON PERFOLIATUS

376 POTAMOGETON OBTUSIFOLIUS

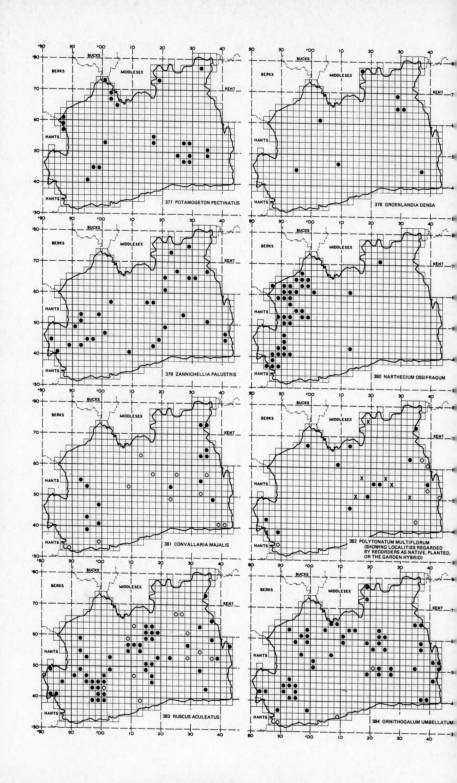

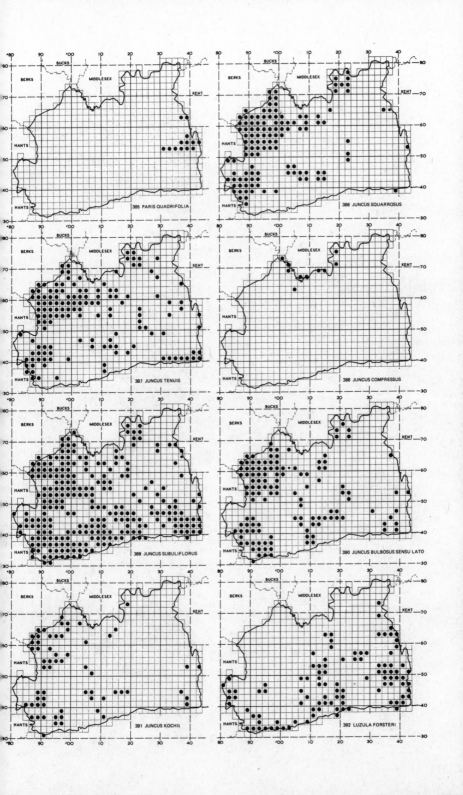

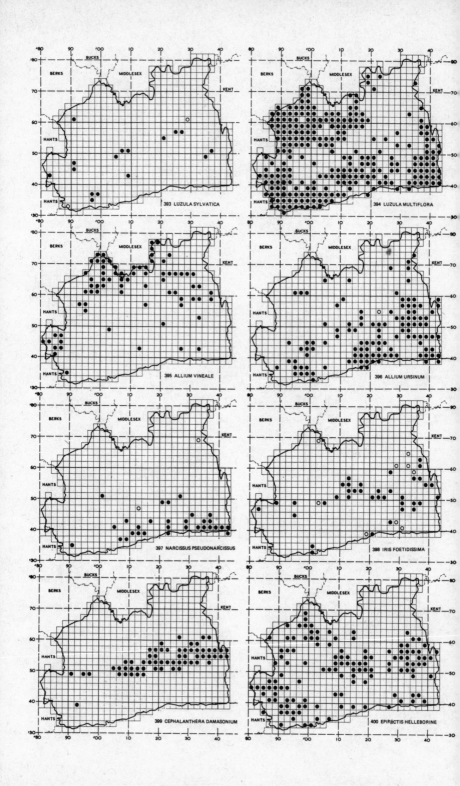

393 LUZULA SYLVATICA

394 LUZULA MULTIFLORA

395 ALLIUM VINEALE

396 ALLIUM URSINUM

397 NARCISSUS PSEUDONARCISSUS

398 IRIS FOETIDISSIMA

399 CEPHALANTHERA DAMASONIUM

400 EPIPACTIS HELLEBORINE

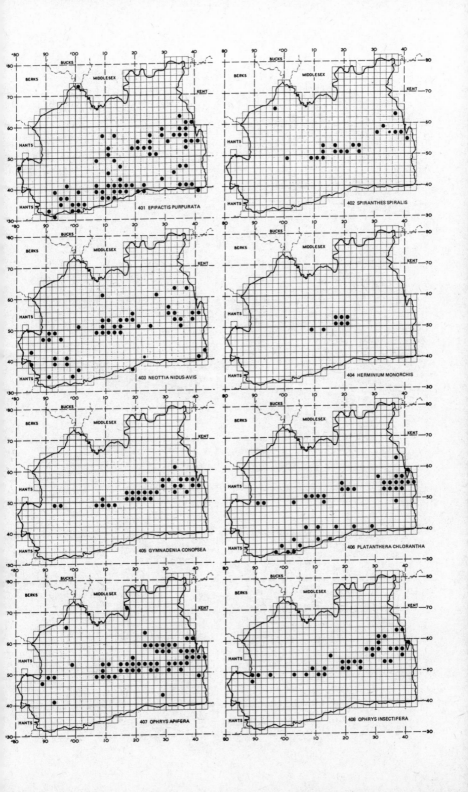

401 EPIPACTIS PURPURATA

402 SPIRANTHES SPIRALIS

403 NEOTTIA NIDUS-AVIS

404 HERMINIUM MONORCHIS

405 GYMNADENIA CONOPSEA

406 PLATANTHERA CHLORANTHA

407 OPHRYS APIFERA

408 OPHRYS INSECTIFERA

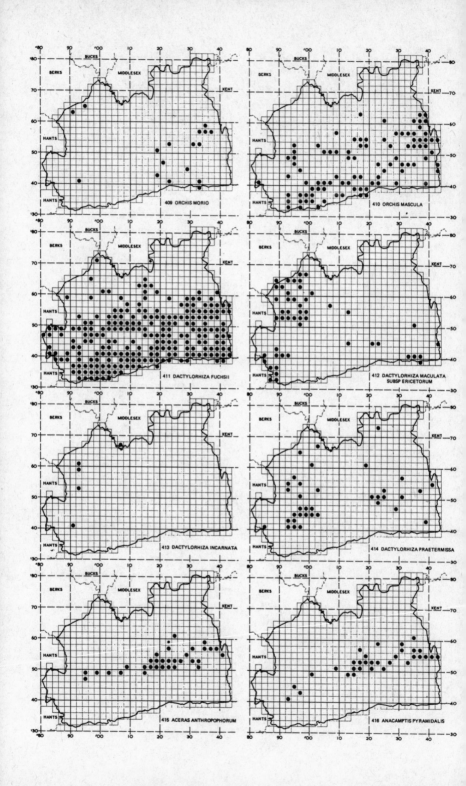

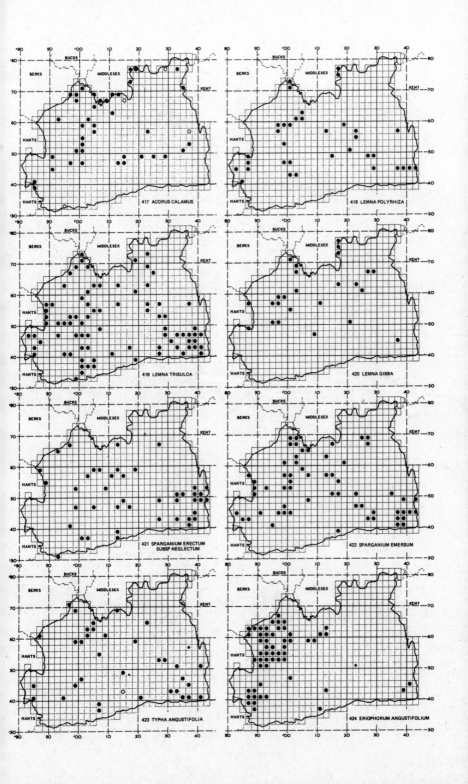

417 ACORUS CALAMUS

418 LEMNA POLYRHIZA

419 LEMNA TRISULCA

420 LEMNA GIBBA

421 SPARGANIUM ERECTUM SUBSP NEGLECTUM

422 SPARGANIUM EMERSUM

423 TYPHA ANGUSTIFOLIA

424 ERIOPHORUM ANGUSTIFOLIUM

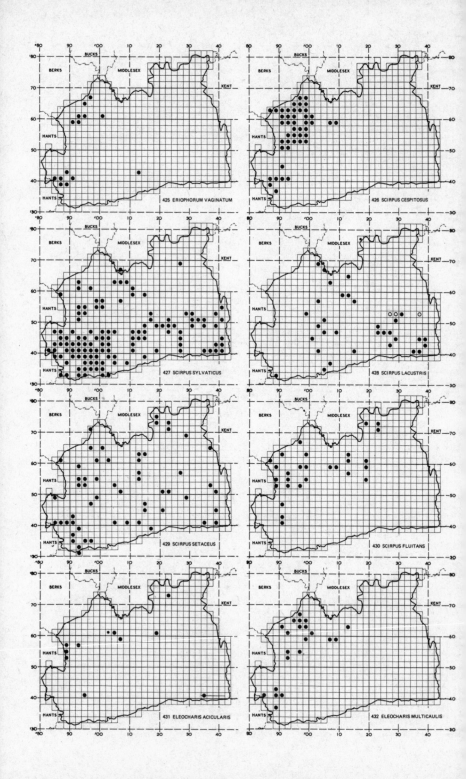

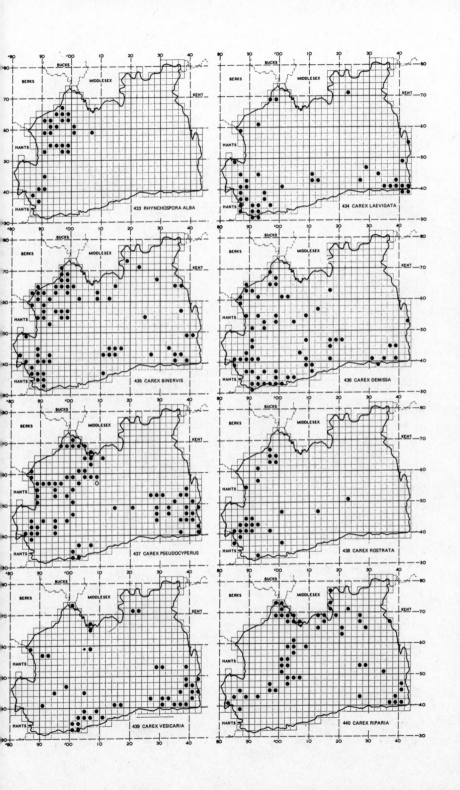

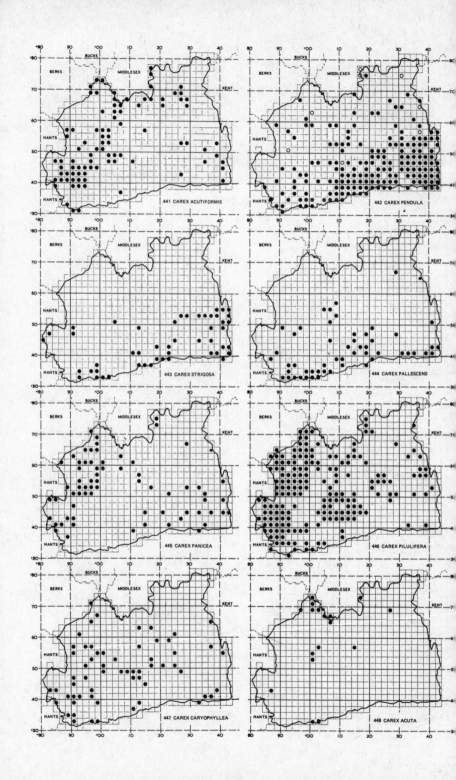

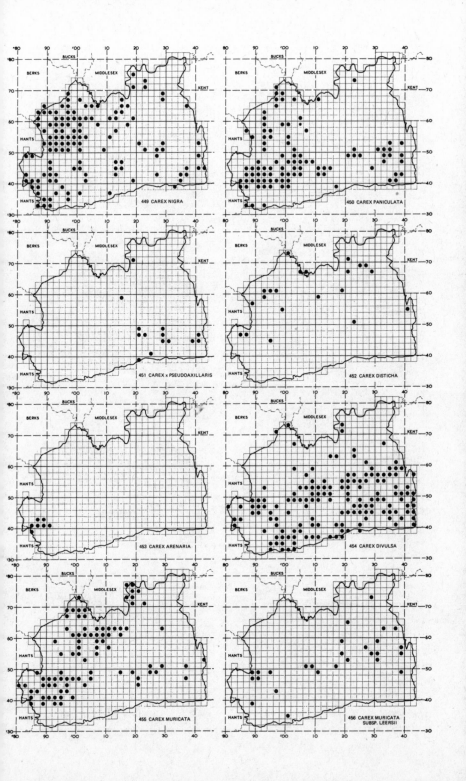

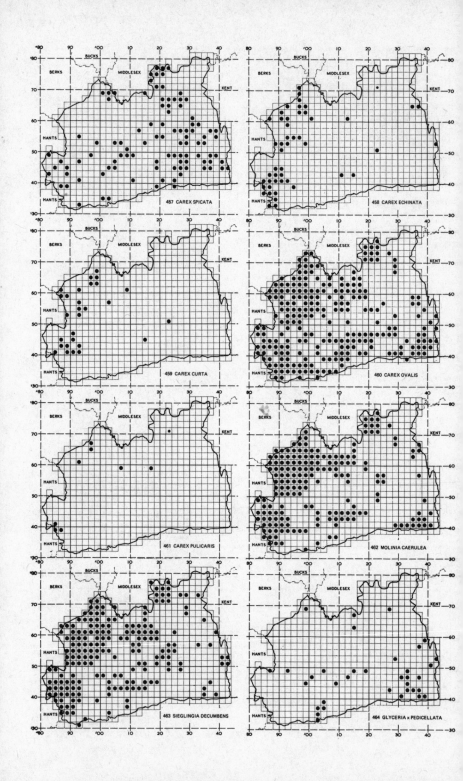

457 CAREX SPICATA

458 CAREX ECHINATA

459 CAREX CURTA

460 CAREX OVALIS

461 CAREX PULICARIS

462 MOLINIA CAERULEA

463 SIEGLINGIA DECUMBENS

464 GLYCERIA x PEDICELLATA

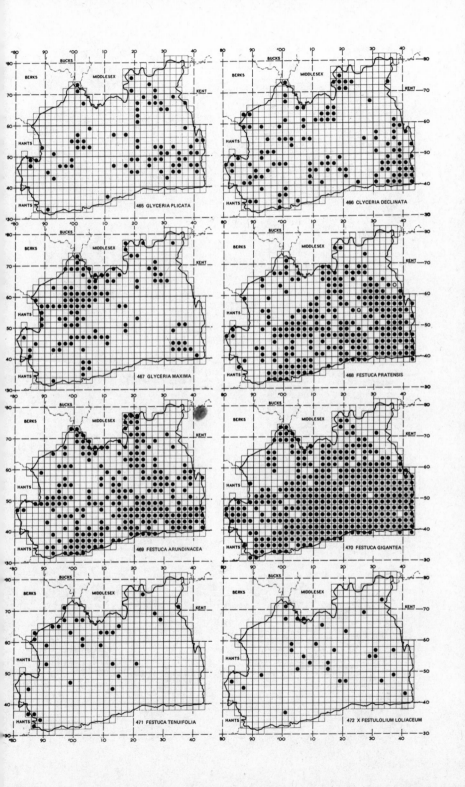

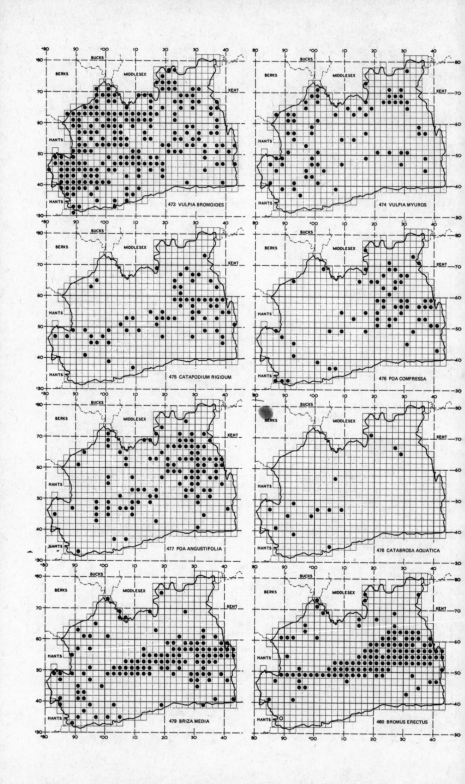

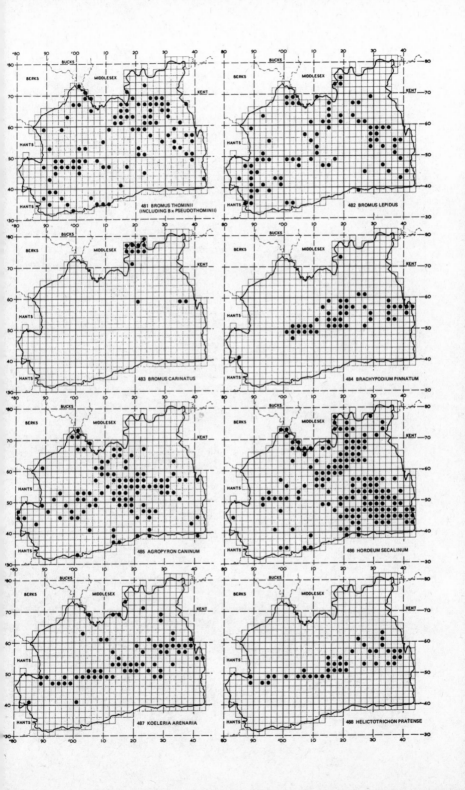

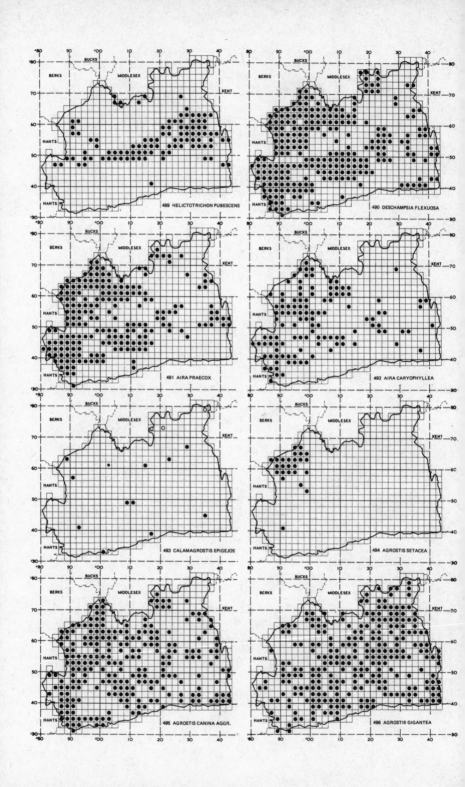

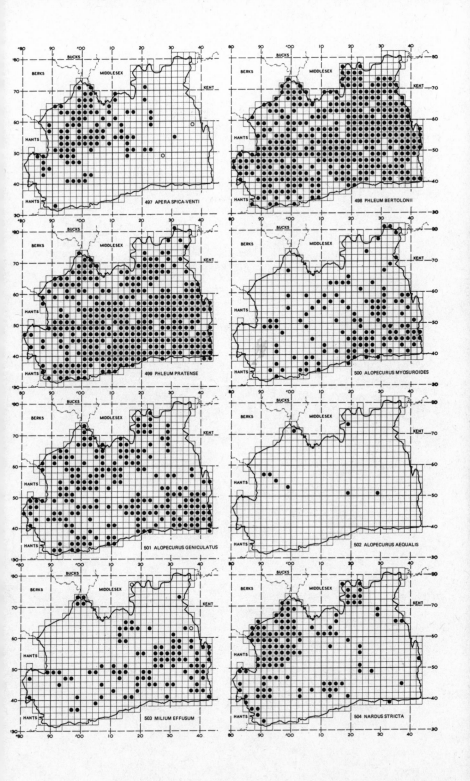

497 APERA SPICA-VENTI

498 PHLEUM BERTOLONII

499 PHLEUM PRATENSE

500 ALOPECURUS MYOSUROIDES

501 ALOPECURUS GENICULATUS

502 ALOPECURUS AEQUALIS

503 MILIUM EFFUSUM

504 NARDUS STRICTA

Bibliography

Most general works on the British flora include references to Surrey plants. These, and other references to individual species, are cited in the text if appropriate. Many references to early works cited in Salmon, 1931, are not repeated here.

Baker, E. G. (1930). 'Charles Edgar Salmon (1872–1930)—obituary,' *J. Bot.*, 68, 50–3

Bangerter, E. B. (1967). 'A Survey of Calystegia in the London Area,' Fifth and Final Report, *Lond. Nat.*, 46, 15–23

Beadell, A. (1927). *Wild Life at Selsdon* (Reprinted from the *Croydon Advertiser*, Croydon)

Beadell, A. (1932). *Nature Notes of Warlingham & Chelsham*, (Croydon)

Beeby, W. H. (1884). 'A New Flora of Surrey', *J. Bot.*, 22, 77–80

Beeby, W. H. (Edit.) (1902). 'Botany', in *Victoria County History of Surrey*, 35–69

Bishop, E. B., Robbins, R. W., & Spooner, H. (1928–36). 'Botanical Records of the London Area,' *Suppl. to London Nat.*, 1927–35, 7–15

Brewer, J. A. (1856). *A New Flora of the Neighbourhood of Reigate, Surrey*

Brewer, J. A. (1863). *Flora of Surrey*. The author also has an interleaved copy heavily annotated by James Britten, J. C. Melvill, A. S. Montgomrey, and others

Bridges, Lord & Sankey, J. H. P. (1969). *The Book of Box Hill*. Ed 2

Britten, C. E. (1932). 'Notes on Surrey Plants', *J. Bot.* 70, 314–37

Carruthers, S. W. (1882). *Notes of the Botany of Surrey*. MS notebook in possession of the writer

Carruthers, S. W. (1883). 'Botany of Dulwich', *Dulwich School Science Society, Fifth Annual Report* (for 1882), 46–8

Castell, C. P. (1950?). *Natural History Records of Wimbledon Common*—Botany. MS in possession of writer

Clapham, A. R., Tutin, T. G., & Warburg, E. F. (1962). *Flora of the British Isles*, Ed 2 (Cambridge)

Clapham, A. R., Tutin, T. G., & Warburg, E. F. (1968). *Excursion Flora of the British Isles*, Ed 2 (Cambridge)

Cocksedge, W. C. (1933). 'The Great North Wood', *Lond. Nat.*, 13, 48–51

Cooper, Daniel (1836). *Flora Metropolitana*, with Supplement, 1837

De Crespigny, Eyre Ch. (1877). *A New London Flora*

Druce, G. Claridge (1928). 'British Plants contained in the Du Bois herbarium at Oxford, 1690–1723', *Rep. Bot. Exch. Cl.*, 8, 463–93

Druce, G. Claridge (1932). 'The Flora of Surrey', *Rep. Bot. Exch. Cl.*, 9, 680–94

Dunn, S. T. (1893). *Flora of South-west Surrey*

Elwes, H. J. & Henry, A. (1906–13). *The Trees of Great Britain*

Ernest, E. C. M. (1932). 'Plant Communities on Croham Hurst', *Proc. Croydon Nat. Hist. & Sci. Soc.*, 10, 152–5

Fagg, C. C. (1941). 'Physiographical Evolution in the Croydon Survey Area and its effects upon Vegetation', *Proc. Croydon Nat. Hist. & Sci. Soc.*, 11, 29–60

Fritsch, F. E. (1927). 'The heath association on Hindhead Common, 1910–26', *J. Ecol.*, 15, 344–72

Fritsch, F. E. & Parker, W. M. (1913). 'The heath association on Hindhead Common', *New Phytol.*, 12, 148–63

Fritsch, F. E. & Salisbury, E. J. (1915). 'Further observations on the heath association on Hindhead Common', *New Phytol.*, 14, 116–38

Godfrey, M. J. (1933). *Monograph & Iconograph Native British Orchidaceae*

Hobson, J. M. (1924). *The Book of the Wandle: the story of a Surrey river*—Plant Life, 158–74

Hall, A. D. & Russell, E. J. (1911). *A Report on the Agriculture and Soils of Kent, Surrey and Sussex*

Howard, Maud (1962). *A List of Flowering Plants and Ferns of Haslemere and District* (Haslemere)

Hutchings, G. E. (1952). *The Book of Box Hill* (Dorking)

Irving, A. (1838). *The London Flora*

Jackson, Rose (1909). *A List of the Flowering Plants & Ferns occurring within six miles of Haslemere*, Supplement (undated) edited by E. W. Swanton (Haslemere)

Jebens, H. Dieter (Edit.) (1968). *Basingstoke Canal—the case for restoration*

Jermy, A. C. & Tutin, T. G. (1968). *British Sedges*

Johnson, W. (1910). *Battersea Park as a centre for Nature Study*—Botany, 67–96

Johnson, W. (1912). *Wimbledon Common, its Geology, Antiquities and Natural History*

Johnson, Walter, (1925). *Talks with Shepherds*

Jones, E. W., (1968). 'The taxonomy of the British species of Quercus', *Proc. Bot. Soc. Br. Isles*, 7, 183–4

Jones, William, & Malcolm, J. (1794). *General View of the Agriculture of the County of Surrey*

Kent, D. H. (1956). 'Surrey Canal wild flowers', *P.L.A. Monthly*, 1956, 343–4

Kent, D. H. & Lousley, J. E. (1951–7). *Handlist of the Plants of the London Area*—issued in seven parts as a Supplement to the *London Naturalist*, 30–6

Lousley, J. E. (1931). 'The Schoenoplectus Group of the Genus Scirpus in Britain', *J. Bot.*, 69, 151–63

Lousley, J. E., (1932). 'Flora of Surrey, a review', *Rep. Bot. Exch. Club*, 9, 595–600

Lousley, J. E., Castell, C. P., & Robbins, R. W. (1939). 'The Survey of Limpsfield Common', *Lond. Nat.*, 18, 53–74

Lousley, J. E. (1945). Botanical Records (of the London Area) annually, *Lond. Nat.*, 24 et seq

Lousley, J. E. (1948). '*Calystegia sylvestris* (Willd.) R. & S.', *Rep. Bot. Exch. Cl.*, 13, 265–8

Lousley, J. E. (1950). *Wild Flowers of Chalk & Limestone* (New Naturalist Series, 1969, Ed 2)

Lousley, J. E. (1951). 'Basingstoke Canal from Byfleet to Woking (Surrey, V-c 17)' *BSBI Year Book*, 1951, 60–5

Lousley, J. E., (1958a). 'A new Surrey Flora', *Countryside*, 18, 183–4

Lousley, J. E. (1958b). 'A revision of the Flora of Surrey', *S.E. Nat.*, 1957, ix–xii

Lousley, J. E. (1959a). Surrey Records for 1958, *S.E. Nat.*, 1958, xi–xiii

Lousley, J. E. (1959b). 'Dulwich Woods; relics of the Great North Wood', *Lond. Nat.*, 38, 77–90

Lousley, J. E. (1960). 'Further notes on relics of the Great North Wood', *Lond. Nat.*, 39, 31–6

Lousley, J. E. (1965). 'Limpsfield Common Revisited', *Lond. Nat.*, 44, 16–7

Lousley, J. E., (1971). 'Mitcham Common and its Conservation', *Proc. Croydon Nat. Hist. & Sci. Soc.*, 15, 35–46

Luxford, George (1838). *A Flora of the Neighbourhood of Reigate, Surrey*

Malcolm, James (1805). *A Comparison of Modern Husbandry . . . of Surrey*

Marsden-Jones, E. M. & Turrill, W. B. (1954). *British Knapweeds*

Marshall, 'Mr', (1798). *The Rural Economy of the Southern Counties,* 2—The Heaths of Surrey, 81 et seq

Martin, Edward A. (1923). *A short account of the Natural History and Antiquities of Croydon* (Croydon)

Merrett, C., (1666). *Pinax Rerum Naturalium Britannicum*

Mill, J. S. (1841). 'Notes on Plants growing in the neighbourhood of Guildford, Surrey', *Phytol.,* 1, 40–1

Ministry of Agriculture, Fisheries and Food, *Agricultural Statistics for England & Wales* (1897 onwards)

Monckton, H. W. (1916). *The Flora of the Bagshot District,* privately printed

Monckton, H. W. (1919). *The Flora of the District of the Thames Valley Drift between Maidenhead and London,* privately printed

Nature Conservancy, 1970. *Conservation in Surrey, An appraisal of selected Open Spaces: Report of a Working Party*

Nicholson, G., Andrews, S. Boyd, et al. (1971). *Walks on Wimbledon Common* (Wimbledon)

Nightingale, D. A., & Lovis, J. D. (1951). 'The New Pond, Merstham,' *Proc. Croydon Nat. Hist. & Sci. Soc.,* 69–75

Payne, L. G. (1939). 'The Crested Buckler Fern, *Lastrea cristata* Presl', *Lond. Nat.,* 18, 29–31

Payne, L. G. (1942). 'The Royal Fern (*Osmunda regalis* L.) in Surrey', *Lond. Nat.,* 21, 12–13

Pearson, A. A. (1918). *The Flora of Wimbledon Common*

Perring, F. H. & Walters, S. M. (1962). *Atlas of the British Flora*

Perring, F. H. (1968). *Critical Supplement to the Atlas of the British Flora*

Prime, C. T. (1972). 'The vegetation of the Blackheath Pebble Beds in the Croydon District', *Proc. Croydon Nat. Hist. & Sci. Soc.,* 15 (4), 47–80

Pugsley, H. W. (1948). 'A Prodromus of the British Hieracia,' *J. Linn. Soc. London (Bot.),* 54

Robbins, R. W. (1932). Wild Flowers, in L. G. Fry's *Oxted, Limpsfield and Neighbourhood,* 124–31 (Oxted)

Robbins, R. W. (1940. 'The Flora of Limpsfield Common (2),' *Lond. Nat.* 19, 25–7

Rogers, D. (1972). 'The Wilderness of Nunhead,' *Illustr. Ldn. News,* 260, 53–5

Salmon, C. E. (1931). *Flora of Surrey*

Salmon, J. D. (1849). An Outline of the Flora of the Neighbourhood of Godalming, in Newman, E. (Edit.) *The Letters of Rusticus*

Schotsman, H. D. (1954). 'A taxonomic spectrum of the Section Eu-Callitriche in the Netherlands', *Acta Bot. Neerl.*, 3, 313–84

Schroeder, F.-G., (1972). 'Amelanchier-Arten als Neophyten in Europa', *Abh. Naturwiss. Verein Bremen*, 37, 287–419

Smith-Pearse, T. N. H. (1917). *A Flora of Epsom* (Epsom)

Stevenson, William (1813). *General View of the Agriculture of the County of Surrey*

Swanton, E. W. (1915). *Vanished and Vanishing Animals and Plants of the Haslemere District*. (Reprinted from *Farnham, Haslemere, and Hindhead Herald*)

Swanton, E. W. (1958). *The Yew Trees of England*

Thompson, H. S. (1930). 'Charles Edgar Salmon (1872–1930)', *Rep. Watson Bot. Exch. Club*, 4, 7–11

Trimen, H., (1864). 'Additions to Brewer's *Flora of Surrey*,' *J. Bot.*, 2, 78–94

Turner, D., & Dillwyn, L. W. (1805). *Botanist's Guide through England and Wales*, 2, 577–95

Watson, W. C. R. (1958). *Handbook of the Rubi of Great Britain and Ireland* (Cambridge)

Welchman, R. H. & Salmon, C. E. (1905). 'The Flora of the Reigate District', *S. E. Nat.*, 1905, 1–8

Whitehouse, Mrs M. (1952). 'The present flora of Barnes Common, Surrey,' *Lond. Nat.*, 31, 17–19

Williams, R. A., (1968). *The London & South-Western Railway—1, The Formative Years* (Newton Abbot)

Young, D. P. (1952). 'Studies in the British Epipactis III-IV,' *Watsonia*, 2, 253–76

Young, D. P. (1958). 'Oxalis in the British Isles', *Watsonia*, 4, 51–69

Young, D. P. (1961). 'A Botanist explores Bethlem', *Bethlem Maudsley Hospital Gazette*, 4, 182–5

Young, D. P. (1962). 'Studies in the British Epipactis VI', *Watsonia*, 5, 136–9

Acknowledgements

I am grateful to so many people for help over the Surrey *Flora* during the last fifty years that it is quite impossible to enumerate them all. My field work in the county with E. C. Wallace has extended over the whole of this period and to him above all I am grateful for generous help from his unrivalled knowledge of the detailed topography and botany of Surrey. I owe a special debt also to the members of the Surrey Flora Committee whose diverse botanical skills and territorial interests have knit together in a remarkable way to produce, with their supporters, many thousands of records from the 555 tetrads as a basis for the distribution maps.

To Barabara Welch, their first Secretary, I must pay a special tribute for the sustained effort and outstandingly high standard of precision which provided such a firm foundation. To her successor, Joyce Smith, I owe much encouragement without which this *Flora* might not have been completed. Donald Young put in much work of a plodding nature, first on the Croydon squares for which he was initially responsible, and then, on his appointment as Editor, on special outings to improve tetrads inadequately reported, and in transferring the records on to maps. That he failed to do more was due to an insidious illness which made sustained work almost impossible. The maps left by Donald after checking and editing were copied ready for the printers by James Stevens. I am also grateful to him for preparing the general maps (figures in the text) and for writing the account of the geology. Most of the contributors of records are given on pages 87–93 and this list also includes many of those who assisted by identifying specimens—the rest are mentioned in the text. To Dr Cyril West, Dr Peter Yeo, Desmond Meikle and Alan Newton I owe much help with the accounts of *Hieracium, Euphrasia, Salix* and *Rubus* respectively.

For superb photographic illustrations, all taken in the county and some taken specially, I am indebted to D. M. Turner Ettlinger (Plates 2, 6, 7, 8, 9, 26, 29 and 30) and Peter Wakely (Plates 4, 5, 11, 15, 16, 18, 20, 21, 22, 27 and 31). The Meteorological Office has provided climatic statistics, including those in Table 1, which are reproduced with the permission of the Copyright Division, HMSO, Crown Copyright reserved.

But for the patience and encouragement of my wife this book would never have seen the light of day.

Index

Page numbers in italic type indicate illustrations. Figures with 'M' prefix refer to the Distribution Maps (pages 405 to 467)